12
SECONDS
OF
SILENCE

How a Team of Inventors,
Tinkerers, and Spies Took Down
a Nazi Superweapon

JAMIE HOLMES

Mariner Books
Houghton Mifflin Harcourt
Boston   New York

First Mariner Books edition 2021

Library of Congress Cataloging-in-Publication Data
Names: Holmes, Jamie, author.
Title: 12 seconds of silence : how a team of inventors, tinkerers, and spies took down a Nazi superweapon / Jamie Holmes.
Other titles: How a team of inventors, tinkerers, and spies took down a Nazi superweapon
Description: Boston : Houghton Mifflin Harcourt, 2020. |
Includes bibliographical references and index.
Identifiers: LCCN 2019045690 (print) | LCCN 2019045691 (ebook) |
ISBN 9781328460127 (hardcover) | ISBN 9780358309512 |
ISBN 9780358313144 | ISBN 9781328459855 (ebook) | ISBN 9780358508632 (pbk.)
Subjects: LCSH: Proximity fuzes — Design and construction. | World War, 1939–1945. |
Military weapons — History. | Weapons industry — United States — History — 20th
Century. | United States. Office of Scientific Research and Development — History. |
Johns Hopkins University. Applied Physics Laboratory — History.
Classification: LCC UF780 .H65 2020 (print) | LCC UF780 (ebook) |
DDC 623.4/542 — dc23
LC record available at https://lccn.loc.gov/2019045690
LC ebook record available at https://lccn.loc.gov/2019045691

Book design by Emily Snyder

Printed in the United States of America
DOC 10 9 8 7 6 5 4 3 2 1

The beginning of every war is like opening the door into a dark room. One never knows what is hidden in the darkness.

— ADOLF HITLER, 1941

CONTENTS

PART III: VICTORY

PROLOGUE

On May 7, 1944, in the deep of night, a train carrying the 130th Chemical Processing Company of the U.S. Chemical Warfare Service steamed toward London. The old locomotive had no speedometer, and the blackout hiding southern England from Nazi bombers made it hard for the engine driver to gauge their velocity using landmarks along the tracks. No lights were allowed to shine or leak from the train. Windows had been screened. Snaking through the dark, they passed stations lit only by shaded gas lamps and quiet blue lights.

Before sunrise, capping a three-week trek, the Americans disembarked in the British capital onto trucks with shuttered headlights. Winds from the east blew cool and damp. Nothing had prepared the soldiers, many of them still teenagers, for wartime London. Not the chemical weapons training at hot and dusty Camp Sibert, Alabama. Not the crossing on the SS *Exceller* accompanied by six destroyers, an aircraft carrier, twelve cargo ships, and nineteen tankers. Not the radio reports.

This London felt otherworldly. Flashlights were dimmed by brown paper, and the bicycle lights permitted were so feeble that many riders pedaled without them. News vendors shouting "the latest" were marked by their cigarette cherries, fireflies in the gloom. On moonless nights it was a real chore reading a public telephone dial. Naturally, road accidents had spiked. Pedestrians were run over unseen. Commuters even stepped out of trains on the wrong side, while others fell down holes or off bridges. Posters recommended carrots for keener vision.

Adapting to the chaos, Londoners had slathered white paint onto curbs, lampposts, car bumpers, and traffic islands. Grand public steps looked like chessboards. Citizens donned white patches, sleeves, and hat brims. Even pets were protectively attired in white. Traffic signals had been shrunk to slivers of red, green, and amber. Policemen in ghostly tunics wielded faint green flashlights.

To the men of the 130th, London's dilemma was too surreal to absorb at once. As their trucks navigated the dark streets, they were awed by the damage: a proud capital in tatters, ruins on nearly every block, bomb craters, stripped façades, charred skeletons of buildings, broken timber beams protruding like compound fractures. They could imagine what it all meant: high explosives, parachute mines, incendiary bombs, firefighters battling raging flames, and rescue squads digging all night for the missing. Those heaps of brick signified great trauma and loss.

When the Blitz came, four years prior, the city had scarcely any protection from the German air force. "He comes when he wants," one Londoner said, of the Luftwaffe. "There's no stopping him." Panic spread like an airborne virus, prompting the British air command to alter its overwhelming reliance on Royal Air Force fighters and to install dozens of antiaircraft gun sites in and around the city. The booming guns helped reassure residents they were not defenseless.

But aside from the psychological boost to morale, the antiaircraft cannons were largely worthless. In the early weeks of the Blitz, the ratio of fired shells to downed aircraft—a metric described in RPB, or rounds per bird—was twenty thousand to one. Taking out a single German aircraft took twenty thousand shots.

The enemy could be hit, but mostly when they flew low and straight during the day. At night targeting was nearly impossible. Planes had to be held in searchlights for primitive aiming devices. American antiaircraft defenses were just as shoddy—a potential disaster for the U.S. Navy. At the start of the war, an American scientist said, "It would be just a sheer stroke of luck to hit anything."

Figuring out how to shoot down airplanes—solving the rounds-per-bird puzzle—was one of the toughest, most urgent scientific tasks of World War II.

By the time the 130th Chemical Processing Company reached England in May of 1944, Londoners were tired—of fear, of rationing, of the blackout. A German "Baby Blitz" claiming more than fifteen hundred

British lives was still winding down. And Allied military planners believed things were about to get a lot worse.

The one hundred and forty-eight American chemical warfare specialists were billeted in brick houses at 4, 6, and 8 Sloane Court East in Chelsea, a residential neighborhood on the river Thames. The apartments were situated between a burial ground and a hospital. Just yards away stood two buildings highlighted on the targeting maps of Luftwaffe bombardiers: the British Army barracks and the Duke of York's Headquarters.

Among the 130th was Samuel Edward Hatch, eighteen, from Massachusetts, with a wide face, large quiet eyes, and dark brown hair. They called him Ed. He was assigned a second-floor room at 6 Sloane Court East with another teenage Bay Stater, Teddy Booras, a skinny boy with an overbite. The sun rose and the pair set about bringing their improvised quarters up to military standards.

On Monday, May 8, hours after arriving in darkness, Hatch scrawled a careful, bland, reassuring note to his parents back home in Springfield, a midsize city in western Massachusetts. "We have been informed by our officers that it is alright to tell you that I am in London and am living in very spacious and beautiful apartments. The food here is the best I've had in the Army."

D-Day was a month away.

Thousands of soldiers, sailors, planes, combat ships, and beaching craft were being readied for the Normandy landing. An invasion of France would pose a direct threat to the heart of Germany, and the Allies fully expected Hitler to retaliate furiously, unleashing an airborne arsenal of secret weapons they knew were in development. If Nazi scientists could use the mysterious weapons on civilians, Britain's capital remained their most plausible target. One Allied intelligence agent even guessed a date for Hitler's counterattack on London: "D+7."

American military analysts were especially concerned that Germany was planning to arm their retaliatory warheads with chemical weapons. In World War I, around ninety-one thousand troops died from chemical warfare, mostly from phosgene, a colorless gas four times denser than air that smells like freshly cut grass. By 1938, German science had invented something far more lethal: sarin gas. And the Nazis had already initiated plans to mass-produce the deadly poison.

Chemical weapons didn't have to be fatal to be useful. Weaponized mustard gas, yellow-brown and carrying the scent of horseradish, was highly effective at incapacitating enemy troops. A wet, heavy gas, mustard is an irritant and blistering agent. It seeps through clothing, causing severe burns and lesions.

The 130th Chemical Processing Company was in London in preparation for the expected counterattack—a guessing game of worst-case scenarios. Should the Nazis use chemical weapons, the 130th would defend soldiers by "impregnating" military uniforms with protective chemicals.

Hitler's answer, as predicted, would come from the air, and present a profound test to the men of the 130th. But it would not be the test they prepared for.

In the first war in history to be decided by weapons that did not exist at the onset of the conflict, the Führer's response to D-Day would initiate the climactic battle between American science and Nazi science. The summer of 1944 saw two secret weapons face off head-to-head over the waters and skies of England.

The V-1, a Nazi drone missile that propagandist Joseph Goebbels dubbed "Revenge Weapon One," was a pilotless aircraft that sounded like a two-stroke motorcycle with strep throat. It was a 4,900-pound "robot bomb" flying on autopilot, and no one had ever seen anything like it. At launch in Nazi-controlled France, odometers in the drones were set to target London. At their destination, engines cut and the bombs silently drifted down to their marks. Londoners came to fear this quiet interval: twelve seconds of silence to wonder if their time had come.

Over eighty long, hot summer days, more than seven thousand V-1s would be catapulted toward England. At first, the antiaircraft guns could do almost nothing. The Royal Air Force did little better; the V-1s were faster than its airplanes. The assault pummeled Londoners' morale to its lowest point of the war.

And then, something changed.

In mid-July, America would deploy its own secret technology in Britain. Its very existence was unknown to all but a few scientists, intelligence officers, military brass, and gunners. It was America's answer to the antiaircraft problem, a secret device that delivered a staggering rounds-per-bird ratio of a hundred to one. American scientists made shooting planes out of the sky radically easier.

What changed was the proximity fuse.

By September, the V-1s were stopped cold.

Known as the world's first "smart" weapon, the proximity fuse (or fuze) was a five-pound marvel of engineering, industry, and can-do spirit.

The gadget, screwed into the tip of an antiaircraft shell, had a brain. It was able to sense nearby aircraft by sending out a radio signal and then listening for the signal to bounce back off the airplane. If it did, the fuse would trigger the high explosives in the shell, unleashing a lethal barrage of shrapnel. American factories would ultimately produce twenty-two million fuses at a cost of over a billion dollars. In the company of the atom bomb and advances in radar, it ranks among the most decisive technological breakthroughs of the war.

Today, few know its story.

The fuse has its accolades: shortening the war in the Pacific by a year, saving many thousands of British lives, winning the Battle of the Bulge. But this book isn't about awards or medals. Nor is it a painstakingly detailed "grand history." It's a story about the spirit and sacrifice of the Americans who made the fuse. They were physicists, engineers, statisticians, and ham-radio hobbyists.

12 Seconds of Silence is the lost tale of a bunch of nerds, tinkerers, and everyday Americans who came together, under horrible pressure and a cloak of secrecy — in a race with Axis scientists — to embrace one of the war's trickiest challenges. It's about the fierce leader of the fuse effort, and the America that shaped him. It's about the triumph of the fuse as a window to the role of science during the war.

It's about the scientists who defended democracy.

The story will take us from London to Washington, D.C., to factories and research labs, to a private farm in Virginia, and to a secret base in the foothills of New Mexico. To tell it, I've drawn from five thousand archival documents, including memoirs, letters, cables, diaries, and exclusive, newly released records.

For the first time — during a war, no less — science moved to the core of military strategy, birthing a strange new field of scientific intelligence and espionage. That subplot will lead us to a Nazi spy ring in New York, a twenty-eight-year-old British physicist called "the father of scientific intelligence" by the CIA, a French spy in occupied Paris known for her beauty, who spoke five languages, and eventually, to a covert meeting in a Manhattan automat with none other than Julius Rosenberg.

Along the way, we'll learn about the forgotten wartime outfit behind the fuse, the Office of Scientific Research and Development (OSRD). Led by the brilliant engineer and inventor Vannevar Bush, OSRD was the research and development hub of the Arsenal of Democracy. Van Bush, "the most politically powerful inventor in America since Benjamin Franklin," in the words of his biographer, saw early on that World War II would be a desperate "battle of scientific techniques."

Within Bush's organization, a group of four men with zero background in weapons development took on the challenge of the fuse. The project was a true long shot. No one, not even Bush, thought that they would succeed. Step by step, as they progressed, the small band of four morphed into a staff of over one thousand.

In honor of their leader, Merle Tuve, Bush called them Section T.

Part I

PEACE

It is being realized with a thud that the world is
probably going to be ruled by those who know
how, in the fullest sense, to apply science, whatever
their other attributes may be.

— VANNEVAR BUSH, 1941

1
ZING-A-ZING!

Merle Tuve's life was ordinary in the best sense.

In September of 1939, Tuve was thirty-eight and happily married. He and his wife were raising two small children. They lived on a pleasant block with grassy lawns in a modest suburban home. He drove a fine Chrysler Airflow known for its streamlined design. And his commute was a plum. It was less than two miles, as the crow flies, from their house on Hesketh Street in Chevy Chase, Maryland, to his research institute in Washington, D.C., proper. In the Chrysler, it took about eight minutes.

While he worked long hours, his home life appeared typical. Neighbors had no reason to suspect that he would alter the course of history, or that his home was about to fill, almost nightly, with hard-drinking Navy officers. If Tuve himself were told that his employer, the Department of Terrestrial Magnetism, would soon be ordering explosives in bulk, he would have been stunned.

Tuve had "Viking" blood running through his veins, one colleague said. Born in Canton, South Dakota, he was a uniquely American product and a mesmerizing scientist, a fiery Midwesterner with a bizarre name, raised by Norwegian immigrants busy inventing a new American lifestyle.

He was the dynamic soul of the fuse project.

The saga of Section T begins with the story of Merle Tuve, a rare childhood friendship, and a singular event that reshaped his character.

Science kept the boy out of trouble.

Tony and Ida's son had energy, a real thirst for life. More than anything, the ten-year-old was *involved*. He wanted to try everything! And Canton, a town of 2,100 people in 1911, got pretty dull in the summertime. The child had to be kept busy.

Like other boys, Merle bought firecrackers and shot BB guns. But he also liked to pry and meddle. He was curious about things that he shouldn't be. When very small, he went through an unfortunate phase of starting fires. It was also true that he once set a trap, caught a skunk and —thinking it was a cat—lifted the animal up to check its gender. This resulted in multiple baths in an outdoor tub.

He was born at home, in a house built by his father that faced the northern winds. With blue eyes and blondish hair that would later turn brown, he was a delicate-looking child. But he was never timid.

According to family lore, he once found his brother in the snow, frostbitten, and to avoid warming the two-year-old up too quickly, he soaked the infant in cold water before placing him in the oven. The toddler was not overcooked.

Tony, a professor, and Ida, a music teacher, ran a loving but strict Lutheran home. Drinking, dancing, and playing cards were not to be tolerated, and they warned Merle and his two brothers and sister never to lie, steal, or get into fights. When Merle was sent upstairs to his room as punishment, he would stay quiet for a while, dutifully, and then begin to scratch on the heating grate above Tony's study, whispering to his father to ask whether he could come down.

Not even his many chores—which included pumping water, beating rugs, shoveling snow, and tending the cows—kept him safely busy and out of mischief. The family calf had the exact same personality that he did. "He's awfully inquisitive," Merle noted, impressed by the new creature. "Butts around everything."

He and his friend Ernie, who lived catty-corner from the Tuves' house, were both curious about electronic gadgets. At age nine, they rigged a telegraph wire between their houses and learned Morse code so they could converse at night. They built their own "contact keys" to tap out the signals. They pestered the maintenance man who came by to replace the old batteries in home telephones.

"Can we have these batteries?" Merle begged, of the discards.

"Sure." They unwrapped the old batteries, hammered little holes in them with tenpenny nails, and soaked them in ammonium chloride in Mason jars to recharge them. They'd link together a few dozen and "get a good shock."

Merle grew fascinated by magnets and electric bells. He stuck his thumb inside little motors to stop them, singeing his skin on the more powerful ones. He gathered cast-off carbon rods from street lamps and made huge sparks. He scoured the shelves of the town drugstore for the latest issues of *Modern Electrics, Scientific American, Popular Mechanics,* and *Electrical Experimenter.*

Playing with gadgets was better than playing pranks.

One Halloween, Merle and Ernie threw trash on a neighbor's porch —while she was still on it. Ernie escaped but the neighbor, Mrs. Sorum, snagged Merle by his pants and pulled him back through the fence hole. Ernie also got up to mischief on his own, once stomping down several rows of a local farmer's corn.

Canton's bullies must have been just as bored: Merle and Ernie were two of their favorite targets. Ernie and a friend were once hitched to a wagon, like a team of oxen, and forced to cart an older boy through the streets. The bullies would ambush Merle and Ernie, grab them by their wrists and ankles, and toss their skinny bodies violently against tree trunks or telephone poles.

Merle was so sensitive a child that when silent movies arrived in Canton, his mother covered his eyes during battle scenes with "Indians" or train holdups. He was "high strung," she said. Ernie's strongest curse words were "Sugar!" and "Oh, fudge!" Lanky, stammering, a brainy child "born grown up," he and Merle made a funny pair. They already knew they didn't want to be farmers. Doctors, maybe.

Somehow the misfits, next-door neighbors in a tiny South Dakota town, would go on to develop the most potent weapons of history's deadliest conflict.

But in 1911 they were just professors' sons.

One day, after trying to avoid a bully by crossing the street, Merle found real trouble. The ruffian followed him, and the taunting and abuse became too much. Merle snapped and attacked back, banging the bully's head on the sidewalk until he was pulled off.

Merle had been ready to kill the boy, literally. After that episode, for the rest of his life, he feared his unchecked anger—afraid of his own temper. It was exactly the kind of trouble Merle's parents were hoping to keep him out of.

Canton was dull because it was newborn. Situated on rich lands above a bend of the Big Sioux River, on the border with Iowa, the town was not far removed from its frontier days—from American Indian raids, droughts, and grasshopper plagues.

It was only thirty-some years before Merle was born that Halvor Nilsen from Gol, Norway, founded Canton, then known as Commerce City. Nilsen struck it rich in the California gold rush, bought over five thousand acres, and thirty families settled there in 1868. They were far from railroads, and had no church, school, market, or streets. Steamboats were rare on the Big Sioux River. Tributaries and marshes were onerous, prairies were wild, and forests were nearly impassable. In this sort of life you tried, failed, got up, fell again, and then started over.

Canton was nearly as new as its home state. The year Merle's father was hired by Augustana College, 1889, was the same year South Dakota was admitted to the Union.

In 1890, with the college in dire financial trouble, Tony became its president, a post he held for twenty-five years. He took on fiscal responsibility for it as well. Five instructors, fewer than a hundred students, and a tiny budget compelled Tony to teach—and learn to teach—a dizzying array of subjects. These included algebra, commercial arithmetic, the Bible, bookkeeping, business practice, English reading, commercial law, civics, elocution, geography, solid geometry, plane geometry, grammar, natural science, orthography, pedagogy, penmanship, psychology, physiology, rapid calculation, religion, and rhetoric.

Tony *was* Augustana.

All four of Merle's grandparents were Norwegian immigrants. Tony was raised on a farm in rural Iowa, didn't learn much English as a youth, and had no college degree. But he studied, leaving dirt fingerprints in textbooks during breaks from plowing.

Now Tony read Shakespeare, Tennyson, and the New Testament to his children as they hoed the peas. He led them in prayers and, sometimes, old Norse hymns. He took them to Canton's Carnegie Library. Merle's father, according to a student, was "a princely specimen of manhood

—handsome, tall, well proportioned, straight, broad shouldered, athletic . . . Morally, he was also very attractive."

But if the days of city building weren't over, the era of wild conquest was. Boredom was a new luxury, an ironic byproduct of sacrifice and hard struggle. Canton's long summer days felt tedious to restless boys. As Merle approached adolescence, his boredom loomed large. Ernie's parents were also worried. Kids all over town were pulling pranks to pass the time. They even disassembled a lumber wagon in the middle of the night and put it back together atop the town courthouse. Canton wasted a thousand bucks in wood scaffolding to get it down.

Tony's solution, like most of his resourceful fixes, was pioneering. The kids would join an inventive little outfit, only a year old, called the Boy Scouts of America.

As Scout leader, Tony enlisted over a dozen neighborhood recruits. The boys took six a.m. hikes, camped for a week at Newton Hills in the bluffs by the river, and managed a three-day trip from Canton to Sioux Falls, a nearly twenty-mile journey.

Merle found it easy to be seduced by the lush wooded hills. He remembered the river "sparkling in the morning sun" and the view of the vast valley from high above, "a giant's checkerboard, in all the colors of late summer."

Tony gave informal talks to the Scouts as they hiked or gathered around the campfire. He told Merle about the trees, and at night he named the stars, instilling in his son a lifelong passion for "the romance of astrophysics."

How far away is the edge of the known universe? Merle asked.

"Five thousand light years, probably," Tony said.

What's out beyond it?

"Nobody talks about ten thousand light years," his father replied. "These are very faint stars. Even the biggest telescopes don't go any further."

The Scouts learned knot tying and the history of the American flag. If any of them accidentally used words that could not properly appear in print, they had to submit to a strict punishment: a cold cup of water poured down the sleeve.

They learned the Scouts' rally: "Be Prepared!"

And the customary reply, as a chorus: "Zing-a-Zing!"

It might have been 1913, shortly after Merle, age eleven, purchased his first Boy Scout Diary for twelve cents and began jotting notes, that his lifelong affinity for exclamation points began: "Sunshine!" "RAIN!" "Buckwheat cakes!!!"

The Scouts held a fundraiser to buy uniforms, and with the leftover cash ordered equipment for an exciting new technology: wireless radio.

The uniforms were spiffy, but Hugo Gernsback's Electro Importing Company of New York was slow as molasses in shipping out the radio gear. Nearly two years passed before the boxes of wires and baffling doodads arrived, and by that time the older Scouts had lost interest in wireless radio, whatever that was.

"Let's see what they've got anyhow, see if we can put some of this together," Merle told Ernie.

For a year, the two boys puzzled over byzantine instructions and struggled to build their own parts. They made an antenna using two-by-fours and fishing poles, stringing wires from the chimney over the hayloft. Tony helped them dig under the foundation of the house to ground the wire in case of lightning.

They set up a receiver just across the river in Beloit, Iowa, and Merle tapped out the first message to Ernie in Morse code: "Have the folks got there yet?"

Canton's former mayor was so impressed that he penned a lengthy column about the episode in the local paper, the *Dakota Farmers' Leader*. "A Great Accomplishment," the headline read. "Local Boys Construct and Have in Operation Wireless Outfit Between Canton and Beloit." The front-page article gushed over the two "young boys without experience or expert knowledge" who'd solved "one of the most delicate scientific problems of the day."

Late at night, Merle put on the Brandes headphones hanging by his bed on a peg—years later, he was buried with them—and listened to the Morse code floating through the night, the signals received by a unit on his desk and then "detected" (converted to sound and amplified) by a metallic mineral crystal.

"Merle, now you stop this and go to bed," his mother warned, when he stayed up too late tinkering. "You won't be any good tomorrow."

When the inventor Lee de Forest released his Audion tubes, which replaced crystal detectors, Merle had to have one. These glass vacuum tubes looked like small light bulbs. Their vacuum seal helped electricity flow. Also like light bulbs, they had wire filaments that glowed when plugged in. Vacuum tubes were so delicate they were hung upside down so the filaments didn't sag.

To buy them, Merle and Ernie spent a summer building sand greens for a local golf course, callousing their hands. With the new tubes, they picked up radio signals from overseas and listened to war news out of Germany during World War I.

Merle grew so expert at transcribing Morse code that he could carry on conversations while he was doing it, a skill that nourished a rude but strangely impressive lifelong habit. Since he could listen and reply at the same time, he'd answer people's questions while they were still talking to him.

Merle's voice turned to bass, and he sang in the glee club. He began a love affair with his voice teacher, a women some twenty years older. He let his grades slip and pondered a singing career. Tony was disgusted: his son had no sense of urgency.

"I haven't observed that you have any natural tendency to overwork," his brother remarked.

In 1918, Merle's complacency was shattered.

On June 2, on the gorgeous lawn of the Tuves' Third Street home, the trees planted fifteen years earlier when Merle was an infant were mature and sumptuous. Rhododendrons bloomed. Merle recalled the "big, long tables, planks on sawhorses, gallons and gallons of ice cream, strawberries and cake."

Tony and Ida were celebrating. It was their twenty-fifth wedding anniversary, and Tony had also just raised a half a million dollars for Augustana College. Church sponsors were thinking of relocating the school, now going strong after years of hardship, to Sioux Falls. But with the extra funds, Tony felt sure he could persuade them otherwise.

The next day, at age fifty-four, Tony boarded a seven a.m. train to Fargo, North Dakota, to discuss the matter with church leaders. Merle never saw his father again.

2
EMERGENCY MODE

I am neither insane nor expectant of death."

Merle Tuve, age twenty-three, scribbled the morbid note at the bottom of his last will and testament. Should he die by sudden illness, the document explained, his insurance would pay seven thousand dollars, and he wanted his mother Ida to have the money after his debts were settled. But if he died in an accident, the payout was higher: a total of twenty-two thousand dollars from Northwestern National, Phoenix Mutual, and Aetna. And if a railroad accident somehow killed him, Aetna would owe even more.

His sister Rosemond and brother Lew would have some of that potential payout. For college, another portion would go toward his brother Richard (the one he baked in the oven). Merle didn't want a costly funeral. The lawns of Princeton University, where he'd just spent a year as an instructor, could have his ashes.

"Don't make a fuss about it, nor ask any permissions," he advised. "Just scatter my few grains of dust where they may rest happily!" Merle's natural tendency to overemphasize things now extended beyond exclamation points: "unlikely but yet <u>possible</u> contingencies," "<u>funeral</u>," "<u>cremated</u>," "<u>greatly desire</u>."

He may not have expected to die, but twenty-three-year-olds don't normally take out three life insurance policies. Young men don't usually worry—or worry so responsibly—about what would happen to their families without them.

Slim, dapper, and handsome, Merle now stood nearly six feet tall, with his short neatly trimmed cut parted carefully in the middle. His fine posture made him look even taller. His smile seemed a little thinner than before, and his blue eyes looked a bit less warm — preoccupied maybe, but deadly serious.

Weakened by despair, Merle's father had died of the flu in 1918, when Merle was seventeen. After the Augustana board ruled to relocate the school from Canton to Sioux Falls, despite his fundraising, Tony fell ill so suddenly that he had to be removed from the train home. For six weeks, he fought for his life at a hospital in Minneapolis.

Here was a wake-up call for a teenager. For political reasons — quickly, unfairly, illogically — his father's life work, and then his life, were snatched away. One moment Merle's childhood was literally a picnic, ice cream, strawberries, cake, and the next it was a cold funeral procession over the Sioux River by the Iowa bluffs.

Suddenly the world had fangs.

For several months, Merle's new reality didn't hit home. The change was too abrupt. Ida squirreled away what she could of Tony's insurance money (after the funeral costs), and prepared to raise and educate four children by herself. For many years, before going to bed at night, she kissed her late husband's photograph.

That fall, Merle still showed little direction. While friends went off to college, he found a job selling dry goods, ready-to-wear clothing, and women's hats. He dreamed of marrying his voice teacher and joining an opera company.

It wasn't until Ernie Lawrence came back from St. Olaf College that first Christmas, spinning tales of his new intellectual life, that Merle understood the gravity of the situation. If he wasn't careful, he'd get stuck selling corsets and diaper cloth.

Just as his father once studied during breaks from plowing, Merle took to hiding a textbook in a bolt of cloth behind the counter. "Well," Ida said, "you can go to college for as long as you can afford it — for as long as we can make it."

Merle enrolled at the University of Minnesota. He scraped through a semester at a time, borrowing money and never knowing whether he could afford to finish. "I am sorry," he later wrote to Ernie's father, to whom he owed money, "but my life has been almost a continuous succession of financial emergencies." He drafted what he titled a "Financial Statement," reconsidered the meaning of that phrase, and then crossed out "Financial" and wrote "Debit."

"Life job, I guess," he noted tartly.

University classes could be just as stressful. Education *hurt*. Merle had to learn how to learn. He bought a manual called *How to Study* that had sections like "Importance of the Questioning Habit." His mother and older brother encouraged him to push through the confusion. "Don't be scared of things," they'd say.

Racking his brain over a baffling assignment felt a bit like assembling a wireless radio. It all looked like gibberish at first. But if you fought the thing and kept at it, the picture slowly came into focus. Merle described a new desire "to <u>learn</u>," took ample and "<u>very valuable</u>" notes, and jotted down lengthy to-do lists.

He began to do everything with urgency.

"Can you come down?" asked Gregory Breit, a physicist at a research institute in Washington, D.C., called the Department of Terrestrial Magnetism, or DTM.

"I want to see if you receive short waves up there in Baltimore."

Breit had phoned Merle Tuve with an experiment in mind. He remembered Tuve, now a doctoral student in physics at Johns Hopkins University, from jogging laps around the athletic track back at the University of Minnesota. Breit needed his help sending radio signals between DTM and Baltimore, thirty-five miles northeast.

Radio waves, both men knew, somehow bounced off the electricity in the upper atmosphere. That had to be, the reasoning went, how radio beams could reach around the curved globe. But no one had proven it. Breit wanted to erect a big dish antenna on DTM's lawn, ricochet shortwave signals off the atmosphere some ninety miles up, and see if Merle could receive the signals at Hopkins.

Tuve didn't think Breit's plan would work. When he had tinkered with short waves in Minnesota, it was hard enough to pick up signals thirty feet across the room. It might be rather difficult to detect them across some 475,000 feet.

"Hold on a minute," he told Breit. "There's another way."

Instead of short waves—which, it turned out, would have gone right through the upper atmosphere—Tuve suggested they transmit long radio waves, and also send the signals out in pulses, which would make them easily recognizable.

Using Tuve's innovative method, he and Breit discovered the first radio wave pulses echoing off the atmosphere—a historic achievement that, it would be widely acknowledged, should have won him a Nobel Prize. He was only twenty-four.

Tuve continued his various echo experiments using advanced Navy equipment, setting up a receiver on the roof of the administration building at the Naval Research Laboratory, in Southwest Washington, D.C., by the Potomac River.

The "only damn trouble" was an airstrip a mile north. When planes came in for landings at nearby Bolling Field (a navy base), Tuve couldn't get a clean bounce-back signal because the radio pulses echoed off the airplanes. Waiting around for four or five minutes every time an aircraft touched down "got to be quite a nuisance." As far as twenty miles away, he could tell when planes were approaching.

Annoying. And also the birth of radar.

A young Navy lieutenant, Deak Parsons (whom we'll meet later), heard about the "problem" and had the idea of detecting aircraft by radio waves classified in 1932.

Tuve's research on the ionosphere was far more than a stellar graduate thesis. It was impressive enough to land him a job at DTM, where he soon developed a reputation as a bright new talent.

By November 1931, Merle was a successful experimental physicist. Park Savings Bank sent him his due congratulations. "Your account with the Bank appears to be overdrawn $0.16," the note said. "Your immediate attention will be much appreciated."

The 1930s were a golden era for American physics, and Tuve was smack-dab in the middle of the excitement. It didn't hurt that the Department of Terrestrial Magnetism's parent organization, the Carnegie Institution of Washington, was flush with research funds. DTM scientists were spoiled rotten.

Five miles north of downtown D.C., DTM stood proudly atop a hill off Rock Creek Park on a seven-acre tract. The rural site was chosen for its distance from "disturbing influences" like electric streetcars. To reduce magnetic interference, its main building, styled in Italian Renaissance brick, employed a minimum of steel. A wooden building allowed for magnetic instrument standardization. No expense was spared to ensure that scientific measurements were precise and pristine.

DTM ran experiments and chased hunches. They *built* stuff.

Tuve had vowed not to become "one of these contented professors" with an easy job and "no realization at all of how little they are accomplishing."

"If I go into academic work I shall go into it to *do* things," he promised.

DTM had no students. What it had were research labs and machine tools. At the time, physicists were theorists *and* mechanics. They learned glass blowing and lathe work. You had to build cutting-edge instruments to run your experiments.

Tuve joined the new frontier of nuclear physics. He was an atom smasher, prying open the secrets of nuclei by launching high-speed particles at them. The more energy a particle accelerator could throw at an atom, the more you could learn. A race for volts was on.

Merle climbed a slapdash ladder to assemble an eighteen-foot Van de Graaff accelerator on the DTM lawn. It looked like an alien water tower. To keep electricity from leaking, he sank a Tesla coil in a drum of oil, wading around waist-deep in the muck to adjust it. He assembled magnets, cogs, pumps and bypass condensers, filaments and batteries and silk support lines.

To photograph atomic particles, he installed an intricate rig with angled cameras, a cylinder "cloud chamber," vacuum pumps, and tangles of plugs and tubes.

Physicists of the era, as one scientist explained, had to pit their brains "against bits of iron, metals and crystals" to force their fickle contraptions to work. An experimental physicist, another said, was a "Jack-of-All-Trades, a versatile but amateur craftsman. He must blow glass and turn metal . . . carpenter, photograph, wire electric circuits and be a master of gadgets of all kinds."

And so Merle squatted atop a room-sized generator, sporting a three-piece suit, holding a wrench. He perched on a desk to adjust a parabolic antenna that would send a beam of light into space to explore the stratosphere. He crouched under the huge particle accelerator that resembled a twenty-foot metal spider boring a drill into the floor, aluminum rings running up the shaft like vertebrae. He leaned over convoluted apparatuses expectantly, like a mother over hatching eggs.

DTM was a place of high adventure.

Tuve's institute was running projects in Australia and Peru. Its scientists studied magnetism under the Earth's surface, the strange energies of the volcanoes above it, and the atmospheric electricity at the edge of space. At one point, DTM even had a sailing vessel, the *Carnegie,* which roamed the globe measuring magnetic currents. (It blew up, as science experiments sometimes do, in Samoa in 1929.)

One of Tuve's closest collaborators at DTM had recently spent three years on an Arctic expedition. He'd been to Baghdad, tried to fly a plane over the North Pole, and caravanned across the Sahara. An explorer in every sense of the word, he'd joined up with Tuve for a simple reason: that's where the action was.

Merle compared scientific research to "gambling all the time."

By 1935, his betting skills were widely known.

By 1939, Tuve had spent over a decade at DTM. Big man on campus, he'd just overseen the construction of one of the most powerful particle accelerators in the world. Encased in a fifty-five-foot-high sheet-metal egg resting inside a concrete cup in the middle of DTM, the contraption resembled a giant's frugal breakfast. Under fifty pounds of air pressure, its generator produced five million volts.

In January, as Tuve and Nobel laureates Enrico Fermi and Niels Bohr attended the fifth Washington Conference on Theoretical Physics, news arrived that German scientists had split the uranium atom by bombarding it with neutrons. Within hours, Tuve had DTM's accelerator confirm the results. *Nuclear fission.* Splitting uranium released some of the tremendous energy holding together the neutrons and protons in its nucleus: equivalent, it turned out, to two hundred million electron volts.

In a photo snapped of seven physicists at DTM after the German results were replicated, Enrico Fermi grins. Only Merle, unsmiling, stares worriedly off to the side.

A media blitz followed, dragging Tuve into the public eye. Coverage went far beyond New York, D.C., and Los Angeles. "Dr. Merle Tuve" showed up in dailies from Tampa, Florida, to Salem, Oregon, Lubbock, Texas, and Butte, Montana.

He was not yet forty.

And he needed a break. Financially, Merle finally had some slack. But he felt overworked, and had a family to think about now. He had married Winifred Whitman, a psychoanalyst and a college crush, in 1927. Merle called her Winnie; she called him Bug. Their boy, Tryg, was five, and Lucy was only seven months.

Neither wanted their children to grow up as Easterners. In small-town America, people just seemed nicer. Raising kids was a community effort.

"The 'feel' of the people in the West is so much more cordial," Merle wrote to a friend, "and the people are so much more like real people."

"We'd like our own kids to know what it's like." Maybe, he imagined, he and his family would buy a second home somewhere in the great West, like New Mexico.

Tuve dumped as many obligations as he could. Too much time away from his family at conferences had worn on him. He cherished the little telegrams Winnie sent him. "All ok but lonesome." And the note Winnie and Tryg wrote together. "Don't go away again! Please!" And the one from Tryg with a sketch of a gentleman with a mustache. "I'll be this big if you don't come home soon."

And the simplest note of all: "I want my papa."

Yet there was another document Tuve treasured. He'd kept it tucked away, underlined and folded in eights, since his childhood in Canton. "The Union Soldier," a speech delivered by Nebraska senator John Thurston, told the story of an aging Civil War veteran. Not "born nor bred to a soldier's life," the veteran was called to arms, like so many Americans summoned from "the plow, the forge, the bench, the loom, the mine, the store, the office, the college, the sanctuary." He cherished peace and "quiet ways" and yet "left good-bye kisses on tiny lips," and "when the war was over he took up the broken threads of love and life as best he could." He answered the call of duty and accepted its burdens.

In September, 1939, Hitler invaded Poland and the United Kingdom and France declared war on Germany. By the end of the year, front pages of newspapers, above quotes from Tuve on atomic physics, brought grimmer news of British and Nazi naval battles, and a desperate fight for air mastery with the Luftwaffe.

The world was imploding with, for Merle, a familiar logic. Tony Tuve's son had never been crazy. He simply understood catastrophe—sudden, unfair, illogical disaster. He knew exactly how to prepare for it, and how to cope with its pitfalls.

"Free speech and the radio," Merle wrote to Ernie, now a physicist at the University of California, "has put me solidly behind England already."

"I'll go into war work as soon as it becomes proper for U.S. citizens."

He was entering the most intense period of his life.

3
SCIENTIFIC SPIES

On October 17, 1939, an emaciated, sickly man named Bill Sebold walked through the doors of the American consulate in Cologne, Germany. His creased face, weathered from years of outdoor labor, looked sunken and haunted.

Born in Germany, Sebold was a naturalized U.S. citizen, and this time he was determined to be heard. A few weeks prior, he visited the very same consulate, with no luck. Since his arrival in Germany in January, as he had politely explained to a doubtful clerk, Nazi intelligence had been aggressively recruiting him, with the full intention of sending him back to the United States as a spy.

He had a background in airplanes and machine science.

"Too bad," said the clerk, who could have been nineteen. "You had better run away." Better get out of Germany as soon as possible to neutral Belgium.

Panicking, Sebold had rushed into the streets of Cologne and flagged down cars with foreign license plates, begging to be smuggled over the border. Failing that, he considered escaping by railway, but got cold feet at the train station. Gestapo seemed to be following him. Strange men stood and stared at him.

This visit, Sebold would not waste time with low-level clerks feeding him haphazard advice. He asked for, and was granted, a face-to-face with the head honcho, the consul general. But Alfred Klieforth, a thin-lipped, bushy-eyebrowed functionary, was skeptical of Sebold's evidence: a letter from his German handler, one "Dr. Renken," asking him to travel to Hamburg for spy training.

"We cannot help you," Klieforth said. "You are in a bad spot. You have to know yourself what you are going to do. We have nothing to do with this. But I have to take this letter and copy it and send it to the State Department."

"Mr. Sebold," Klieforth wrote in his dispatch to Washington, "claims that through a strange set of circumstances he is now at the mercy of certain German secret organizations interested in the production of American military aeroplanes."

"He is in fear of his own life."

Bill Sebold was not spy material. He'd just wanted to visit his mother in Germany. But to Nazi intelligence, he appeared to be an expert airplane engineer. Maybe it hadn't been such a good idea for him to blab so freely to that group of Hitler Youth on the SS *Deutschland* during the sea voyage over from New York City.

It was true that he worked briefly for the Consolidated Aircraft Corporation of San Diego, a military contractor that made seaplanes for the U.S. Navy. But he was little more than a "bumper" at Consolidated, shaping aluminum parts. He'd dabbled in invention, worked as a junior engineer on an oil tanker, and knew his way around diesel engines. But he wasn't some sought-after expert. Since arriving in the United States in 1929, he had toiled as a gold dredger in Alaska, a maintenance man, and a porter and elevator operator on Manhattan's Upper West Side.

Sebold's Nazi handler, Dr. Renken, did not think much of him. "He did not particularly impress me," Renken noted. "He looked quite ordinary, was of average intelligence, and obviously came from modest means." After Renken had Sebold undergo a brief physical, the examiners diagnosed the frazzled, overwhelmed mechanic with similar disdain. "This guy can't hurt a fly."

The day after his physical, Sebold had a nervous breakdown. He was admitted to St. Marien Hospital in Cologne, and remained there for nearly two weeks.

Sebold had a weak constitution and a shaky temperament. There was very little reason for the Nazis to believe they could trust him. He had served in the German army in World War I (not uncommon among new

Americans of German heritage). But why would he stay loyal to his birth nation over his adopted home? His engineering expertise was limited. He had a criminal record and once even assaulted two police officers. He had spent time in a mental hospital.

Even his Nazi code name, Tramp, reeked of disappointment.

That Dr. Renken and his superiors were recruiting a dud like Bill Sebold, aggressively and carelessly, spoke to a terribly urgent fact: in the brewing military conflict, whichever side won the battle in the air would win the war.

Renken, a plump, pampered man with a prominent gold tooth, was actually Nikolaus Adolf Fritz Ritter. He was a chief of air intelligence for the Abwehr, a clandestine service that gathered data on foreign militaries. To steal secrets for the German air force, he was organizing a spy ring inside the United States.

Ritter's job was to supply the Luftwaffe with information about American breakthroughs in bombing and antiaircraft technology.

After meeting with Consul General Klieforth, Sebold waited. So did England and France. While the German army ruined Poland, a smaller Nazi force held a fortified defensive barricade running along the French border. Manning the other side and the border with Belgium, French and British troops readied for an attack. England dropped propaganda leaflets over Germany, hoping for peace. Sporadic dogfights and fracases punctuated the uneasy calm. The British press called it the *Sitzkrieg*, the "war of sitting." The French dubbed it the *drôle de guerre*, the "funny" or "bizarre" war. To Americans, it was known as "the phony war."

On December 6, Sebold awoke in darkness, in his hotel room north of Düsseldorf, to find a stranger in the shadows. "I am your uncle Hugo," the man said.

The Luftwaffe needed him in America as soon as possible.

Still quite ill, Sebold was shepherded to the spy school in Hamburg. He studied Morse code and ciphers, was trained to shrink secret blueprints into microphotographs, and learned how to adjust a microscope to read microdots—miniature specks of text about the size of the period at the end of this sentence.

Renamed "Harry Sawyer," Sebold was booked on an ocean liner out of Genoa, Italy, and given instructions to find work at an American aircraft factory.

Crossing the Atlantic, on his way back to New York, the Nazi recruit pondered bad options. Through the American consulate, Sebold had put in a request to be met in the United States by FBI G-men. But maybe, instead, he should kill himself? Or what if he ran away with the Nazi espionage money? He gave his spy materials to another passenger, reconsidered it, and then snatched them back.

On February 8, 1940, Sebold's ship docked at a quarantine station in sight of the Statue of Liberty in New York Harbor. As health inspectors made the rounds, FBI special agent Albert Franz searched for Sebold among the four hundred and twenty-seven passengers — three hundred of whom were Jewish refugees escaping from Nazi Germany.

Franz asked Sebold to accompany him to an FBI office in Foley Square, near City Hall. They disembarked quietly and shuffled past reporters waiting for the Irish novelist Liam O'Flaherty. Two FBI cars trailed Sebold and Franz downtown.

FBI director J. Edgar Hoover would personally brief President Franklin Roosevelt on the story Sebold shared with interrogators over the next two days. The mechanic spilled the entire bizarre tale of his recruitment, spy training in Hamburg, and mission in New York. In his possession were nine hundred and ten dollars in small bills and a book, *Hawkins Electrical Guide,* for operating a clandestine shortwave radio.

Inside the back of Sebold's watch, FBI investigators discovered the crown jewels. Microphotographs, with detailed instructions, were to be delivered to three German agents already in New York: Fritz Duquesne, on West Seventy-Sixth Street; Lilly Stein, at East Fifty-Fourth Street; and Everett Roeder, in Merrick, Long Island.

The instructions listed the military secrets the Luftwaffe craved:

- The U.S. Chemical Warfare Service may have devised a new method of protecting military clothing from chemical attack. Find out if the chemical solution is still being tested. If it is being produced, find out by whom. "What is the chemical make-up of the new protection?"
- "Find out everything possible about new developments in the line of anti-aircraft guns."
- "Is there any-where in the States an anti-aircraft shell with so-called 'Electric Eye' being manufactured?" If so, find out everything you can about its testing, the shell, caliber of the round, and how it is fired. "How has the complicated inside mechanism reacted to the firing off shock?"

- Obtain several sales catalogues from the General Electric Company and RCA for their radio sending and receiving vacuum tubes.
- Get a copy of the espionage law.

Poor Bill Sebold agreed to be the first FBI "counterspy," or double agent. He pledged to help dismantle Nikolaus Ritter's New York spy ring.

Sebold had exposed vital military secrets. Whoever selected the targets inside his wristwatch had been carefully advised by German scientists. The Nazis, the instructions showed, sought a fuse with an "electric eye"—a proximity fuse—that could be fitted inside a shell and shot from an antiaircraft gun. They were researching chemical warfare. And, clearly, they had no qualms about placing spies in U.S. factories.

If America entered the war, any group coordinating science and industry for the military would have to do so with utter discretion.

While the Germans conspired to steal American technology, and Merle Tuve prepared to join the war effort, British intelligence tried to appraise the current state of Nazi military science.

At 54 Broadway Street in London, a brass plaque identified the occupant of the dingy building near Buckingham Palace as the MINIMAX FIRE EXTINGUISHER COMPANY. Inside, behind frosted-glass windows, Reginald Victor Jones glared suspiciously at a mysterious cardboard box on his desk. Fred Winterbotham, head of air intelligence for Britain's foreign intelligence service, MI6, had delivered the parcel still wrapped. Jones knew quite well the package could be a bomb.

"Here is a present for you!" Fred told him.

R.V., as friends called him, was a twenty-eight-year-old harmonica enthusiast. Son of a postman, with a muscular face and wavy hair, he looked less like a physicist than a poster boy for Savile Row bespoke tailors. He had gone to Oxford on scholarship, graduated with top marks, and was a crack shot with a pistol.

He was also a prankster who indulged in telephone hoaxes. At Oxford, he once posed as a telephone repairman and convinced a graduate student—in order to "test the line"—to sing loudly into the receiver, stand on one leg, dip his hand in a bucket of water, and then submerge the receiver in the bucket.

R.V. was playful enough to take some pleasure in it when the shared secretary, Daisy Mowatt, intentionally inserted mistakes into his typed reports out of her own sense of mischief. On one occasion, R.V. answered the phone to discover that Daisy had been indulging her quirky sense of humor with a total stranger.

"Is that really Dr. Jones?" an upset voice asked. "I have just been talking to a most extraordinary lady who asserted that you had just jumped out of the window!"

"Please don't worry," R.V. replied, without skipping a beat. "It's the only exercise that we can get."

The sealed box on his desk was accompanied by seven typed pages, in German, along with a translation added by the British embassy in Oslo. Inside the box, according to the letter, was part of a Nazi secret weapon.

The provenance of the package was not comforting.

Captain Hector Boyes, British naval attaché in Oslo, had received an anonymous note asking if new German weapon developments would interest him. If so, the tipster requested that the BBC broadcast into Germany begin, in place of the usual preamble, with the words "Hallo, heir ist London." The BBC made the change and the mysterious item followed by post in early November, 1939.

Why on earth, Jones wondered, did the source go through the whole charade with the BBC? If the intelligence was useful, why not simply send it along? No one knew who the source was, if someone was playing a trick, or what was in the box.

Gingerly, he opened the parcel.

Inside, after noting with pleasure that he had not been blown up, R.V. found a little glass tube. It was a vacuum tube, a prototype meant for an antiaircraft fuse.

"A proximity fuse is being developed for anti-aircraft shells," the accompanying letter said, in Jones's summary. The new fuse "required the development of an electronic trigger tube, a sample of which is enclosed in the accompanying package."

Jones didn't know it, but his tipster had just confirmed Bill Sebold's Luftwaffe wish list. No wonder the Nazis wanted to learn how "the complicated inside mechanism reacted to the firing off shock." The Germans were *actively* developing a new prototype that could explode by sensing a nearby aircraft. But they had not been able to figure out how to make its electronic parts strong enough to withstand the tremendous blast of an antiaircraft gun.

Not yet, anyhow.

R. V. Jones was not a prominent figure within British intelligence circles. He was a junior adviser on loan from the Air Ministry, a humble liaison with a shared secretary. He was so far down the totem pole that he didn't meet Stewart Menzies, the new chief of MI6, for a year. He did meet the vice chief, who sought him out, for a favor, after hearing that "there was now a scientist in the office."

There is now a scientist in the office. Apart from the code-breaking mathematicians, Jones was the only research scientist working for MI6. The little glass tube and the anonymous letter, which came to be known as the Oslo Report, were dumped on his desk because no one else there was qualified.

By default, Jones had become MI6's go-to analyst on Nazi military science. Earlier that year, when Hitler publicly bragged of a secret weapon that would blind and deafen its victims, Jones was the one assigned to figure out what it was. (He concluded it was a false alarm. By blind and deafen, Hitler meant "render thunderstruck." And by weapon, *waffe*, Hitler meant a military branch, not a specific weapon.) Jones also sorted out, from wild rumors, the Nazi threats to take seriously: bacterial warfare, new chemical gases, and "pilotless aircraft."

Jones was in desperate need of help. His purview was too vast and complex. Intelligence work was tricky enough without mixing in science. Analysts already had to parse maddening ambiguities to judge a morsel of data. Is the source credible? What are their motivations? Is it a diversion? Is the intelligence stale? Is the threat important enough to devote resources to?

Add to this unstable pot physics, chemistry, and advanced engineering, and the uncertainties amplified. Even assuming a source was trustworthy and well placed, they might not grasp the minutiae of cutting-edge science. There was simply no reliable way for nonscientists to evaluate highly technical intel. And the scientists who consulted with MI6 were not trained in intelligence analysis.

The "first watchdog of national defense," Jones wrote later, should be a "constant vigil for new applications of science to warfare by the enemy." In 1939, the field of scientific intelligence did not exist.

In December he submitted a bold proposal to create a "Scientific Intelligence Service." He asked for analysts with technical backgrounds to be placed in the intelligence services of the army, Royal Navy, and Royal Air Force.

Any new weapons the Nazis were developing would have to move from small trials, to manufacturing, to large-scale field tests before they were put to use. The British, he argued, could cultivate sources within each stage of production. Among Nazi scientists, they'd certainly find babblers, drunken braggers, mistresses, non-German nationals from occupied nations, and generally sympathetic ears.

He distributed his proposal, and waited.

John Buckingham, deputy director of scientific research for the Royal Navy, opposed Jones's plan. Buckingham preferred, as analysts of technical intelligence, the existing scientific specialists who were working on British weapons.

Jones would not receive any of the staff he requested. Nor would he be allowed to retain a single assistant or even to hire his own personal secretary.

He got even less traction with the Oslo Report.

The anonymous letter that accompanied the boxed glass tube warned of many other frightening Nazi projects. The letter's details alarmed Jones, and some of its claims had been verified. Germany, the source reported, had developed new forms of radar. The Nazis were allegedly building a pilotless aircraft, the FZ-10. And the German navy, it said, was designing remote-controlled gliders, FZ-21s, at an experimental station —a place called "Peenemünde, at the mouth of the [river] Peene, near Wolgast, in the vicinity of Greifswald."

Whoever wrote the Oslo Report was a true expert, someone with advanced and wide technical knowledge—suspiciously wide, maybe. Its signature read, simply, "a German scientist who wishes you well."

Again, John Buckingham pushed back, dubbing the report a "plant." Its seven typed pages detailed a plethora of secret plans. No single source, Buckingham reasoned, could have access to such a broad range of projects.

The Air Ministry, the Admiralty, and the War Office thought so little of Jones's analysis and warnings that they had their copies of the Oslo Report disposed of.

Nobody took him seriously.

For the time being, Jones kept the Oslo Report in his own files. And somewhere in the back of his mind, he tucked away the name of the alleged experimental research base that no one in MI6 had ever heard of before. "Peenemünde."

4
THE WIZARD

Vannevar Bush was not supposed to testify about Nazi airplanes in 1939. He was only the vice chairman of the National Advisory Committee for Aeronautics, or NACA, which conducted aeronautics research for the military. But the chairman was sick in the hospital, and Bush was next in line to lobby the Senate Committee on Appropriations for cash. NACA desperately needed a new research facility, in California, to compete with the rapid advances of the Luftwaffe.

The threat of airpower was not to be taken lightly. Bush saw that new lethal aircraft were a product of advancing science. But he also felt that while "science has produced a weapon, so also can it produce in time a defense against it."

Forty-nine, elegant, wiry, with rimless glasses and gray accents in his thick black hair, Bush was not well suited for the hoopla of Capitol Hill.

His entire life, he had been prone to anxiety and illness. Raised in Massachusetts by a minister who "hit a hard punch," Bush had a childhood punctuated by unfortunate maladies: rheumatism, which left him dragging his left leg for a period, typhoid fever, a ruptured appendix. In college, he was too frail for football so he managed the team. He ran track poorly, by his own admission. Bush was not only a bit of a hypochondriac, he was a prima donna.

As president of the Carnegie Institution of Washington, Bush was Merle Tuve's boss's boss. He was also among the most famous scientists in the country, and was certainly one of the best connected. Bush was "emperor," *Time* wrote, of the "biggest scientific empire under one management in the world."

The Carnegie Institution had an endowment worth, in today's currency, over seven billion dollars. In 1939 alone, Bush oversaw investigations of cosmic rays, gaseous rocks, the evolution of early humankind, and the rotation of the galaxy. His outfit was running projects in Guatemala and Java. Bush himself had just returned from a romp through the Yucatán, where he and his wife were met with flowers from Governor Humberto Canto Echeverría. Carnegie funded Mayan excavations, and floated observers down the Amazon. Bush's scientists peered into deep space to study fast-moving nebulae hundreds of millions of light years away.

Naturally, Bush was accustomed to a particular lifestyle. He commuted to the Carnegie Institution's D.C. headquarters by chauffeur from the Wardman Park Hotel, where he complained about the bathroom light, too dim for shaving; the erratic water temperature, which made showering "almost impossible"; and the refrigerator which, his cook reported, struggled to freeze ice cubes.

Intellectually, Bush's résumé was eye-popping. A former vice president of MIT and a founder of Raytheon, he was an inventor with a quirky collection of patents. Bush designed and built one of the first reliable machines, replete with gear shafts and motors, to solve complex mathematical equations. (Today we call them "computers.") He once constructed a land surveyor using bicycle wheels. He was a master of electrical circuits who carved his own tobacco pipes.

In his free time, Bush tinkered with color photography. He would ponder improvements to engines, fishing rods, archery bows, and even to a bird feeder with a spring for dumping off thieving pigeons. He accepted honorary degrees and glad-handed university presidents and captains of industry. He hobnobbed at top social clubs and societies, chatted up famous pals like the aviator Charles Lindbergh, and talked gadgets with Orville Wright, of the Wright brothers.

Five American presidents would call him Van because, he joked, they couldn't figure out how to pronounce his full first name. (*Vannevar* rhymes with *achiever*.) He signed letters with an iconic *V*. To a pal at MIT, the inventor was, simply, "Wizard."

Already in 1939, he was a revered figure.

Bush's testimony before the Senate, in April, was not the first time that he had dealt with politicians who seemed numb to the danger of German science. Two months earlier, in February, he had testified before the House of Representatives on NACA's proposed expansion. At the time, Virginia congressman Clifton Woodrum had peppered him with a barrage of single-minded questions.

Woodrum, a trim, sour-mouthed pharmacist who often led fellow congressmen in singing the tune "Carry Me Back to Old Virginny," did not think much of Bush's request for funds. Woodrum was sure NACA already had a research facility. It was called Langley Field, and it was located in his home state of Virginia.

"Germany is building better planes for military purposes than we are," Bush told Woodrum. "Germany has now at *one* station," he said, "more personnel and more facilities than we have in this country for aeronautical research."

"They have five major stations."

"What is the reason," Woodrum asked, "for thinking that the facilities can be better increased by the building of a new station" so far from Langley field?

There wasn't enough land at Langley field.

"There is plenty of land available," Woodrum bluffed.

Bush and the other NACA reps had kept at it. Not only was there not enough land at Langley—no, *really*—there wasn't enough power for the new wind tunnel. Besides, the military preferred two research sites, in case of attack. And the heart of the aircraft industry was in California, which made liaison work easier. And there was already an airfield at the proposed site. And the weather allowed for year-round flight tests, with nearby mountains protecting the area from fog.

"You are figuring on war," a congressman said.

No, NACA explained, but Germany was manufacturing planes that flew four hundred and twenty miles per hour, far faster than U.S. aircraft. And while the existing American wind tunnel could only reach speeds of five hundred miles per hour, the Germans' could reach seven hundred. A single research station in Berlin employed sixteen hundred people, while Langley Field had only four hundred and fifty employees. Germany was spending between four and six times more on aeronautics research than the United States. And the Germans had done away with engine carburetors and switched ingeniously to direct fuel injection.

Future wars, Bush felt—scientists at NACA felt—would be won in experimental labs long before military technologies ever reached the battlefields.

"Why could you not do that work here as well as out in California?"

Bush's reputation as an engineer seemed to count for very little.

Now, on April 5, 1939, Bush discovered that his scientific credentials apparently meant even less to Woodrum's colleague Carter Glass, the Democratic senator from Virginia. Nor did Tennessee senator Kenneth McKellar appear to appreciate that Bush, a self-described "Cape Cod Yankee," was asking for millions of dollars for a facility in *California.* After all, what about Langley Field?

Bush met the senators at a narrow, thirty-foot table in the Capitol Hill committee room, under a chandelier and frescoed ceiling of Roman gods and goddesses, surrounded by murals of classical maidens in flowing robes.

"Why is it impossible to extend the research facilities at Langley Field?" Senator Glass said.

Langley Field didn't have the acreage, Bush said.

"Why cannot they be located at Langley Field?" asked Maryland senator Millard Tydings, just moments later.

What followed was a maddening, head-spinning, tail-chasing attack on NACA's plan, Bush's motives, and even his scientific expertise. Wasn't having two facilities inefficient? Why was this so urgent? Why was NACA warmongering? Why not put it *near* Langley Field? Wasn't Bush overestimating the Nazi threat?

"Do you think," Glass asked Bush, that Nazi airplanes "could go three thousand miles across the ocean and attack this country with their aircraft?"

"Not today, Senator, but it may happen tomorrow," Bush said.

"It will not happen in a thousand years from now," Glass replied.

Did Bush really think Nazi airplanes could reach America? McKellar asked.

"Senator, when I was a boy," Bush said, trying to explain the nature of scientific discovery, "we did not even think it possible that a man could ever fly."

"I am not talking about your boyhood," McKellar spat.

Senators Glass and McKellar prattled on in circles, insulting, teasing, and provoking Bush until he lost his temper. Their petty bickering was infuriating.

Nazi science. Uh-huh. But what about Virginia? Bush accused the senators of bad faith, and banged his fists on the table. Committee members claimed to be offended. The back-and-forth grew so heated that Bush thought the senators might actually throw him out. After things cooled down, according to Bush, the committee had the bitter exchange struck from the permanent congressional record.

Congress denied him the money for the California facility. Bush returned home to his wife Phoebe to report that he'd lost NACA some five million dollars.

"I made a mess of it," he said, "and I took a beating."

Later that year, after some delicate maneuvering, NACA did get the money. But Bush never forgot his first hard lesson in American politics. In matters of national defense, the United States Congress was flat-out "asleep on the technical end."

During the First World War, defensive weapons were far stronger than offensive ones. Machine guns guarding trenches and bunkers easily mowed down charging troops, turning battles into stagnant, grueling affairs. Trench warfare meant static clashes of attrition between enemies who weren't all that mobile. Airplane engines were feeble, tanks unreliable, and soldiers were stuck in the mud. Cities were relatively safe, and civilian deaths were 14 percent of total losses.

That dynamic had now changed. Hitler's forces were *maneuverable.* Germany's Blitzkrieg or "lightning war" on Poland proved devastatingly effective. Rapidly advancing armored vehicles, backed by air support and motorized infantry, gobbled up foreign lands with mammoth, swift, mechanized pincers.

Germany reached Warsaw within a week. Poland's air force was pulped in two days. Sad shards of old biplanes littered unused runways. Polish troops on *horseback* were left to defend against Nazi tanks.

Offensive weapons were suddenly more powerful than defensive armaments. Luftwaffe bombers could carry heavy payloads over vast distances. In late September, in the largest air raid in history yet, the German air force dropped 560 tons of high explosives and 72 tons of incendiary bombs on Poland. The terror bombing of Warsaw wrecked much of the capital and cost thousands of innocent lives. Civilians, it was clear, were newly exposed: civilian losses during the Second World War would soar to a stunning 67 percent of all deaths.

The Soviet Union and Germany were busy carving up the globe. Adolf Hitler had already annexed Austria, and following the Munich Agreement, he assumed control of western Czechoslovakia. Then he invaded the rest of that nation unopposed and "retook" part of Lithuania lost after World War I. Meanwhile, the USSR brutalized Latvia, Estonia, the rest of Lithuania, and over a long winter battle, Finland.

As the conflict stretched into early 1940—while Hitler proposed "peace initiatives" through December, January, and February—Van Bush found himself surrounded by fellow scientists increasingly worried about the war.

Bush was now chairman of NACA, among his many duties. He also sat on the Committee on Scientific Aids to Learning, an advisory group that promoted the classroom use of technologies like radio, motion pictures, calculating machines, and microphotography. Serving with him were the president of Harvard, the chairman of the National Research Council, and the head of Bell Telephone Laboratories.

The men agreed that America would be drawn into the war, that the United States was shamefully unprepared, and that the existing military labs could never produce the new technologies now in reach given the current state of science.

While the Army and Navy research laboratories did perform experiments, they were poorly funded and unfocused. No one was investing in the kind of basic research that could lead to radical innovations. In 1938, the Army devoted only 1.5 percent of its budget to research. The Navy treated radar like a hobby project.

"When military men are queried," Bush complained, in 1939, "they usually reply that 'the answer to a plane is another plane.'"

Military traditions and protocols were stubborn, and the brass weren't always inclined to listen to civilian scientists. President Roosevelt observed wryly that trying "to change anything in the Navy" felt like "punching a featherbed."

Bush envisioned a new organization of American scientists, recruited from universities and industry, to make the military weapons and devices he saw were possible. He understood quite well that his biggest obstacles were political.

Military skeptics could slow down scientific research as easily as congressional partisans. If scientists were distracted by, as Bush put it, "the mysterious ways in which one operates in the Washington maze," valuable time would be lost. The problem could not be left up to military luddites or politicians who didn't believe that bombers would ever, not "in a thousand years," be able to fly three thousand miles.

In April 1940, Germany overran Denmark and seized Norwegian ports. Leaflets fell from the skies over Copenhagen calling for Danes to peacefully accept the Nazi occupation. Swastikas unfurled in town squares and outside requisitioned buildings. German infantry marched through burning Norwegian villages.

Within weeks, Bush drafted a four-paragraph plan to marshal science to the war effort. Using his carefully tended contacts, he arranged to meet with FDR. He first passed his memo to Frederic Delano, a Carnegie trustee and Roosevelt's uncle, and secured a sit-down with Harry Hopkins, a former social worker and Roosevelt's closest aide. Hopkins immediately saw the promise of Bush's plan.

On May 10, Germany invaded Holland and Belgium. Hitler then hurled his mechanized army north of the Maginot Line through the thick Ardennes forest—a route thought to be practically impassable with armored units.

France was caught completely off guard.

Militarily, the French were disorganized. They were also outdated in their use of technology. When French supreme headquarters had to contact the northeastern front commander, the general would drive to communicate *in person*. Command was using *motorcycle* messengers, instead of radios or telephones.

It could take two days for an order to reach the front.

The "strange defeat" of France, in the phrase of one historian, was all the more puzzling because France and Italy did in fact have the tanks and airplanes to counter Germany. On paper, it seemed like an even match. Yet France swiftly fell, and Italy announced that it was joining the war on the Axis side.

Two days before the Nazis occupied Paris, at four thirty p.m. on June 12, 1940, Van Bush met President Roosevelt, for the first time, in the Oval Office.

Harry Hopkins joined them. Under the title "National Defense Research Committee," Bush presented his one-page scheme, a plan for scientists to help the military quickly devise new breakthroughs in technological warfare.

The president had already made up his mind. He scrawled "O.K.— FDR" on Bush's single sheet and the discussion was over. The meeting lasted ten minutes.

Employing an old legal authority left over from World War I, Roosevelt's cabinet quickly approved the proposal, creating the National Defense Research Committee, or NDRC (later called OSRD). Bush was named chairman.

Unusually, the NDRC would report to the White House, and not through the customary military channels. And Bush's organization would also be virtually immune from congressional oversight, drawing its cash through the executive branch.

On June 14, as Paris fell, the president quietly announced the new agency. Within days, Nazis would be goose-stepping down the Champs-Élysées.

"I do not think I have very much news this morning," Roosevelt told the press corps. NDRC, as he explained, was chartered to enlist academic and industrial scientists in the national defense. Led by Bush, they would "handle practically all research problems with the exception of the problems of flight," which was still overseen by the National Advisory Committee for Aeronautics.

"Does Dr. Bush remain the head of [NACA]?" a reporter asked.

"I suppose so, yes," Roosevelt replied.

With that, American science had an eighteen month head start to prepare for war. Bush would oversee the entire business of placing bets on scientific projects of military value. He would have to answer the most urgent military needs, predict what the Germans and Italians (and, soon enough, the Japanese) might produce, anticipate counter-weapons, revamp American industry, and convince skeptical military brass to embrace new technologies on the fly.

The creation of NDRC was greeted with apprehension. Within the military, Bush later said, many saw it as "an end run, a grab by which a small company of scientists and engineers, acting outside established channels, got hold of the authority and money for the program of developing new weapons."

"That, in fact, is exactly what it was."

In his office at the Carnegie Institution, Bush mounted a bronze plaque. Under a coat of arms, a faux-Latin motto read ILLEGITIMIS NON CARBORUNDUM, or "Don't let the bastards wear you down." The motto, a visiting scientist said, was dedicated to D.C.'s "Senators, Congressmen, and other pretentious people."

To celebrate his victory, Bush spent a late night out at the Cosmos Club, an all-male hub by the White House. Membership was limited to those "distinguished in a learned profession or in public service": statesmen, educators, lawyers, businessmen, and scientists. Bush held court past midnight.

But he was too loose with his words and opinions. Jerome Hunsaker, an aeronautics expert, pulled him aside. "Van, you talk too goddamn much," he said. "In your new position you can't talk. . . . You don't expose your thoughts."

Hunsaker was right. The old rules were gone.

5
SECTION TUVE

Eight blocks north of the White House, W. T. Ensign marched determinedly up the steps of 1530 P Street, past the Italian vases and titanic Ionic columns of the portico, under the limestone balustrade, and through the giant bronze doors.

Very soon, the inlaid marble floor under his feet would be cluttered with desks and cubicles. Square feet were scarce as government agencies expanded rapidly across the capital, and all available space within the baroque halls of the Carnegie Institution of Washington, including its great rotunda, would be needed for Van Bush's new scientific initiative. Bush had already abandoned the dark oak panels of his old office above the rotunda in favor of the mahogany boardroom, with its marble fireplace, distinguished oil portraits, and relative calm.

Ensign was an inventor who dreamt of a new method of flying. He had phoned Bush's office already, with no luck, asking for a thousand dollars for a test flight. There was a model of his device nearby on G Street, and if someone cared to examine it, surely they would approve the funds. He had worked previously with the Smithsonian Institution's Samuel Langley, he said, on Langley's flying machine.

Bush's secretary had typed up Ensign's pitch. "Mr. Ensign calls his apparatus a 'gravity eliminator' and he claims it will be of the utmost importance in national defense . . . he asks for only three minutes" to explain his design.

Now Ensign demanded to speak to Bush in person. Staff were trying to steer the stranger away, to palm him off on someone else, when Bush himself walked down the corridor and quickly found the would-be pioneer in his face.

Ensign explained his financial needs, his "gravity eliminator," and its military value. Of course, it was a harebrained scheme. Afterward, a staffer caught up with him.

"Then you were taken care of all right," the assistant said.

"It was not all right," Ensign barked. "I told Dr. Bush that the Germans had offered me 500,000 dollars for it. Bush said he didn't believe it. . . . I told him I would sell out to Hitler. And God help this country if Hitler gets it!"

Nothing beats treason to liven up the morning.

Ensign's "gravity eliminator" wasn't exactly typical of the wild proposals that, inventors claimed, could help beat back the Nazis in Europe. Neither was the extortion attempt. But Ensign's pitch did reveal what Bush was facing in June of 1940. Every tinkerer in the country had some half-baked idea.

Up to then, Bush discovered, there was no real system for considering military inventions. The Navy's Bureau of Aeronautics referred ideas to NACA. Army proposals were sorted, helter-skelter, through the Adjutant General's Office. Many Army departments seemed to have no formal process at all.

Now, with the president's announcement of NDRC, which made news nationally, Bush was a prime target for serious proposals *and* for all sorts of quack ideas.

He was mailed fantastic letters. One inventor claimed his new propeller could make airplanes fly 70 percent faster. A Boston doctor suggested that huge electrically charged balloons could somehow protect against falling bombs.

Military leaders had urgent questions for Bush. They asked for data on how ships surge, sway, roll, pitch, heave, and yaw, how to banish weeds from jungle runways, and keep their pilots from passing out at high accelerations. They needed better radar, night vision, defense against chemical gases, improved fuels and explosives, more accurate aiming devices, rockets, and new portable bridges.

Then there were projects the military didn't know they needed.

Bush had to sort through all the proposals, pipe dreams, and could-bes and determine which ideas were scientifically feasible and which were implausible based on, as he phrased it, the underlying "physics of the situation."

The Yankee had a simple recipe in mind for NDRC projects. Step one: Hire the smartest scientists in America. Step two: Give them full authority.

The problem, he saw, was that his hires "must distinguish the really practical idea from the thousands of screwball proposals" that always abounded.

"If they are a sound group," he felt, "they will."

On July 2, 1940, at the Carnegie Institution, NDRC met for the first time.

To help run it, in addition to representatives from the Army and Navy, Bush recruited his friends from the Committee on Scientific Aids to Learning. As chairman, he supervised the show. Under him, each running a division, were Richard Tolman, James Conant, Frank Jewett, and Karl Compton.

Tolman, a dean at the California Institute of Technology, was a dapper mathematical physicist who shared Bush's love for pipe tobacco. He would head Division A, focusing on new armor and ordnance. Tolman had been so eager to contribute to the war effort that, weeks earlier, he informed Bush that he was hoping to enlist in the Army Reserve. He was fifty-nine years old.

Conant, president of Harvard University, was a chemist. He had helped develop chemical weapons in World War I, and met many of Germany's chemical experts. Slender, stooped, with a natural diplomat's broad smile, he had butted heads a few times with Bush when the latter was running MIT. He would lead Division B, and supervise the cutting-edge science of chemicals, bombs, and fuels.

Division C fell to Jewett, president of the National Academy of Sciences and of Bell Telephone Laboratories. A bald electrical engineer with pince-nez glasses, Jewett was an entrepreneur educated in a one-room, one-teacher school in California who spent his youth hunting, fishing, and laboring in apricot groves. He would oversee experiments in communications and transportation.

Compton, president of MIT, was a physicist by training. During the First World War he worked in the Signal Corps, and would now handle innovations in radar and related instruments under Division D. Athletic, tough, jovial, captain of the football team in college, Compton had always been both more amiable than and a step behind, intellectually, his Nobel Prize–winning brother Arthur.

NDRC would also supervise a notable "Committee on Uranium," a small group of physicists that already included a Bush employee: Merle Tuve.

Bush had his coaches. Now they needed star players.

Merle Tuve spent the early months of 1940 riling up his colleagues at the Department of Terrestrial Magnetism. He prowled DTM's grassy campus—suddenly indifferent to the ovoid Atomic Physics Observatory and the extraordinary experiments it had hosted. He implored colleagues to put aside their dreams of protons, geomagnetism, cosmic rays, and atmospheric electricity.

"Let's not do any more research, if the Germans are going to inherit it," Tuve told them. "If those Nazis are going to inherit it, what's the use."

"We've got to find out how we can contribute to stopping this."

He found it difficult to focus on the marvels of the natural world with the kinds of headlines now feeding the American imagination. "Nazi Drive on Britain Seen: Reich Looks to New Air Weapons." "50 Finns Killed, 100 Hurt in Wide Russian Air Raids." "The German Plan: Attack on Scandinavia May Be Either a Diversion or Major Blow at the British." "Nazis Stress Plan of 'Total Warfare.'"

In May, the British sent a scientific attaché, Archibald Hill, to the United States to explore exchanging knowledge in experimental military science. Tuve invited Hill for dinner at his home in Silver Spring, and found the attaché in poor spirits. Hill was anxious that America wasn't taking the war very seriously.

"Don't kid yourself," Merle said. "If you'd sense the temper of people in this country, you'll know that there are all kinds of people just like me that are determined that this thing has to be stopped, and we'll put our shoulders behind it."

On the day of NDRC's first meeting, at Van Bush's request, Tuve wrote down a list of scientists and experts who might make valuable recruits. He sketched out a quick ranking of candidates in his typical, high-energy fashion.

Physicist Robert Brode, at the University of California, Berkeley, was "very able. Thorough. Honest. A quiet leader," he wrote. Francis Bichowsky, a physical chemist from Ann Arbor, was an "odd person, but a <u>shark</u> <u>technically</u>." John Strong in Pasadena was a whiz at building various instruments. Ed Salant, of New York University, had great ideas, wide knowledge, and was "<u>anxious</u> to serve." John Frayne, a Hollywood soundman, was a first-rate electrical engineer.

Merle dissected dozens of personalities from across America:

- Dowdell, University of Minnesota: "Lively. Knows metals."
- Poulter, Armour Institute: "Makes mistakes but is a pusher. Engineering."
- Sanders, Naval Research Laboratory: "Thermodynamics. Heat engines. Knows men."
- Beams, University of Virginia: "A laboratory genius. Finest type of personality. . . . Perhaps too modest and mild. Handles men will get results."
- Dill, National Bureau of Standards: "Heat insulation. . . . Stronger man than he seems."
- Turner, Princeton University: "Good personality. Brains. Trifle short on steam."
- Van Lear, University of North Carolina: "Party hot air, but evidently can direct things. Hydraulics."
- Hitchcock, George Washington University: "Has horse sense. . . . Knows concrete, cement, structures."
- Condon, Westinghouse Research: "Wide contacts. Theory. Slightly erratic."
- Berkner, Carnegie Institution: "Radio engineer, Navy man. Good leader."

He delivered his thirteen-page list to Bush three days later, on July 5. Many of the recommended names went on to work for NDRC, or for Tuve himself.

In addition to serving on the Committee on Uranium, Merle would be working under Richard Tolman in NDRC's Division A, on some ordnance project. Before he could begin, he needed security clearances from both the Army and Navy. In the meantime, he did some reconnaissance. He picked up a navy ordnance manual and met with Navy captain

William Blandy, soon-to-be chief of the Bureau of Ordnance, and Mike Schuyler, head of ordnance research and development.

Blandy and Schuyler knew very well what their biggest problem was: the Navy was really bad at shooting down aircraft. They just didn't know what to do about it.

Was there any way, they asked, that Tuve might develop an "influence fuse" that could trigger near a target? It would be a lifesaver, quite literally, if Merle and his group—"who knew something about electronics"—could figure out how to get a Navy antiaircraft shell to blow up in proximity to an airplane.

In late July and August, military intelligence began to process the clearances. Merle went to Minnesota with his wife and kids to visit her family. All he could do, he said, was to "chew nails and read [about] ordnance and wait."

"As soon as your clearance comes through," Bush assured him, "I'll ask you to do some things." Merle fidgeted, restless, a cat on a hot tin roof.

The Navy's request for some kind of smart fuse had a new and urgent basis: airpower now posed an unprecedented risk to American vessels.

It used to be that the only way to sink a ship from the air was a massive torpedo attack from slow-flying bombers. Level bombers could also drop heavier ordnance from a great height, though it was much harder to aim. But with stronger airplane engines, you didn't need sluggish bombers flying in predictable formations. Agile, fast, single-engine dive-bombers—brutal targets for naval gunners—were poised to transform the traditional dynamics of maritime warfare.

In 1937, Germany had tested the new method of bombing against the old. Diving bombers scored a 40 percent hit rate against a target ship. Level bombers scored 2 percent. American gunnery officers, after their own naval exercises, reported that only airplanes could possibly protect the Navy against dive bombers.

Along with the revolutionary development of aircraft carriers, breakthroughs in fighter planes, bombers, and airborne weapons had the potential to fully overwhelm the antiaircraft guns on battleships and other naval vessels.

Both sides' carriers could now fill the skies with planes. By 1940, pressurized catapult systems lining carrier decks could fling seven-thousand-pound aircraft up to seventy miles per hour over just fifty-five feet. To land, the pilot had to aim precisely so that the airplane's hook snagged a wire on the deck and engaged hydraulic braking gear.

The Navy had a large air arm, and was pushing to squeeze more airplanes onto the decks and into the hangars of its carriers. But carriers operated behind battle lines. The vessels in the most danger from new airplanes were escort ships, light and heavy cruisers, destroyers, and battleships. Forward vessels would be in serious trouble if they had to defend against attacking aircraft by themselves.

In the Pacific, Japanese naval power was on the rise. Japan had the world's largest, most modern carrier fleet and had just introduced, in July of 1940, the most advanced carrier-based fighter, the Mitsubishi Zero, an agile long-range dogfighter with a twelve-to-one kill ratio (meaning twelve enemy airplanes would be destroyed for every Zero lost).

Yet in the U.S. Navy, ack-ack guns were dangerously inefficient. And you could cram only so many heavy guns on the decks of a naval vessel.

Stuck in Minnesota, studying up on ordnance, Merle Tuve was learning just how big a problem the Navy had. The 1938 specifications for the Navy's standard mechanical time fuse, in particular, were no less than shocking.

Naval gunners manually set fuses on explosive rounds. The technique was archaic. First, you had to calculate with telescopes where an airplane might be in, say, fifteen seconds, measuring height, bearing, and range. Then you twisted a metal ring on the fuse, which formed the tip of the shell—very much how you might twist an old kitchen timer—and loaded the shell, fired, and hoped the thirty grains of explosive black powder sitting inside in a silk bag blew enough shards of brass shrapnel into the flight path. Fire the shell an instant too early or too late, and even if you missed your aim by only a foot, the round would be thousands of feet off.

Shells had to explode within a window of one-fortieth of a second.

No wonder it took thousands of rounds to knock one "bird" out of the sky. No wonder every ack-ack gunner dreamed of a shell that could automatically explode near a target. The Navy's Bureau of Ordnance, in fact, was already being "bombarded and barraged" by inventors claiming to have devised a proximity fuse. Patents were being filed by hopeful engineers in the United States, England, and Sweden. The rounds-per-bird problem was a clear and lethal threat.

None of the existing ideas were practical.

On a Sunday in August 1940, Merle Tuve ran into his colleague Philip Abelson in the laboratory of the Department of Terrestrial Magnetism.

Tuve was thinking about shooting down airplanes.

The Luftwaffe had executed a massive raid on England, and he had just been listening to the radio reports. Among others, journalist Edward R. Murrow was filling American ears with morbid details: the red fires in the London night, the "unreality" of the attacks, ambulances "weighted down with an unseen cargo of human wreckage," windows blocks away from explosions eerily shattering. Murrow reported the official advice for dealing with a falling bomb: lie face-down on the ground, cover your ears, and hold your mouth slightly ajar.

Abelson listened to Tuve sound off passionately about "the need for defensive measures." Merle had his security clearance, but he was the only one at DTM who had been vetted. He could not tell Abelson or anyone else what he'd learned from the Navy. But he could discuss the physics at the core of the rounds-per-bird puzzle.

Any sort of proximity fuse, Merle knew, had to be smart enough to arm itself after firing, sense a change in its environment that would indicate an aircraft, and then detonate a primer within a split second of receiving that signal.

Devising some kind of sensor was easy enough. The problem was that antiaircraft rounds, unlike bombs or rockets, endure extraordinary pressures. And the electronics of the day were very delicate. One component in particular, electronic vacuum tubes, seemed far too fragile. The same glass bulbs Merle had pined after for his wireless set as a boy—the brittle tubes hung upside down so the filaments didn't sag—would have to withstand the shock of cannon fire.

Imagine shooting a light bulb out of a Colt .45.

To offer a few comparisons, a space shuttle during launch does not exceed three times the force of gravity. Bombs are simply dropped. Rockets had to tolerate up to one hundred g's. But the pressure from an antiaircraft gun reaches up to twenty thousand g's. A five-pound fuse fired from a Navy gun would effectively weigh fifty tons.

When the Navy asked Van Bush's outfit for an antiaircraft fuse, on August 12, they knew very well that it "had little or no prospect of real success."

"Here is something the Navy badly needs," as Captain Blandy put it. "We don't know how you're going to do it. If we did—we'd have done it!"

The "how" was left to Richard Tolman's Division A.

Section T—Section Tuve, as Merle occasionally called it—was formally established on August 17, 1940. Much was still up in the air. For several weeks, Van Bush had been sick and was in and out of the hospital with dysentery. The organization of NDRC hadn't taken full shape. Harold Urey, a Nobel Prize–winning chemist, petitioned Tolman in Bush's absence to recruit twenty or thirty young scientists as "scouts." If the scouts were well connected, the idea was, they could match American scientists with specific Army and Navy projects.

Merle had a different idea. He proposed to set up, at DTM, a "sorting" lab for promising military innovations. Quick experimental tests of feasibility could help separate workable from impossible proposals. NDRC badly needed such a lab. And Bush, president of the Carnegie Institution, already oversaw DTM.

On September 3, Bush wrote to Tuve formalizing his agreement with Merle's plan. Attached was an oath of allegiance to the United States, to be taken before a justice of the peace or notary public. Bush reminded Tuve of the "need for the utmost secrecy." The assignment was on a "part-time volunteer basis."

The "duties of Section T," Bush wrote, "will be Preliminary Investigations."

Merle couldn't get the fuse out of his mind.

6
CHOICE OVERLOAD

W ell," Tuve figured, "let's shoot them straight up and they'll come
down again, and the chances of them hitting us are very small.
We'll find out what happens."

An old howitzer, from 1916, resembling a small Napoleonic cannon,
fired a one-pound round vertically, high into the Northern Virginia air.
With the faintest of nods to safety procedures, Tuve and two junior Section T scientists, Dick Roberts and Henry Porter, held thin panels over
their heads. The boards may have helped psychologically, but they were
flimsy shielding. The men stood a hundred yards apart, listening intently
for the shell's whizzing descent.

They stood and they waited. It was late 1940.

A friend of Tuve's, an economist named Gardiner Means, had offered
the space on his seventy-four-acre farm as a makeshift testing ground. A
country estate of rolling knolls and steep hills, the tract enclosed a forest,
a nineteenth-century farmhouse, hay fields, sheepdogs, and flower gardens. Means's farm was big enough and it was close to the capital, and
Merle was in no mood to waste time. Based on the physics, from preliminary tests, he believed a "rugged" fuse might be possible. He couldn't be
bothered to ask permission, endure needless delays, or convince skeptical
military officers to find him a firing range.

He was on his own. Even Van Bush was doubtful.

"Quite frankly," Bush said, when Tuve pitched his plan to build a
smart fuse for antiaircraft guns, "I think this is an impossible task, given

present-day techniques, in any reasonable time." But he believed in Merle's competence.

"In spite of my qualms," Bush said, "I'll back you up while you try it."

Early on, Tuve's team consisted of just three young physicists who were good with their hands: Dick Roberts, Larry Hafstad, and George Green. It didn't help matters that their security clearances were taking longer than hoped. Even as Tuve recruited his DTM colleagues to defense work in late August and early September of 1940, he couldn't tell them exactly what they were working on.

When Merle first asked Roberts, in mid-August, to join the project, he merely tasked the physicist with testing the strength of vacuum tubes. Roberts, twenty-nine, was a cocky upstart from Titusville, Pennsylvania, with a perpetually eager face. He was close enough with Merle to ask him to be godfather to his child (and to innocently request, for the baptism, some lightly radioactive tritium water).

Could a tube possibly be built to withstand 20,000 g's?

At nine p.m. on August 17, Roberts ventured down to the Atomic Physics Observatory's "target room," the circular bunker where fission had been confirmed the year prior. Reinforced by thick concrete walls, the room was buried under the egg-shaped particle accelerator. You could reach it by a zigzagging, underground tunnel from the main lab or, more directly, by a thin metal ladder that disappeared in front of the accelerator into a small dark hole in the concrete floor.

In the target room, Roberts attached a vacuum tube to a lead bar and hung the metal block from a string. Then he pulled out a .22-caliber pistol and fired a bullet at the lead block. The small glass tube seemed to be intact. But when he calculated the force of the bullet's impact with a slide rule, he found it was only five thousand g's.

The next day, he tried another approach. He put on his metalworker hat and melted some lead into a "nice hemisphere." Then he attached a short section of pipe to the top, filled the pipe with Plasticine putty, and pushed in a vacuum tube. With his strangely knobbed half ball of lead, he marched up to roof of the Cyclotron Building and tossed the device at a heavy steel plate thirty feet below.

Again, the glass tube looked fine.

All of which posed an annoying security dilemma for Merle when he heard about it. Because he couldn't explain why he wanted the fuse, Roberts was running the wrong tests. A bullet's impact, like the sphere hitting the steel plate, lasts only an instant. But rounds from an antiaircraft gun endure high g-forces for over a dozen seconds. If Roberts could have fit a vacuum tube inside of a bullet, it would have been a more appropriate experiment.

Now Merle had to carefully make clear to his tiny volunteer staff that "the duration of the acceleration was too short to be a meaningful test." The physicists were far from dumb. Vacuum tubes were used in a range of military equipment, but the only application associated with such prolonged and intense pressures probably involved artillery guns intended to shoot down airplanes.

By the time Merle asked his small staff to purchase some explosive blasting powder, the ruse was formally up. Not that he could confirm it.

Tuve hoped to "put a vacuum tube in a shell."

Go over to Georgetown, he told them. "There's a place that sells raw powder."

If they couldn't simulate the forces inside an antiaircraft gun, they would have to test tubes inside an antiaircraft gun. For that, they needed blasting powder. So the men looked up *Dynamite* in the Yellow Pages.

When Larry Hafstad, Tuve's bespectacled, slick-haired confidant, reached 3209 M Street NW, the thirty-six-year-old physicist found a flower shop. The florist, Linden Shenk, ran a side business selling "Blasting and Sporting Powder, Caps, [and] Exploders." Shenk kept the powders at home, and could bring them into town in four-pound batches, the maximum allowed by law.

On September 4, Shenk became Section T's first official supplier. His powder selection was rather baffling, but Hafstad chose two canisters of DuPont black blasting powder. To Shenk, it might have seemed odd for a man like Hafstad—clearly not the sporting type—to buy explosives. After all, the physicist knew little about propellant and couldn't explain why he needed it. The florist's receipt, for $3.60, vigilantly noted Hafstad's address, age, and hair and eye color.

Now Tuve needed an antiaircraft gun. But, he soon discovered, he couldn't buy an artillery piece or easily procure one from the Army or Navy.

How was he supposed to build a radically new type of bullet without a gun?

No matter. The physicists built their own "cannon" at DTM's machine shop out of Shelby steel tubing, a bored-out chunk of metal, and a spark plug. They needed "shells" so they made them from smaller pipes. It wasn't safe to run tests at DTM, so they set up their improvised gun on Gardiner Means's farm, burying it in a pit in case the hardware exploded when the powder ignited. They potted some vacuum tubes in wax inside their homemade shells and fired them straight up.

They shot rounds high up into the air, and listened intently as the pods descended, whistled, and thumped back to earth. Hunting the pasture for the holes in the ground, they dug up the capsules and examined the insides.

Failure. The tubes were "powdered glass."

By the time Merle finally wrangled a real howitzer from the Navy and set it up on the farm, weeks later, Section T's problems had multiplied.

A fuse had to be "smart" and sturdy. It also had to be small.

A common Navy gun barrel was nearly sixteen feet long, with a five-inch bore. Its rounds weighed fifty-four pounds apiece. Normally a clock fuse, twisted onto the round like the top of a flashlight, would form the nose of a shell. Clock fuses were three and a half inches long with a conical tip and a short, narrow lower body. Any new design would have to conform to the same basic shape and width.

To work inside a fuse, a vacuum tube had to be both strong enough to absorb the twenty thousand g's *and* "subminiature." It had to be smaller than a paper clip. Yet the smallest conventional vacuum tubes were double the needed size.

Tuve also had to decide what technology to focus on. There were a number of ways to sense an aircraft. You could try to use a plane's shadows or its glare, and design a "photoelectric" fuse that went off with a change in light. Or you might try to design an infrared sensor to home in on an engine's heat signature. Or you might build a listening device to explode near the sound of an airplane propeller. Or you could make a fuse with a little "radar ear" using radio waves. You might even design a sensor to trigger near an airplane's low-intensity magnetic field.

In early September, Tuve asked his team to assemble the circuitry for a light-sensitive model. He still didn't tell them what it was for. They knew anyway. Within a week, they'd built several crude mockups using old coffee cans. Their design wasn't rugged—the prototypes would have been pulped inside of a gun—but the sensors could effectively detect a 1 percent change in light.

On September 17, Section T received an early assist from London. Professor John Cockcroft, part of a British delegation sent to share secret military technologies with the United States, met with Army and Navy brass and Division A's Richard Tolman to discuss their own progress on a smart fuse. The British were pretty well along, he told them, in testing photoelectric fuses for two-hundred-and-fifty- and five-hundred-pound bombs. These could be dropped, in theory, on Luftwaffe aircraft from above.

Bombs smaller than two hundred and fifty pounds, Cockcroft said, weren't practical because the fuses wouldn't fit. The British had made fainter efforts toward acoustical fuses that used microphones, and least developed of all were radio fuses for bombs and rockets. Developing fuses for shells, Tolman reported, had "of course" occurred to the British, but the g-forces and the size constraints were a problem, as were the firing vibrations, which were "much worse than in rockets and bombs."

Two days later, at Tuve's home in Chevy Chase, Maryland, Cockcroft showed the Section T chairman a simple design for the circuitry of a radio fuse. Developed by an Australian physicist, the wiring was nearly identical to the photoelectric and acoustic designs, except that instead of a light sensor or little microphones, it used an oscillator, a radio circuit that can broadcast and receive signals.

Each design was essentially the same: some kind of sensor connected to amplifier vacuum tubes that boosted the signal and "told" the trigger to detonate.

There was a reason that the Luftwaffe wish list, smuggled in inside Bill Sebold's watch in January, hadn't asked about bomb or rocket fuses. Nor were the Nazis nervous about the wiring of various designs. Their assets in New York were asked about "the firing off shock" in shell fuses because *that* was the puzzle.

The circuitry may have saved Section T time, but it didn't help Merle answer the key question: Which type of fuse could he make small, smart, and tough?

Acoustic fuses might work. But what if other battle noises set them off? According to the British, if you shot two rockets, the sound of the second one firing sometimes blew up the first. Photoelectric fuses, the focus of British efforts, also had disadvantages. They wouldn't work through clouds or at night, when most Luftwaffe raids would occur. Glare from the sun or sea could falsely trigger them. Radio fuses, closest to Tuve's own expertise, could probably be jammed.

Yet another idea pursued by the British was a "pulse" fuse detonated by remote. But how could you select, from the ground, one fuse among many? And within a one-fortieth-of-a-second window? The British delegation informed NDRC's Division A that they were also interested in a fuse trigged by static electricity.

Merle, exactly as Van Bush predicted, had too many options. He faced a buffet of what-ifs and wouldn't-it-be-nices. And each idea—a fuse that reacted to sound, or heat, or light, or radio waves, or magnetism, or remote control, or electricity—could prove to be a dead end, wasting money and time they didn't have. In late 1940, Section T ran preliminary tests on all of them.

Their budget was twenty-five thousand dollars.

Merle urgently needed more staff and greater control.

He had Van Bush's full support, but John Fleming, the director of the Department of Terrestrial Magnetism, was still his immediate boss. While England suffered, Merle still had to maintain, Fleming reminded him, a "reasonable balance" between defense work and his usual responsibilities, which included overseeing the construction of DTM's latest particle accelerator, a cyclotron.

Fleming ran a tight ship. Sixty-three, aloof, with features that were somehow both soft and scolding, Fleming was wary of defense work taking over DTM. He initially asked Merle to request prior approval for "any additional assistance" he required from DTM scientists or machinists. All Merle had at the beginning was his own time, a fourth of Green's time, and Hafstad and Roberts half-time.

On September 25, Fleming did finally agree to allow Tuve himself to work "full time, if required," on defense projects. But even then Fleming put him on a short leash. His authorization was for only two weeks, until October 10.

Merle made a wish list of 31 scientists he wanted cleared by military intelligence, and scribbled grouchy to-do lists in a small aqua notebook.

"Our own men (all) can use mach tools in . . . shop at any time," he wrote, evidently because, as things stood, they couldn't. William Steiner, a Fleming administrator, seemed to be hogging the machine shop. "Complete freedom in allocating machinists time. . . . Without asking Steiner." Then, just a few lines down: "Machine tool time allocation by Tuve, not by Steiner (i.e. we may interrupt work)."

By October, Merle was "busy days and nights and Sundays with Defense research," he told a friend. In November, Fleming allowed him six men, full-time. But only until the end of the year. And should Fleming need them back, he had priority.

Section T, in short, was hamstrung by a small staff, lack of shop time, secrecy protocols, a tiny budget, a makeshift gun, shells made of pipe, an overabundance of options, and a firing range that consisted, after all, of a friend's backyard. And yet there was an even more fundamental obstacle they faced.

They had no experience making weapons.

They discovered that one reason the early tests destroyed the glass tubes was that Hafstad had selected the wrong kind of powder. Although Dick Roberts's wife, Adeline, lovingly portioned out the explosive into convenient, cheesecloth bags, the stuff burned too hot. The pressure in the gun was too high. Even when the tube survived, the electronics inside looked like a "tiny wad of tinfoil."

Roberts and Henry Porter, a hulking, friendly-looking engineer and Tuve's first outside hire, tried out a different propellant to fire the shells, a "smokeless" powder grain that came in thin, one-foot sticks. On the farm, Roberts lay on the ground beside the gun, ready to check the instrumentation to measure the g-forces.

A safety fuse burned down, crackling like on a stick of dynamite.

Bang! Above the men, suddenly, hung fiery rods — burning sticks of explosive propellant—fixed in the air as if suspended on a devil's mobile, and then, quickly enough, descending all around them. Porter howled. Both men sprinted. Roberts set a "record for a fifty-yard dash from a prone start."

The shell went barely fifty feet in the air.

The next morning, Porter went back to their trusted florist in Georgetown and bought four different kinds of powder to test.

Even after the physicists found the right propellant, vertical firing was a tough challenge. You had to strain to hear where the rounds landed. When Tuve, Roberts, and Porter first fired the obsolete navy howitzer, the shell landed in a briar patch. Normally, the rounds went straight up and landed base-first. But when it was windy, you had to tilt the gun into the breeze, and if you misestimated the angle, or the wind shifted quickly, the rounds could "turn over" and soar off.

Division A, meanwhile, needed Section T to run preliminary tests on unrelated projects. Within three months, Merle had thirteen projects, with fourteen more queued up. Even though he was already passing off some of them to other labs, he still desperately needed more hands on deck, immediately and full-time.

Progress felt slow and grinding. Merle and his team experimented with fuse circuitry, spun tubes under pressure in centrifuges, and shot them from the howitzer.

They did manage, by way of a rumor, to find a possible solution to the tube-size problem. In Massachusetts, a Raytheon engineer named Percy Spencer had built what, in 1940, might have been the greatest children's toy in the world. Spencer devised a flying, remote-controlled airplane for his kids. To control it, he adapted a glass vacuum tube from its smallest application, in hearing aids.

Raytheon's midget tubes, built for two-piece hearing aids (an earpiece and a battery-powered microphone unit the size of a large wallet), were small enough for the fuse. With the new blasting powder, some of the tubes even survived being shot from the howitzer. But many broke or were warped into artistic shapes.

Tuve reached out to experts at Raytheon—and, for good measure, at Bell Telephone Laboratories and Hytron—to begin modifying the tiny vacuum tubes.

He dashed off cryptic telegrams to Ivy League professors. "Important use your services Defense Research stop your secrecy clearance completed yesterday permits proceeding stop Please inquire possibility obtaining leave of absence three months or more. . . . Visit us in Washington our expense Friday Saturday or Monday."

Merle bought a safe, and had DTM's technicians fingerprinted.

Fleming *was* losing control over his precious domain. Van Bush gave Section T the authority to make full use of DTM's scientific and machine tools, five thousand square feet of its instrument shops, and up to fifteen thousand square feet in the Cyclotron Building, the Atomic Physics Observatory, and the main labs.

With Bush's backing, momentum was starting to shift Tuve's way. Merle arranged to borrow a more powerful centrifuge capable of spinning tubes at high enough g-forces to simulate an antiaircraft gun. He hired machinists and started interviewing ham-radio operators and electricians.

He penned a fiery missive to Bell Labs over delays: "Structurally special tubes . . . rapidly approaching urgent status stop can come to New York almost any day if this will expedite attack stop test apparatus available here." He scrawled instructions to Raytheon about rugged components, warning them about "vibrations."

The British were right to worry about the vibrations from artillery cannons. Even if Tuve could modify subminiature tubes, the fuse was dead on arrival if he didn't fix the microphonics problem. The issue was that, with pressurized shells rotating at such high speeds — some four hundred revolutions a second — tube filaments would bend and vibrate, interfering with the electrical signals. Outside experts thought the microphonics problem was so bad that a smart fuse for shells was impossible, full stop.

By December, Merle had split his fifteen or so scientists and technical staff into five teams. They tested fuses sensitive to sound, shadows, radio waves, and remote control. The fifth team worked exclusively on rugged components. Tuve was not afraid to delegate arduous tasks. If an important question required the exhausting method of trial and error, they would succeed by trial and error. Renowned theorists were assigned menial chores. One physicist tasked with the monotonous testing of microphonics inside various vacuum tubes apparently felt that the important assignment was more suitable for a clever chimpanzee. He quit, leaving behind a couplet that echoed in the annals of Section T:

If M. A. Tuve is so goddamn tough,
let him test tubes, I've had enough.

7

UNCANNY DAYS

R. V. Jones expected to be back in London before nightfall.

Bletchley Park, home to British codebreakers working on encrypted German communiqués, was less than two hours from the capital by car.

Bletchley was feeding Jones valuable intelligence, and Edward Travis, deputy head of the Government Code and Cypher School (the cryptographic headquarters of MI6), had asked R.V. to come give a talk. Codebreaking was exhausting work, and it would buck up the troops to hear how their efforts were paying off.

On the morning of November 28, 1940, Jones drove north out of London. With him was Charles Frank, an old friend from Oxford and an exceedingly clever physicist. Frank was the first scientist R.V.'s boss let him hire, finally, after months of sounding alarm bells over Nazi science. Jones called Frank his "deputy."

Beyond heavily guarded ornate gates, Bletchley Park's main building was, by one description, "irretrievably ugly," and by another account, the "ugliest country house in the whole of England." A British millionaire had converted a farmhouse into a mansion, resulting in an odd mixture of architectural styles. But the rural plot also held, within barbed wire, plentiful trees and a pond with ducks.

Bletchley's work was subdivided among different "huts"—spare, whitewashed shacks intended to be temporary. Hut One housed Alan Turing's deciphering machine. Hut Three translated German. Hut Four was for naval intelligence. And so on.

Hut Six was doing the actual code-breaking of German army and air force messages. A modest structure sixty feet by thirty, the building had eight rooms, linoleum floors, and bare walls. Inside its decoding room, a staffer wrote, "operators [were] constantly having nervous breakdowns on account of the pace of work and the appalling noise" of decrypting machines. They would crack thousands of cipher keys monthly, revealing a web of tiny details: names, units, leaves of absence, supply requisitions, transfers. Every day was a marathon of concentration.

Jones had been putting their hard work to good use. With Bletchley's help, he had begun building a scientific intelligence service.

He was gaining a reputation.

Earlier that year, Bletchley Park codebreakers passed on to R.V. a deciphered message that included the word *Knickebein*. Literally, the term meant "bent leg." Following a series of vague clues, including parts from a downed Luftwaffe aircraft and a bugged conversation between Nazi prisoners of war, Jones concluded that the Germans had devised a deadly new method for aiming bombs.

According to his theory, the Germans had modified Lorenz, a radio-beaming system originally designed for landing aircraft under poor visibility. The scheme used a radio beam projected into a plane's approach path. If the pilot veered to one side of the beam, a radio operator in the plane heard dotted signals. If the plane veered to the other side, the operator heard dashes. In the center of the beam was a continuous signal that narrowed as a plane approached the runway.

But Jones suspected that instead of using Lorenz for blind-landing, the Luftwaffe was directing the beams to *tell* planes where to drop their payloads, allowing bombers to locate targets, at night, with far greater accuracy than the British could.

Under the new system, two beams from two German cities were projected over a target in England. A Luftwaffe bomber could fly along the first beam and drop its bombs upon crossing paths with the second. Like half of an X marking the spot, two beams jointed over a target looked like a bent knee. *Knickebein*.

R.V.'s theory, like his warnings about German science, was treated with disbelief. The Lorenz blind-landing system only reached thirty miles, and a top British scientist doubted the beams could extend some ten times that range, as Jones claimed. Professor Frederick Lindemann, Winston Churchill's science adviser, also didn't buy it. At the radio frequency required, Lindemann said, the beams wouldn't bend over the curvature of the Earth. Even the chemist Henry Tizard, who led the scientific exchange with the United States, was a skeptic.

"One cannot possibly get accurate bombing on a selected target in this way," Tizard wrote.

But R.V.'s theory gathered enough evidence that he was summoned to present the clues to Churchill. When the secretary left him a note to proceed at once to 10 Downing Street, R.V. checked to make sure it wasn't one of her practical jokes.

He was nearly half an hour late to the meeting. Inside the cabinet room, he found the prime minister with nine advisers, including Tizard, Lindemann, the air minister, and senior officers in the air force. In a low voice, Jones laid out his case.

"For twenty minutes or more," Churchill recalled, "he spoke in quiet tones, unrolling his chain of circumstantial evidence, the like of which for its convincing fascination was never surpassed by tales of Sherlock Holmes or Monsieur Lecoq."

Jones's real breakthrough had come during the inspection of a Lorenz receiver on a downed Heinkel bomber. At R.V.'s prodding, an investigator noticed that the Lorenz radio receiver was far more sensitive than would be needed for blind landing. Bingo.

According to Churchill, "a general air of incredulity" filled the room after R.V. spoke. Nevertheless, a scout plane was dispatched to search for a beam over one of its suspected locations. The beam was exactly where R.V. said it would be.

The episode marked a victory for scientific intelligence and proved that the newborn field could save lives. Bletchley Park's codebreakers helped Jones locate the German beams, which allowed the British to interfere with them.

But the Nazis had been developing another system of beams for precision bombing. As they now controlled northern France, their beaming stations were closer to England and thus more accurate. Days earlier, in mid-November, using the latest method, the Luftwaffe had delivered a brutal bombing raid on Coventry.

Jamming the beams, or sending counterfeit signals, was the only hope. British night fighters didn't have good enough airborne radar to counter the attacks. And after the Blitz began, it quickly became clear that the antiaircraft guns stationed throughout London and its suburbs—on playing fields, dotting the banks of the Thames, on a railway wagon on Blackfriars Bridge, and in the flower garden in Greenwich Park—weren't of much help either.

The result, one general said, was "largely wild and uncontrolled shooting."

The Luftwaffe's free rein was flat-out terrifying.

Churchill had warned that the world might sink into an "abyss of a new Dark Age made more sinister, and perhaps more protracted, by the lights of perverted science." Now his warning took on a new and urgent meaning. Planning for an invasion and government in exile, England had transferred to Canada some five billion dollars in gold bullion, foreign securities, and foreign exchange reserves.

As protection from low-altitude attacks, the Royal Air Force's Balloon Command inflated thousands of large blimps, sixty-three-foot-long hydrogen-filled inflatables known as "barrage balloons." Steel cables, anchoring the balloons, were designed to detach if snagged by an aircraft and unfurl two parachutes to bring their catch down. The method was nearly futile: the blimps destroyed only sixty-six Nazi airplanes.

During the Blitz, Germany lost just 1.5 percent of sortied bombers.

While some English families evacuated cities entirely, others opted to leave each night—a practice known as "trekking." Many affluent Londoners decided it was safer to sleep in the suburbs, making it impossible to find a hotel within seventy miles of the capital. In the port cities of Plymouth and Southampton, citizens slept in tents in the countryside. Dockworkers were bused out daily. Thousands from Clydebank slept on hillsides or marched nightly to a railway tunnel in Greenock.

In London, fourteen hundred shelters were declared unsafe, and the thousands of brick shelters under construction couldn't withstand a direct hit. Nor could the small corrugated-steel Anderson shelters dug into gardens throughout the city. In central London, public shelters were used by just 3 percent of the population. Droves flocked to the subway stations to sleep on dirty floors, platforms, and escalators. Many shelters were horribly overcrowded, had no latrines, lacked cots, and could generally be characterized, in the words of Churchill's wife, Clementine, as "cold, wet, dirt, darkness and stench." Soiled floors were littered with used condoms.

While over a hundred thousand people sheltered in London's Underground stations, many carried on in relative normalcy—whether from stoicism, fatalism, or a sense of romantic excitement. Twenty-somethings played a strange game called "No Man's Land," hopping from one party to another while bombs rained down around them. Twenty thousand children had been evacuated early in the Blitz, and some were now returning. Signs warned parents: CHILDREN ARE SAFER IN THE COUNTRY . . . LEAVE THEM THERE.

Nazi bombers had turned London into a killing field.

R. V. Jones, as he briefed Hut Six on November 28, was acutely aware that a Luftwaffe raid on London was expected that very night, and that streams of packed cars would soon be exiting the capital in observance of the now-normal ritual. He was all the more anxious because his wife, Vera, was alone and six months pregnant.

He had planned to drive straight home after his talk. But he ran into a friend from Hut Three. The huts at Bletchley tended to form their own little subcultures, and as it turned out, Huts Three and Six had something of a rivalry brewing. As long as R.V. was there, his friend felt, Hut Three deserved the same update that Hut Six got. Couldn't Jones give the briefing again? R.V. was caught. It was four thirty p.m. when he wrapped up his second speech, and sunsets came early in November.

R.V.'s headlights were already weak from the requisite shutters. Facing the London exodus now pushing the opposite direction, as a precaution, he dimmed his lamps further to avoid blinding the oncoming drivers.

He squinted at the headlights of the passing caravan.

Outside of St. Albans, still over an hour from London, Jones and his "deputy" Charles Frank met a string of vivid lights. R.V. could barely see at all.

Jones had been authorized to hire Frank in part as a reward for his success with *Knickebein* and the ensuing "Battle of the Beams." But the main reason for adding Frank was that Jones was carrying around in his brain a good deal of secret information that would be lost if he was killed in the Blitz. Frank, in essence, was a backup hard drive. And yet both men, the whole of MI6's scientific intelligence service, were now in very real danger.

Jones did not have time to brake. The truck had been parked carelessly, overlapping the road, while the driver went off to look for more gas. By regulation, the rear of the truck should have been painted white and been clear to see. Instead it was covered with mud. Jones and Frank were thrown right through the windshield.

The stream of Londoners fleeing Luftwaffe bombs, which had now begun to drop on the capital, continued apace beside the scene of the crash.

The men took their bearings. Directly across the street was St. Albans Hospital.

R.V.'s car was wrecked. But Frank's worst injury was a cut over his eye, and Jones's was nothing more than a gash on his forehead. They were stitched up at the hospital, and R.V. called his wife from the police station. He knew of a friend nearby the two men could bunk in with. The pair arrived at MI6's offices in London the next day, "bandaged and with pronounced headaches," at the normal time.

They had intelligence to sort through. Data was pouring in from prisoners of war, from cracked German ciphers, and—increasingly—from French spies.

One of the war's most effective operatives, Jeannie Rousseau, whose reports would prove critical to R. V. Jones's and Merle Tuve's efforts, was only twenty-one years old when the Nazis invaded France. It was July of 1940 when she graduated at the top of her class from the Paris Institute of Political Studies, the famed Sciences Po. Abstract lectures on political theory must have contrasted harshly with Nazi troops lining the border, crossing into France, and infesting the capital itself.

When the war began, Jeannie and her parents, Jean and Marie Rousseau, moved from Paris to the Brittany coast. Dinard, a seaside resort village in northern France, seemed to be far enough west to be safe from the invading army.

Lively, beautiful, with sly arching eyebrows, Jeannie knew quite well how men saw her. She was viewed as a "*gamine*," a harmless child, a "charming young woman" who couldn't be "responsible for anything nasty."

But Jean and Marie's daughter was far from just a pretty face. She was an A student who spoke five languages. She had a near-photographic memory.

The Rousseaus believed that Dinard was safe, but their illusions were shattered just days after the Germans entered Paris. The hospitals in Dinard were evacuated to make way for wounded men from the front. Like would-be vacationers dressed for summer on a frigid fall day, revelers at the resort town didn't want to accept the truth. They filled the gorgeous beaches, swam under the bright blue sky, and tanned under a beaming hot sun while, on the water, in plain view, naval vessels began to evacuate twenty-one thousand British troops to England.

On balconies overlooking the crowded sands, wounded Frenchmen watched the strange spectacle unfold, and sometimes sang or whistled.

Days later, Allied planes woke the residents in the early hours of the morning. Radio stations began to ring with Germany victory songs. French families gathered their valuables and buried them in their gardens. The post office stopped delivering, and stores in Dinard ran empty as townspeople stocked up on supplies. On the same day France signed the armistice agreement with Hitler, partitioning the country into an occupied zone and a collaborating "free zone," some four hundred German soldiers poured into Dinard. The troops promptly took over L'Hôtel Royal, a plum, expansive, beachfront property overlooking Saint-Malo Bay.

The invaders scoured the Rance River estuary, searching for boats that the German military might want to commandeer for a naval assault on England.

Clocks in Dinard were advanced one hour, to match German time. Outside L'Hôtel Royal, Nazi officers stared audaciously at passing women.

Gas tanks were forcibly emptied. Driving was banned without a special permit, and a curfew was established at eleven p.m. French citizens had to watch what they said in public. Cafés and bars couldn't play British broadcasts. If you wanted to hear radio news from London, you had to lower the volume and close the windows. Ration cards for sugar, bread, and other foodstuffs were distributed. Sardines, coffee, and rice were suddenly hard to find. You couldn't buy laundry detergent.

"We are becoming 'Nazified' every day," a resident said.

Soon German airplanes flew overhead with black-and-white crosses on their wings. Locals living around Dinard airport were evacuated. English bombers flew over the resort town while German antiaircraft guns fired from French soil.

In nearby Saint-Servan, British hand grenades lost during the evacuation floated to shore, wounding three children when they tossed them against the rocks.

The post office began to deliver again, now under Nazi control. Young German men were everywhere—relaxing on the beach, racing across Dinard in high-speed cars, snacking at Le Bras, the most delicious pastry shop in town. Troops knocked on doors of local homes and demanded room and board.

German high command arrived. Even the head of the Luftwaffe, Hermann Göring, was said to be visiting the belle époque villas and soft beaches.

The mayor of Dinard, of course, needed a translator to communicate with the German army. There had been rumors that the Nazis required unmarried women between the ages of eighteen and thirty-five to report for work details. Perhaps because Jean Rousseau felt that his daughter would be safe alongside the mayor, he volunteered her as a linguist.

"She doesn't want anything but to serve," he insisted.

Jeannie donned a crisp white shirt and smart blue suit, and began translating the various demands of German high command. The Nazis, she found, "wanted to be liked." Her German was fluent—she could pass as native—and the army officers were happy to have someone like her to chat with. Quite naturally, she earned their trust. She also overheard tidbits of intelligence: plans, numbers, names.

She was perfectly placed to pick up valuable secrets. Dinard was the headquarters for Hitler's Sixth Army, commanded by Walther von Reichenau. Under orders from the Führer, Reichenau and others had begun to plot an invasion of Britain as part of Operation Sea Lion. Reichenau's unit was to embark from Cherbourg, France, land in Dorset, England, and take Bristol.

In September of 1940, Jeannie was paid a visit by a member of the French Resistance from the nearby town of Saint-Brieuc. He asked her to join them and to report on the secrets of the German officers she worked with.

She immediately agreed.

A self-described "human tape recorder," she honed the tools of her craft. She learned to be both patient and cunning, attended to details, asked seemingly unrelated questions, and connected the dots. She made for an excellent spy.

Perhaps too good. The British, the Germans saw, appeared to be receiving detailed reports out of Dinard. The Nazis believed they had an informant.

In January of 1941, Jeannie was arrested by the Gestapo and taken to a prison in Rennes, France. Jacques Cartier Prison, forty miles south of Dinard, had not been a nice place to be incarcerated *before* the Nazis took over. Months prior, authorities publicly guillotined a man in front of the prison door.

She maintained her innocence. The German Wehrmacht officers from Dinard, testifying as witnesses, defended the young Frenchwoman. Their playful translator simply couldn't be a spy, they insisted. In the end, the Germans had no clear evidence against her, and an army tribunal released her.

Her father asked what she'd done in the first place.

"Nothing, Papa."

She had no intention of telling her family more than she had revealed to the Gestapo under interrogation. But although the tribunal freed her, the Nazis in Dinard still had their doubts; she was ordered to leave the coastal area for good.

But she had a taste for espionage work now. She planned to go right back "into the lion's den," she said, closer to Nazi military secrets. Her plan was to get a job with German officers—in essence, to orchestrate a counterintelligence operation by herself—and then somehow pass along what she learned to the Resistance.

She would risk her life, again. All the while, she wouldn't know, as she put it, "whether the dangerously obtained information would be passed on—or passed on in time—or recognized as vital in the maze of the couriers."

"It is not easy to depict the lonesomeness, the chilling fear," she wrote.

There was only one place to go. Home, to Paris. If there were loose-lipped officers who knew about Hitler's secret plans, that's where she'd find them.

8
PEENEMÜNDE

As Van Bush and NDRC were just getting underway, and Section T was fashioning an antiaircraft "gun" out of steel pipe, the Third Reich was expanding the most advanced weapons research and development center in the world.

The facility was hidden on Usedom, a secluded island on the Baltic Sea in northeastern Germany. Known for its reedy beaches, ducks, and Pomeranian deer with dark antlers, the site offered obvious geographic advantages.

Usedom, surrounded by the Baltic, Swine Channel, Stettin Lagoon, and the Peene River, was connected to mainland Germany by three bridges, and those could be tightly controlled. A nearby islet allowed for even greater secrecy. Along the coast, a natural firing range of more than two hundred and fifty miles extended over the water.

Since 1936, colossal efforts had gone into erecting the covert military base. Before the Luftwaffe's construction department razed the terrain — before some ten thousand German military engineers and hard hats invaded the sparsely populated land — the northern tip of the island was, in the words of a historian, "a forgotten paradise — a secluded place, utterly remote and totally unspoiled."

Swaths of oaks and pines were felled. Miles of roads were paved and railway tracks strewn. Sewage pipes were buried in sandy dirt, and a coal-fired power plant was built. Usedom's northeastern harbor was dredged.

On the islet of Greifswalder Oie, six miles north, landing docks were installed. Tents were pitched on newly leveled land. Generators powered a maze of purposeful wires. Submarine cables connected Greifswalder Oie to Usedom.

Northern Usedom was home to a small fishing community called Peenemünde, a moniker that would lend the secretive facility its name. The village's four hundred and forty-seven inhabitants were evacuated, and most of their homes were demolished.

In 1937, the first three hundred and fifty rocket experts and military technicians arrived from Kummersdorf, a weapons development center south of Berlin.

By 1940, Peenemünde was well established—and well guarded. To reach it by train, visitors passed a string of resort towns before disembarking at Zinnowitz, a beach getaway on southern Usedom known for its spa hotels. A short walk led to a second train station, this one private: a plain, flat-roofed wooden building that required a special pass to enter. From here, a modern electric train, a minor marvel of the era, with a fine interior, large windows, and automatic double doors, transported visitors four miles north, alongside the Baltic through a forest, where a second line of security separated the top of Usedom like a thumb's knuckle.

The scale of the Peenemünde facility was jaw-dropping. At its height it had elegant modern housing for three thousand scientists, engineers, and their families. To reach the residences, one passed through high double columns under a chiseled stone swastika. Wide enough to drive a car through, the entryway looked like the imposing front of a national bank. Peenemünde offered bakeries, cafés, a grocery shop, a bookstore, a beauty salon, a butcher, and four movie theaters.

The military science colony had a school, a sports field, and tennis courts. Two-story row houses complemented freestanding family homes. The "Garden City"–style layout, a writer noted, sought "to spiritually unite German families with their native soil." Residents tended gardens and picked blueberries in the woods. Wednesdays, as "morning sport," they swam and played handball and soccer.

Peenemünders put on plays.

A technician described the settlement as a "clean, perfect town."

By 1940, Peenemünde hosted seventeen hundred scientists, technicians, engineers, and other employees important enough to merit civilian draft exemptions. Over the next years, the number of exempted employees would increase dramatically.

Workspaces included state-of-the-art laboratories, assembly halls, huge production facilities, and administrative offices. Along the northeastern coast, the army built test stands for experimenting with rockets. At the northwestern tip, the Luftwaffe paved an airfield to research groundbreaking aircraft designs.

Barracks and mess halls housed thousands of forced laborers. Joining a German construction force of forty-eight hundred workers, one thousand conscripted Polish prisoners, or "guest workers," had been forcibly transported to Usedom, where they were encamped and made to work as slaves to Nazi military science.

The secret facility cost an estimated five hundred and fifty million reichsmarks to build and get up and running, a sum equivalent to over three billion dollars today.

Secrecy rules were enforced under penalty of death. Beginning in late 1940, the Gestapo placed informants among the scientists and construction workers. "One never knew when one was being watched," an engineer said.

Signs warned employees: BE CAREFUL WHAT YOU SAY — THE ENEMY IS LISTENING! Between 1939 and 1945, at least one, and perhaps as many as twenty civilian workers at Peenemünde died by hanging for violating secrecy rules.

The site was, many of its staff imagined, a model of Aryan genius, the leading facility for elite weapons designers and the gem of the Wehrmacht.

Among Peenemünde's scientific stars was Wernher von Braun, a twenty-eight-year-old rocket expert who walked around with a leather jacket and a pheasant feather in his cap. Tall, handsome, a certified pilot, he maintained a steady string of girlfriends and liked to relax by riding horses in the woods or sailing to the islet. A cello enthusiast, von Braun was a member of a string quartet at Peenemünde.

He was also a member of the Nazi Party and, since 1940, of Hitler's SS — the Führer's dreaded paramilitary terror organization. Among Peenemünde scientists, membership in the Nazi Party ran higher than the German average.

Von Braun's passion for rockets had an adventurous flare. If he lived to be seventy, he said, he would be the first man to walk on the moon. In early tests at Peenemünde, rockets were painted with a beautiful "girl" perched in a crescent moon.

Albert Speer, highly regarded by the Führer and Hitler's inspector general of construction, was ordered to supervise the completion of Peenemünde after war broke out. Speer recalled the facility with stars in his eyes.

"The work . . . exerted a strange fascination upon me," Speer said, of his visits. "It was like the planning of a miracle. I was impressed anew by these technicians with their fantastic visions, these mathematical romantics. Whenever I visited Peenemünde I also felt, quite spontaneously, somehow akin to them."

And yet, despite von Braun's and Speer's poetic gestures, Peenemünde had one purpose, from its conception: to invent new ways to end human life.

It was the ideal site for testing Hitler's "superweapons."

The Paris Gun was the heaviest piece of German artillery prior to World War II. Designed for sieges and used in the First World War, the two-hundred-fifty-ton-plus cannon, with a barrel over a hundred feet long, could bombard enemy cities from afar. In 1918, hidden in a forest in northeastern France, the gun shelled Paris with explosive rounds from seventy-five miles away, bewildering Parisians who heard neither airplanes nor cannon fire.

Walter Dornberger, a German army commander at Peenemünde, hoped to surpass the siege gun. An artillery man and engineer who had fought in World War I, Dornberger dreamed of an even deadlier weapon of long-distance bombardment. He proposed to develop a "rocket with a range twice that of the Paris Gun," capable of delivering "one ton of explosives, 100 times that" of the old cannon.

It was the German army's big hope from Peenemünde: a liquid-fueled rocket that could carry a massive bomb payload from a hundred and fifty miles away.

To devise a stable aerodynamic design, von Braun pressed Dornberger to invest in the most advanced experimental wind tunnel in the world. At an estimated cost of over one and a half million dollars, in today's currency, the tunnel when operational would set a world speed record not exceeded until after the war.

Forget the seven-hundred-mile-per-hour German wind tunnel Van Bush and NACA warned Congress about. Peenemünde's supersonic tunnel would reach speeds of Mach 4.4, over four times that of sound, or nearly thirty-four hundred miles per hour.

To fuel the test rocket, German chemists had devised a method of producing high-concentration hydrogen peroxide, which when catalyzed creates propelling steam with temperatures of over 850 degrees Fahrenheit.

Not to be outdone by rocket scientists funded by the army, Hitler's Air Ministry was considering its own weapon to rival the Paris Gun: a pilotless aircraft.

Aware of Peenemünde's interest in drones, Argus Motoren Gesellschaft, a Luftwaffe contractor in Berlin, submitted a plan to the Air Ministry for a remote-controlled aircraft capable of carrying a one-ton payload. In February of 1940, Fritz Gosslau, an aeronautical engineer at Argus, personally presented his plans for a drone weapon to the Air Ministry.

Forty-one and ambitious, Gosslau had already developed a type of pilotless aircraft for the Luftwaffe, a tiny radio-controlled drone with a three-horsepower engine and an almost-twelve-foot wingspan. It was used for antiaircraft target practice.

Gosslau's new proposal was far more daring.

Fernfeuer, or "Deep Fire," had a larger wingspan and a flight ceiling of over 19,000 feet. The new drone could be maneuvered to its target with a guidance beam and then "told" to drop its payload using a radio signal from a nearby aircraft. Deep Fire wasn't suitable for precision bombing. But, Gosslau argued, for "the fight against area targets, it is to be regarded as sufficient."

Engineers at the Air Ministry weren't convinced. Employing a remote signal wasn't practical for battlefield conditions. An even bigger problem was the relatively low max speed of Deep Fire. Fully loaded, powered by an Argus As 410 piston engine, the pilotless craft had a top speed of only 280 miles per hour. In 1940, British Spitfires could already reach 370 miles per hour. Attack drones would be of no use if the British Royal Air Force could easily catch them and shoot them down.

Rudolph Brée, an Air Ministry engineer, returned with the verdict: Deep Fire would not be approved for development. Even after Gosslau's boss lobbied the Luftwaffe's director-general of equipment to change his mind, the project remained stalled.

Gosslau was not deterred.

Knowing that speed was the major concern, he focused on finding a new engine. As it turned out, he didn't have to look far. Gosslau had already been devising an innovative engine for the Luftwaffe that might fit the bill: a so-called pulsejet.

The pulsejet was a peculiar and completely novel design. Some twelve feet long, calling to mind a huge rifle sight, the steel-tube engine drew air from a forward duct into a combustion chamber. Nozzles inside the chamber sprayed fuel, and a spark plug ignited the fuel-air mixture. The explosion caused a fiery propulsion, closing the front duct from inside and generating thrust out the back. Each blast also sucked the oxygen out of the chamber, reopening the duct and drawing in air for ignition. The process repeated itself fifty times a second, in rapid pulses.

The shutters on the forward duct, resembling steampunk venetian blinds, clapped open and shut in a flurry. The end result was a terribly "loud audio signature" that sounded like a broken lawn mower or a car with a blown gasket.

An aircraft with a pulsejet was projected to fly over 430 miles per hour.

A contract to develop the strange engine was first assigned to Argus by the German Air Ministry in 1939, and Gosslau had already tested a series of prototypes. In 1940, with the help of another German engineer, Paul Schmidt, he perfected the air-valve system. Their design was unlike anything that existed in aviation.

In 1941, Argus began flight tests of the pulsejet by strapping it to a biplane.

Horrific, loud buzzing filled the skies over Peenemünde.

9

DON'T SLOW DOWN

March 15, 1941

TO: Director

FROM: M. A. Tuve

SUBJECT: Prowlers

At 2:45 a.m., March 15, Mr. Price the night watchman noticed two men in front of the Cyclotron Building. After watching them a few seconds he saw them go down the steps at the side of the building toward the driveway. He called the police then, Eighth Precinct, and they arrived some four minutes later and scouted the grounds and the first floor of the Cyclotron Building but could find no trace of the men. Mr. Price searched the top floor of the building but found nothing unusual. He notified Mr. Tuve at approximately 2:55 a.m.

The strangers caught snooping around Section T's hub at the Department of Terrestrial Magnetism, the Cyclotron Building, had no good excuse to be anywhere near it, let alone in the quiet, dark early hours of the morning.

Tuve was anxious about security in general. "Confidential and secret information is being too widely dispersed in our organization," he warned his team.

Beginning in late March, Section T staff were only allowed to discuss test results, project status, or the potential military use of the fuse on the second floor, where access was severely restricted. Fuse prototypes had to be stored there also, and covered when transported. If technicians required a special tool, they had to put in a purchase order rather than borrowing it from another group.

Tuve's message was clear: "Keep out of other's [*sic*] rooms."

Any public discussions of their classified work—at, for example, the local Hot Shoppe restaurant—were to cease immediately.

"The oath of <u>secrecy</u> means <u>silence</u>," Tuve reminded them.

As a general rule, workers were to assume their telephones were tapped. Staff had to present salmon-colored IDs to enter the Cyclotron Building, and visitors could enter only through the south door, and only after signing a register.

Trash marked *secret* or *confidential* had to be incinerated. Lab notes, drawings, and blueprints were locked up nightly, and shades were drawn.

The Cyclotron Building, a functional, two-story structure, today resembles an aging preschool or a high-school science building. But in 1941, the brand-new thirty-six-room construction was home to a particle accelerator meant to appease biologists who wanted to radiate fruit flies, frog eggs, and rats with neutrons.

The accelerator was incomplete, and the building poorly furnished.

Acoustics in the Cyclotron Building weren't designed for keeping secrets. Narrow slits between the panels of glass bricks embedded throughout the edifice—the kind you might see striping the walls of an indoor swimming pool—allowed staff in one room to clearly hear chatter in the next. Tuve had separate fuse projects divided into groups, and scientists tackling one challenge couldn't hear classified details of another. The ceilings had to be soundproofed, and the holes in the walls plugged.

To conceal their workbenches and experimental parts from visitors, the scientists needed some one dozen tarps, roughly fifty square feet each.

At first, Section T staff had no place to hang their coats. They needed chairs and stools. Simple metal chairs would do, Merle urged, in one of a series of pleading memos to the director of DTM, John Fleming.

Ceiling sprinklers—emergency nozzles for chemical fires—leaked.

There weren't enough trashcans.

Merle, rather embarrassingly, was compelled to write Fleming to request "adequate supply of pencils, pads, and other stationary [*sic*]." A DTM administrator, he remarked, "requested a formal memorandum from me to you before he can do this."

Merle wanted six Sears, Roebuck workbenches, three CONFIDENTIAL stamps, ten poles for opening high windows, and some linoleum-topped tables.

One of his team supervisors had no desk, so he held group meetings around a lathe. Even as Fleming dutifully fielded Merle's requests, it was difficult for the DTM director to keep pace. Tuve was adding new recruits to the team weekly. By February of 1941, he employed some thirty physicists, radio engineers, and technicians, three typists and stenographers, and two watchmen.

Equipment to build and test fuse parts poured in: tool sets, tool grinders, drill presses, lathes, milling machines, blowtorches, safety goggles, calipers, clamps, soldering irons, and soldering pencils. Merle needed microscopes and micrometers for precision measurements, a "Vacuum Tube Voltmeter" to check the electrical characteristics of tubes, resin and beeswax to cushion the tube parts inside shells, electric hot plates and coffeepots to melt the wax, thermometers, batteries and battery testers, binoculars and a range finder for field tests, a radio receiver unit, squibs, resisters, sockets, plugs, cables, spools of wire of various tensile strengths and electrical properties, and many pounds of explosive powder.

Merle's office was in room 101 by the entrance. Directly across the hall was the "Powder Room." He notified his staff to "<u>KEEP OUT</u>" of it.

"Don't handle explosives without proper instruction," he cautioned. "Experts are never careless. Beginners must not be careless even <u>once</u>."

When building prototypes, he reminded them, they needed goggles to work with any equipment "liable to throw particles or blow up."

He asked Fleming to order a first-aid kit "of considerable size."

Among the memos and warnings posted on the bulletin board were more mundane messages: apartments for rent, a request not to park by the neighbor's house, and an edict from Director Fleming not to play baseball on the lawns.

Fleming would soon change his mind. If the Section T staff working on the secret project agreed to use the grass at the top of the hill, Tuve relayed, "a <u>not too violent</u> form of baseball will be perfectly all right."

"P.S.," Merle added, "Please don't knock down the weather man's equipment."

Section T needed an airplane.

If the British were right, and some kind of "electrostatic" sensor might detect an aircraft by its electric charge, the first step was measuring the charge.

Tuve outsourced the job to a new consultant: head of the physics department at the University of New Mexico, the aptly named Jack Workman, a scientist in his early forties with a plump face, buzzcut, and streaks of gray over his ears.

Through a military liaison, Richard Tolman arranged for an Army Air Corps monoplane, an O-47 — an aircraft just bigger than a Japanese Zero — to be loaned from Fort Sill, Oklahoma, along with a pilot.

In a series of experiments at the Albuquerque airport, Workman had the O-47 pilot do flybys over an "electrometer," a procedure similar to — but more expensive than — waving a balloon near your hair after rubbing it on the carpet. The metal plane's electric field, Workman found, varied dramatically. Snow and ice lent the aircraft a negative charge; in nicer weather, it held a positive one.

"In one test on a cumulus cloud," Workman reported to Tuve, "the plane was negatively charged after passing through the base . . . and positively charged after passing through the top and returning through clear air." After Workman installed a voltmeter on an aircraft wing, he discovered that the static charge on a plane over fifteen thousand feet was too weak to trigger a sensor.

So ended the electrostatic fuse idea.

Section T's "acoustic" fuse required Merle to order — or ask Fleming to order — microphones that could withstand high heat and humidity. (Imagine the conditions in a Pearl Harbor cargo hold or a tropical munitions lot.) Initially, Tuve's engineers tried to copy the British acoustic design for bombs based on "two photographs of poor quality," one drawing, and "a meager description of a few tests." That plastic prototype was abandoned, and another cylindrical design was constructed of metal, wood, and sponge-rubber rings.

Merle described it as a "flying toilet bowl."

When the scientists attached the listening devices to falling dummy bombs, they learned that the rushing wind, the bombs' "self-noise," was too loud for the microphones to distinguish the roar of a nearing aircraft.

The "acoustic" fuse was dead, too.

Tests on a "photoelectric" fuse that reacted to small changes in light, and experiments on a radio fuse design, on the other hand, showed promise. Through early 1941, Section T assembled crude prototypes of both designs on Cyclotron Building workbenches. Constructing basic models that didn't have to withstand the punishment of antiaircraft guns wouldn't solve the rounds-per-bird problem, but it would allow Section T scientists to field-test the fuse circuitry.

James Van Allen, a twenty-six-year-old physicist, worked on both designs.

Shy, handsome, with a remarkably kind face, Van Allen was a farm boy from Mount Pleasant, Iowa. Like Merle, he was captivated as a child by radios and mechanical and electronic gadgets. He tinkered with old car parts and built a Tesla coil, horrifying his mother by demonstrating how the foot-long electrical surges caused his hair to stand on end and shoot out little sparks.

After high school, he'd hoped to enroll at the U.S. Naval Academy. But he failed the physical due to flat feet, bad eyes, and poor swimming.

When Van Allen was recruited by Merle in late 1940, he was a promising young Fellow at the Carnegie Institution with a stipend of some two hundred and fifty dollars a month. He lived in a single upstairs room in a house just down the hill from the campus.

In the early months of Section T, as the Luftwaffe pummeled London, he noticed that his colleagues were drifting off to other areas of DTM to work on some secret project that seemed exciting, challenging, and important.

He felt, he said, that he was "fiddling while Rome burned."

At first, he worked on a new photoelectric design. Like solar panels, the "photocell" sensors converted light to electricity. If the light shifted, the change in illumination translated to a jump in electric output — which could be used as a signal to trigger.

His trouble was wiring a sensor so it could detect a 1 percent change in light when it was dark *and* during the day, when it was two thousand times brighter. A logarithmic circuit, which reacted only to relative changes, solved the problem.

"If we had the whole place full of bright lights," Van Allen said, the circuit triggered if you simply "waved a fan." When he turned the lights off, and you "could barely see in the room," he explained, "I'd wave the fan again and it triggered."

The "rudimentary" radio circuitry was just as sensitive. All kinds of reflective surfaces would bounce the signal back to the device. The prototypes would trigger, Van Allen said, "if you brought a fly swatter near the machine."

The radio model was "a funny kind of circuit." Normally, a radio that sent and received signals was wired to be insensitive to the objects around it, but with such a simple design, Van Allen found that it was almost impossible to build a prototype that *didn't* have a strong sensitivity to nearby objects. "You could have someone open the door and the thing would kick off."

By the summer of 1941, field tests of the two fuse types made it clear which had the most promise. Fitted into dummy bombs and tested at a nearby air base, the light-sensing model proved inconsistent. The lens opening needed to be of different sizes depending on the weather, and the flight tests had to be flown "away from the sun" to avoid glare catching the lens and triggering them early.

The radio fuse was the most promising design.

And yet, it remained impossible unless Tuve's group could produce tiny electronics capable of surviving a cannon blast. Just as problematic, they hadn't yet solved the "microphonics" riddle of keeping the whole damn journey quiet enough for a miniature flying radio to send and receive its signals clearly.

For those puzzles, Merle needed a specialist, an expert not in radio circuitry but in advanced structural engineering: a scientist who understood how materials react under great stress, how they bow and bend, and when they break.

Ray Mindlin, a thirty-four-year-old assistant professor at Columbia University, had a taste for Chopin, dry humor, and sports cars. He was a specialist in materials science. Confident, stylish, with dark bushy eyebrows, he looked like the type of professor who engaged in profound discussions, in hallways, with his hands in his pockets. He ran track, and was fast off the blocks. According to a friend, "If there had been such a thing as a five-meter race, he would have won it all the time."

Asked how things were going, he would reply: "Fair to Mindlin."

The engineer had studied stress analysis and problems like the "torsion of structural beams" and the weight distribution around tunnels.

He was recruited in January of 1941, by Western Union.

According to a Section T staffer, Mindlin had just finished a consulting job to help explain why the Tacoma Narrows Bridge collapsed in late 1940. Battered by forty-two-mile-an-hour winds, the bridge had wobbled surreally, twisted like a trampoline, and flung chunks of concrete into the air "like popcorn" before crashing dramatically into the waters of Puget Sound nearly two hundred feet below.

Engineers blamed the failure on "excessive oscillations."

At first, Mindlin didn't have security clearance. So Section T lied to him about the tiny, rugged tubes that he was to help build. The vacuum tubes were meant for meteorological balloons, they said, and had to withstand a long fall to earth.

Led by Henry Porter, the bulky Chicagoan, Section T's "rugged electronics" team now had a bevy of centrifuges at their disposal that could simulate thousands of times the force of gravity. As revised hearing aid tubes arrived, Mindlin tested their electrical properties, spun them in centrifuges, and had samples fired in shooting tests. He also pulled engineering tips from strong tube designs, including a Canadian tube that protected its thin wires, which ran vertically up the narrow tube, by threading them through small platforms made of mica.

The real trouble was the fine electronics themselves.

In early 1941, Mindlin produced a sequence of diagrams, "design curves" that broke down how tough each tube part was. It wasn't so different from how engineers approach bridge design. The tubes even had miniature cantilevers inside, like bridge supports. Mindlin worked out the warping and final "yield" points of dozens of minute components: grids, grid wires, and grid posts, getters, which kept the vacuum tubes sealed, "presswire welds," and mica support "spacers." He laid out "the bending moments at the supports and at the midspan and the twisting moment at the supports" for a "grid lateral." He pored over blueprints of tube parts and compared yield points to the muzzle velocities of anti-aircraft guns.

Or, as Porter put it, "Mindlin appeared one morning with a series of graphs."

Working closely with the subminiature tube contractors—Hytron, Raytheon, and Bell Telephone Labs—the rugged electronics team tested, refined, and retested a series of prototypes to fix the myriad design weaknesses. One of the most daunting roadblocks was the filaments. Like carbon filaments in a light bulb, these wires heated up when electrons flowed through them. Each an inch long and 0.00075 of an inch wide, they were the centerpiece of the tubes and the most delicate component. At Hytron, the filaments had to be mounted using magnifying glasses by teams of women hired for their "nimble fingers."

When tubes were spun at high velocities, as they would be when fired from antiaircraft guns, four bad things tended to happen: the electrical characteristics changed, the tubes short-circuited, the filaments broke, or the filaments bowed and vibrated so badly that the microphonics problem was untenable.

To keep the fine wires from bending, the miniature "cello strings" would have to maintain a very high tension. Designing a filament support to keep that tension would pose problems, but the first step was to find a material that had a naturally high tensile strength even at hot temperatures.

Bob Brode, a forty-year-old physicist (who was one of the top experts on Tuve's original list of recruits), reached out to manufacturing companies in Chicago and Newark. He requested samples of tough, fine wires made of silver-clad platinum and nickel and chromium alloy.

Mindlin and Brode found that tungsten (or wolfram, the W on the periodic table) fit the bill. Tungsten has the greatest tensile strength of all metals and the second-highest melting point of all elements, at over 3,000 degrees Celsius.

With Mindlin's design curves, and tungsten filaments, and a ton of trial and error—bit by bit, test by test—the tubes were becoming quieter and stronger. After Raytheon's president called regarding tube production, Merle recorded his reply in his notebook, referring to himself in the third person.

"Tuve said don't slow down."

The tube design that emerged was stacked, with platforms like floors in a high-rise that were strengthened by miniature columns. Mindlin didn't seem to care how minute Section T's devices were. To him, these weren't glass tubes the size of paper clips. They were narrow, three-story buildings.

Gardiner Means was worried about his sheepdogs.

The cannon fire on his Virginia estate frightened the animals and rattled their nerves. Several times, the canines "lit out for parts unknown" and did not return for some time. Also, as Means informed DTM's Director Fleming, he was expecting a hay crop soon in the test field, and depending on how frequently Section T planned to use the site, the men might trample the hay or, in any case, "They may have some difficulty in recovering whatever it is that they are shooting."

The truth was, the old howitzer was too big for Means's farm. Even one-pound rounds, falling from fifteen thousand feet, were quite dangerous, and Section T had already moved on to firing larger ammunition at military firing ranges.

For a short time, Tuve's group utilized a Navy range in Virginia Beach. But the test rounds, shot vertically, kept drifting unpredictably in the breeze. The Navy decided that the shells were landing too close to its torpedo station.

At a proving ground in Maryland known as Stump Neck, Section T shot six-pound rounds out of a 57 mm gun. Accessed by a narrow causeway, Stump Neck was easy to guard, and "drifters" landed in the water. For protection while the sizable metal capsules plummeted back to earth, they erected a four-inch plate of steel on posts and covered the top with sandbags.

The procedure for test shoots was the same as before. They potted various fuse parts in metal cylinders using beeswax and resin, fitted them into hollowed-out shells, fired the rounds straight up, and tried to hear and see where they landed. Using clamshell diggers normally used to make fence-post holes, they excavated the pods for analysis. It took nearly a minute for a shell to speed thousands of yards high, slow, apex, plunge, and finally bury itself four feet in the dirt.

Stump Neck's caretaker had a Chesapeake retriever, Curly, who was not at all frightened by the gun blasts. When the rounds hit the ground, the dog would run out madly and dig into the earth or splash hopefully into the water. Curly seemed to imagine that the gigantic gun would bring down many ducks.

For some early tests, to make sure that the landing impact wasn't damaging the test parts, they used modified signal flares. A "Star Shell," shot from an antiaircraft gun, normally released a parachute holding an illuminating "candle" that could light up the night sky. Tuve's scientists simply replaced the candles.

At an Army proving ground in Aberdeen, Maryland, Ray Mindlin watched as the miniature parachutes floated his test pods gently back to earth.

Mindlin's tubes were withstanding gun blasts at an encouraging rate. In a March test shoot of Bell Labs samples, two-thirds of the prototypes survived.

On April 20, Jack Workman, in from New Mexico, fired a very special fuse part at Stump Neck: an oscillator for sending and receiving radio signals. The transmitter used a tube similar to the one developed by Mindlin as an amplifier, only wired differently and connected to an antenna. Workman turned on a radio receiver, and heard, through headphones, a live radio signal coming from a shell in flight.

A radio fuse was possible.

On May 8, Merle traveled down to a Navy proving ground in Dahlgren, Virginia, to witness the breakthrough firsthand. Accompanying him was the head of NDRC's Division A, the pipe-loving Richard Tolman.

Dahlgren was home to one of Section T's strongest advocates, Lieutenant Commander Deak Parsons; years earlier, he had recognized Tuve's work as a basis for radar. With sharp brown eyes and a wrinkled brow that suggested an abnormally large brain, the thirty-nine-year-old wasn't a typical military man. He was, his biographer wrote, "a new breed of officer." An admiral later described him as "the finest technical officer the Navy has had in this century."

Parsons was deeply interested in — and well informed about — radio engineering. He also happened to be a former gunnery officer and ordnance expert.

At Dahlgren, the artillery was a marked upgrade from the outmoded guns and one- or six-pound shells Section T was used to. The base housed the massive five-inch guns common on Navy ships, guns that weighed two tons apiece and that were in the most desperate need of smart fuse ammunition. These were the kinds of gun that could actually take down enemy aircraft from a great distance. The firing range here was aquatic, stretching twenty miles down the river toward Chesapeake Bay. Parsons often helped Merle arrange shoots toward the target area five miles downstream. During firing, barges and fishing boats were cleared out of the danger zone.

To view that day's test, Tuve and Tolman boarded an observation boat and drifted down the Potomac. The river bottom was littered with shells. Seagulls squawked overhead. Over two miles away, a gun blast echoed and a shell whistled past above their boat as the physicists listened intently to their radio receiver for a signal from inside the round. Thousands of yards downstream, tall white plumes of frothy spray shot up like geysers as the shells splashed down one by one.

They heard signals in three of seven rounds.

For Merle—and for the U.S. Navy—the test was a "go" sign. Van Bush informed President Franklin Roosevelt of the results: "Success is now in sight."

The little tubes weren't quite tough enough yet, and the microphonics puzzle wasn't fully solved, but the basic components of the fuse were in place. "It is now clear that such devices can be made," Merle wrote, on June 11. Urged on by Parsons, he drafted a plan to expand Section T from a staff of sixty-five to a hundred and twenty, and to double his monthly budget to sixty thousand dollars.

DTM's Director Fleming had to order more desks and aluminum chairs.

10
PRYING EYES

"Where are the mics?"

The gentleman asking smiled a slight, fixed grin. Sixty-three, wearing a three-piece suit, he clutched his lapels in a folksy, pleasant manner. Resembling both an alert bird and a farmer holding his suspenders, he proceeded to scour the office for listening devices. He glanced back toward the door he had just entered. He scanned the ceiling and each corner. He wandered by the hat tree and coat rack, and peeked inside the mirrored cabinet over the sink.

A 1941 tear-off calendar on the office wall read *June 25*. Both the calendar and the table clock on the desk were strategically positioned to be captured by the motion-picture camera rolling quietly in the next room, where FBI agents were recording and filming the two men from behind an "x-ray mirror."

In late 1940, the FBI had rented the sixth-floor office, at 152 West Forty-Second Street and Broadway in New York City's Times Square, for an elaborate sting. Painted letters on the office door revealed its main protagonist: WILLIAM G. SEBOLD.

Bill Sebold, aka Harry Sawyer, was now an FBI double agent and a "diesel engine consultant" steeped in tradecraft. He was on the verge of pulling off the largest successful espionage case in the history of the United States.

Satisfied at last that the room wasn't bugged, the visitor, Fritz Duquesne, sat beside the end of Sebold's desk. Nicknamed the "Duke" by the FBI, Duquesne was a Nazi spy described in his 1932 "biography" as "the most adventurous man on earth in the nineteenth and twentieth centuries." As a guerrilla fighter in the Second Boer War, Duquesne apparently escaped a British penal colony in Bermuda by swimming through shark-infested waters. He was a big-game hunter who met Teddy Roosevelt at the White House in 1909. Once declared dead in the pages of the *New York Times,* the Duke had also been a secret agent for Germany during World War I, and was credibly to blame for a bombing that claimed the lives of three British sailors.

In the New York City office, at around six fifteen p.m., the spy put on his glasses, rolled up a pant leg, and pulled U.S. military secrets from his sock—diagrams and photos of a Navy mosquito boat, a Garand semiautomatic rifle, a light tank, and a grenade launcher.

Three days later, on June 28, FBI agent James Ellsworth—a Mormon who learned German during his missionary service in the Weimar Republic—picked up Sebold on Forty-Second Street. The day before, unbeknownst to Sebold, some two hundred and fifty Bureau agents met in Manhattan to finalize plans to dismantle the Nazi spy network.

To avoid reprisals, Ellsworth drove Sebold to a safe house in Long Island.

For over a year, FBI agents had followed, photographed, recorded, and filmed the conspirators, intercepting and altering valuable intelligence meant for Germany. In Long Island, following Nazi orders given to Sebold, the Bureau set up a shortwave radio to transmit fake messages—with fake intel—to Hamburg. Over fourteen months, they sent 301 ciphered messages in Sebold's name, and received back 167.

Across America, German operatives had their cover stories rudely interrupted. In Long Island, a musician was pulled off a hotel bandstand. In Milwaukee, a waiter was nabbed in a hotel room. FBI agents arrested ship stewards and cooks, a bookseller, a restaurateur, a mechanic, a carpenter, and a power-plant engineer.

Lilly Stein, an "artists' model," was entertaining a guest that Saturday night. FBI agents listening in waited for the man to leave before arresting her.

"Well, I'll say one thing," Stein told the federal investigators, after learning that her apartment was bugged. "You sure got an earful."

Everett Roeder was celebrating his son's marriage when G-men knocked. They reportedly found one hundred thousand rounds of ammo in his basement.

Duquesne himself was arrested at his West Seventy-Sixth Street apartment by a Bureau agent who had been posing as his upstairs neighbor. Joining the Duke in cuffs was his Nazi-sympathizing American girlfriend, an artist some twenty years his junior who claimed to dabble in playwriting, sculpting, and toy design.

On June 30, "Harry Sawyer" and the FBI relayed a coded message to Hamburg over the shortwave radio. "Newspapers report arrest here of twenty nine German agents. I believe everything still all right here. Believe Harry Safe."

The spy network was astounding in its scope. Sebold had initially been put in contact with four operatives. The FBI sting netted thirty-three arrests.

The trial was in Brooklyn, New York.

"Nazi Spies Told to Get Secrets of New Weapons, Jury Hears," one headline blared. The Associated Press reported on the operatives' interest in chemical and biological warfare, range finders, "radio sending and receiving devices," airplane production, and, of course, the "anti-aircraft shell with an 'electric eye.'"

In D.C., Tuve filed away a clipping of the article.

One of the spies disclosed, in a letter meant for Germany, that while a photoelectric fuse was discussed in scientific magazines, he was "unable to find anyone who knew anything about the actual experimental use of such a device."

The Sebold episode came at a delicate time, just as Section T was looking to hire outside contractors to assemble their experimental fuses. It seemed likely that other Nazi agents inside the United States still remained at large.

Either for infiltration or sabotage, German intelligence was clearly targeting American companies involved in national defense. Among those arrested were a draftsman, a radio engineer, and an inspector for Westinghouse Electric. One Nazi agent, Hermann Lang, had already delivered details of a highly classified Norden bombsight, an intelligence coup with potentially disastrous results.

Duquesne himself hoped to sabotage the General Electric plant in

Schenectady, New York—a factory that was producing fuse parts for Section T.

After the massive espionage ring was dismantled, FBI director J. Edgar Hoover sent Section T a confidential report entitled "Suggestions for Protection of Industrial Facilities." The booklet advised plant managers engaged in defense work on exactly how to keep foreign agents from infiltrating their factories.

One of the spies arrested, the FBI discovered, already had a copy.

On August 7, 1941, at ten o'clock a.m., American consular officers from Germany and Italy gathered in room 300 of the Carnegie Institution of Washington headquarters on P Street. Among the diplomats was none other than Alfred Klieforth, the consul general whom Sebold had approached for help in Cologne in 1939.

In fact, Klieforth was the star witness.

Presiding over the discussion was Harvard president James Conant, a natural diplomat himself. Conant was now chairman of NDRC.

In May, NDRC had been restructured. Van Bush's organization was now called the Office of Scientific Research and Development, or OSRD, with NDRC incorporated underneath it. Bush's expanded duties included overseeing medical research of military value. Drawing funds directly from Congress, he also now had the authority to actually produce small batches of weapons.

Several other representatives of OSRD were present at the meeting, as were two members of the Office of Naval Intelligence. The diplomats had been invited to reveal everything they knew about German military science. If Conant could find out what Nazi scientists had devised, and what they were working on, OSRD would know which projects to prioritize. New information—let alone fruitful intelligence—was scarce. But even faint clues could help.

According to the diplomats, the Nazis were working diligently to produce synthetics that could replace scarce raw materials and allow the Germans to keep up production in the face of Allied blockades. Chemists had devised synthetic rubber, and synthetic fibers were being made from "potato tops." German researchers were trying to build a motor that ran on coal dust, and a battery that didn't contain lead.

All radio technicians in Germany seemed to be engaged in Hitler's ambitions, and the government was so fearful of running out of scientific talent that they were frantically training more scientists and technicians. University laboratories had been closed to students and redirected to military research.

Old radio towers had been torn down unceremoniously and replaced by peculiar "new structures with movable cross-arms at the top."

Rumors warned of large stores of poison gases outside of Milan.

In Cologne, Germany, more hearsay. Klieforth's cook had a fiancé who worked at a poison gas factory. As a deterrent against leaks, two or three men from the plant apparently "vanished" every month. According to Klieforth, the men were likely innocent, but that did not diminish the effect of their disappearance.

At an air-raid shelter Klieforth visited, the walls were painted in glow-in-the-dark paint that was activated by light. For two hours after the lights shut off, the paint remained luminous enough to read a book by.

He had seen rockets fired near Cologne.

On a trip to Essen, Germany, he visited the Krupp Industrial Works, a steel and weapons manufacturer at the heart of Hitler's rearmament program once described as the "pride of Germany." The Krupp Works remained mostly undamaged, hidden from British aircraft by what Klieforth called "mist projection." The Nazis were using artificial fog, which emerged from hundreds of "fountains" or pipe nozzles and extended some five hundred to eight hundred feet above the factory.

Klieforth compared to it a dry London fog, only "slightly yellowish." The mist didn't smell like smoke, didn't stain clothing, and carried a strong enough foul taste that he kept his mouth closed. The artificial fog was being used to mask production facilities that couldn't be blacked out, like blast furnaces.

Conant, the other OSRD representatives present, and the naval intelligence officers made it a point to question the diplomats on topics of special interest. But the consular officers had no information on new medical treatments for gas burns, or special clothing that might protect against chemical attack. Nor did the foreign service officers have any information regarding where individual German scientists might now be located.

Conant and Bush, in other words, needed information they couldn't get.

At the time of James Conant's meeting with the diplomats, Conant and Bush were considering building a nuclear weapon. The whereabouts of Germany's physicists might have revealed how actively Hitler was pursuing a bomb.

Throughout the summer of 1941, physicists in OSRD's "Section on Uranium" and others were wrangling with that very question. On one side of the debate was Merle Tuve, who saw a nuclear weapon as morally repulsive, a far cry from a defensive weapon to protect against airplanes. An atom bomb, Tuve also believed, was a larger, longer-term project than Hitler could seriously invest in.

"The Germans can't afford to do it in a big way," he argued. The war with Germany would be over, Merle believed, before a bomb was ready. Advocating the other side was his childhood friend from just across the street in Canton, North Dakota, the physicist Ernest Lawrence. Lawrence wanted OSRD to commit a massive budget and substantial human resources toward building a nuclear weapon.

At thirty-nine, Lawrence looked very much the picture of a physicist. Dignified, solemn, he wore small glasses and kept his hair so firmly parted and combed that it suggested a marble bust. He retained his ever-polite manners. He still said "Sugar!" and "Oh fudge!" when angry. But he had influence now, having won a Nobel Prize in 1939 for inventing the cyclotron (the type of accelerator being assembled at DTM).

Earlier that year, behind Bush's back, Lawrence lobbied Conant and others at OSRD, including MIT president Arthur Compton, for a large-scale nuclear program.

On March 17 at MIT, Lawrence startled Compton with the news that he was prepared to produce U-235, the uranium isotope capable of a chain reaction: neutrons splitting uranium atoms, releasing neutrons to split more atoms, and so on, in a colossal chain of atomic firecrackers. Lawrence planned to convert the powerful electromagnet of a cyclotron to separate the isotope.

On March 19, in New York, Bush gave Lawrence a tongue-lashing, telling him in no uncertain terms that he, Bush, "was running the show," and if Lawrence didn't like it, he could exit the debate within OSRD over producing the bomb.

Four days later, Lawrence visited Tuve at the Department of Terrestrial Magnetism. Merle signed his friend into the Cyclotron Building register.

The pair had come a long way. As Merle put it, a lot of water had "gone under so many different bridges since the old days" in Canton, since they'd hiked as Boy Scouts through the wilderness, peered eagerly at unopened boxes of wireless-radio gear in Merle's basement, and toiled on the sand greens for vacuum tubes. Underneath their feet, in the Cyclotron Building basement, stood the incomplete replica of Lawrence's brainchild. A week prior, Lawrence had proposed converting that noble scientific invention, devoted to uncovering the secrets of the atom, into a machine that could enable the most destructive weapon ever produced in human history.

There is no record of what the two men discussed.

Tuve had no desire to build a bomb. He wanted to play defense, to protect Allied cities and navies with a revolutionary "smart" weapon—not offense, with the world's most catastrophic bomb. As Lawrence's faction gained leverage within OSRD, Merle abandoned the "uranium affair" to focus on the fuse.

Section T's Bob Brode would join Lawrence in the New Mexico desert.

When Jeannie Rousseau, the Sciences Po graduate, was released from prison in Rennes after enduring the torments of the Gestapo, she returned to Paris and quickly found work that gave her renewed access to valuable Nazi intelligence.

Despite being suspected of espionage in Dinard, she was hired, once again as a translator, by a group of French industrialists who sold commodities (including strategic goods like steel and rubber) to the Germans.

She worked out of offices on rue Saint-Augustin, not far from the Louvre Museum. Her duties included frequent trips to the old Hôtel Majestic, just a short walk down the Champs-Élysées toward the Arc de Triomphe. Previously a vast palace, the hotel had once housed a duke and queen in exile. Converted in 1936 by the French government for the Ministry of Defense, the regal building became, after the invasion, the headquarters of the German military high command.

Rousseau — whip-smart, charming, undeniably beautiful, with a sharp memory and a taste for tradecraft — had found the heart of Nazi power in France. She began to gather details about the German war machine's industrial engine. She became the top liaison of the industrialists' group, visiting the Nazi offices nearly every day and storing tidbits of information about German needs, inventories, purchase orders, supply lines, and other vital details gleaned from conversations.

She knew the information had military value. But she did not know, until a chance encounter, how to pass it along to the right contacts. And she could not have known that soon enough, by yet another stroke of luck, she would stumble upon secrets that could alter the war.

11
ESCALATION

On May 27, 1941, President Franklin Roosevelt delivered a bleak, haunting broadcast from the East Room of the White House.

With his elbows firmly planted on a desk cluttered with microphones and an American flag at his back, Roosevelt warned of a global future dominated by Nazism and proclaimed an "unlimited national emergency." The Axis Powers' "dreams of world domination," he told Americans, were no wild fantasy.

Hitler's navy had extended its combat zone westward across the Atlantic. In response, Roosevelt planned to establish military bases in Greenland, and send troops to Iceland. Just six days earlier, a steamship crossing under the American flag was sunk by a German submarine. The ship's crew was still missing.

Americans were conflicted over their proper role. Roosevelt's dilemma, as British ambassador Lord Halifax put it, was "to steer a course between . . . the wish of 70 percent of Americans to keep out of war" and "the wish of 70 percent of Americans to do everything to break Hitler, even if it means war."

At rallies on American street corners, fights erupted between "interventionists" and "isolationists." Policemen on horseback broke up scuffles as true believers taunted their opponents, branding each other "Jews" and "Nazis."

The U.S. Congress barely passed a continuance of the Selective Service Act to extend the draft. In the House of Representatives, the vote was 203 to 202. The legislature even reduced the White House's emergency budget, a move that forced Van Bush to secure a stopgap loan from a private citizen, John D. Rockefeller, so that OSRD scientists could continue their vital work uninterrupted.

On June 22, Hitler invaded the Soviet Union, sending more than three million troops into a battlefront extending from the Baltic to the Black Sea.

Americans prepared for war even as they hoped to avoid it. Entire swaths of American industry were repurposed. A shirt and pajama factory won a contract for military uniforms. An agricultural equipment company in Milwaukee geared up to build Navy turbines. Car factories in Detroit were revamped to fashion parts for B-17 and B-26 bombers. GM's assembly lines, once dedicated to Cadillacs, readied to roll out M5 light tanks. Aiming to produce one bomber an hour, Ford began constructing a massive airplane factory that would employ some one hundred thousand workers. At shipyards, freshly welded hulls splashed into harbors. Smokestacks and swiveling cranes beckoned all laborers, mechanics, engineers, and technicians, regardless of their recent employment history.

Suddenly, it was difficult for Tuve to find qualified staff.

By the spring of 1941, many of the sharpest minds in the country, including fourteen hundred physicists, were working to protect the nation. Talent scouting became a frenzied vocation. Graduates were swiftly plucked out of university physics and engineering departments. Radio engineers were in desperate need.

Wrestling over recruits, leaders of rival project's leaders bickered. One MIT physicist even called back an employee of Section T without notifying its chairman.

"Who are you to telegraph Harris," Tuve snapped, "a member of my emergency crew busy on urgent defense work to report to your laboratory for work. Raiding is a capital offense unless done by the polite rules of war. Some regards."

Not everyone inside the pressure cooker of Section T could handle Tuve's sharp elbows. He had "very, very few diplomatic instincts," a coworker said. Of all of Bush's top scientists, Tuve provoked the most strife.

Merle, Bush wrote in a memo, was "very brilliant, not inclined to follow rules, and with a peculiar slant which sometimes makes him very difficult."

More than one associate called Tuve the most "dynamic" person they had ever met. One hire said he was "scared stiff of him." He was hated and loved, admired and feared. Many named him a "driver": he drove staff "unmercifully."

Section T's chairman had no time for wounded feelings.

In May, the goal in sight, he selected his top workers to focus on the final push to create the radio fuse. Then he relegated "Group B"—literally his B team—to designing bomb and rocket fuses on the second floor of the Cyclotron Building.

Smart fuses for AA guns were a military priority deemed nine times more valuable than bomb and newborn rocket applications. Just like that, Group B was locked out of the real action. Chairs from Group B's main office suddenly disappeared. They lacked shop equipment. Making matters worse, the temperature on the upper floor of the Cyclotron Building was ten degrees hotter than on the first. Group B staffers were forbidden to enter the cooler offices below that probably contained their furniture.

A hot spell in D.C. peaked at 93 degrees. On the second floor, sunbeams poured unmercifully through the glass bricks embedded in the walls.

In July, to make way for Tuve's planned expansion, Group B was removed from Section T entirely and relocated to the Bureau of Standards.

By the summer of 1941, American and British fuse researchers—each side pursuing different approaches to the problem—were doing their best to coordinate and share findings. OSRD had an office in London, and its counterpart in D.C., the British Central Scientific Office, was steps away from the White House.

But liaison with the British was not going well.

When Larry Hafstad, Tuve's vice chairman (described as the "balance wheel" to Merle's fiery intensity) visited England in June, he discovered that the British model of the radio fuse for bombs was far different than the American design. Reports had been exchanged, but prior to Hafstad's trip, no single lab had tested the latest British and American fuses side by side. "If this works," British researchers told Hafstad, of Section T's radio fuse circuitry, "we have a lot to learn about fuses." The British had been pushing hard for a final design, while Section T had been working through the problem in steps. One result was that British circuitry was significantly more complex than Section T's, even though both tested equally well.

Nor were British and American scientists on the same page about the trigger mechanism. Section T had in mind a unit that contained the radio transmitter inside the round. But British researchers were focused on the remote control or "pulse" fuse. Unlike a true smart fuse, the pulse design didn't send out a radio signal. That came from the ground to the shell as it flew near an enemy airplane.

Tuve was also pursuing a pulse fuse—as a backup. But he would soon abandon the design, which turned out to be impractical in battle.

Various fuse-related misfires and mishaps didn't help Anglo-American relations. Section T sent to England some of its innovative batteries, but the samples were badly packaged and lost their juice on the journey. American Hytron tubes failed in British centrifuges. English-made tubes sent to Section T likewise tested poorly. Hafstad witnessed a demonstration of British rocket fuses in which every single rocket missed the target. Churchill himself was in attendance.

Then there was Dr. Mark Benjamin's accident.

Chosen as British liaison to Section T, Benjamin arrived in Buffalo, New York, on August 17, 1941. A hard-working father of infant twins, he was a researcher at the General Electric Company outside London and an expert on vacuum tubes.

Hosting the Brit was Section T's Luther Grant Hector, a physicist from Buffalo in his forties. After dinner, Hector led Benjamin on a summer-evening scenic tour: Across the Peace Bridge, down the Niagara River on the Canadian side to the whirlpool rapids, then back to Niagara Falls. From the bottom of the American Falls, they took in the monumental power of the rushing torrents, the barrage of cascading water slapping the rocks, and the vast clouds of floating mist.

Over the next few days, Hector and Benjamin visited several tube manufacturers, including Raytheon in Boston and Bell Telephone Labs in New York.

On August 20, they set out for uptown Manhattan, home of Columbia University and Ray Mindlin's "Microphonics Laboratory."

In room 102 of Philosophy Hall—an edifice that boasts the original cast of Auguste Rodin's *The Thinker* on its front lawn—Mindlin had his own little Section T satellite lab. Today the building houses the English, Philosophy, and French Departments. At the time, it was home to the engineering school, which had generously helped Mindlin set up the space and acquire equipment, free of charge. The department even offered the services of one of its engineers, John Russell.

To tamp down the microphonics, Mindlin's team mounted tubes onto an electric "driver" machine and performed "shake tests." Vibrating the components, they measured the magnitude and frequency of the microphonic disturbances. Then, with an observing microscope and a stroboscope—a rotating device used to examine spinning objects—they would pin down the sources of the tiny noises, locating which miniature tube parts were vibrating the most.

Mindlin demonstrated the techniques to Benjamin, ran sample British tubes through their paces, and discovered the main culprit: the tiny springs inside.

In Washington, D.C., the next day and again that Friday in the Cyclotron Building, Benjamin met with Tuve, who had just returned from a two-week vacation with his family in Bethany Beach, Delaware. They discussed rugged tubes, sharing fuse components and data, and the rounds-per-bird problem.

Ten days later, heading home, Benjamin boarded a British military plane in Montreal. In his possession were special prototype tubes made by Raytheon. The aircraft was originally supposed to land in Prestwick, Scotland, but Prestwick was closed for maintenance. Under heavy, low clouds, after trying an alternative airport, the plane crashed into the top of a hill, littering a quarter mile of Scottish countryside with secret tubes and killing all ten passengers on board.

It took five days to find Benjamin's body.

Tuve was driving to lunch with the president of Raytheon, Laurence Marshall, when he put in Section T's first order for mass production. Ten thousand dollars, he figured, should buy about two thousand rugged "hearing aid" tubes.

"You know, Tuve," Marshall replied, "I am going to have to try to introduce you a little to the facts of life in manufacturing. Whenever we make a tube we throw away the first hundred or two hundred thousand dollars' worth of those tubes. We grind them up. They don't meet the specs. You have to run a procedure and develop a quality control, but the individual technicians make mistakes until it's gotten into a routine . . . especially when you are handling small things like these hearing aid tubes."

Prototype fuse parts that worked, it turned out, didn't guarantee all that much when it came to getting those prototypes produced in bulk.

As vice chairman Larry Hafstad quickly learned, the research methods Section T knew well were "just the opposite of what you do in the production cycle." While experimenting, scientists struggle to get their contraptions to work once. But in production, the job is to prevent a device from ever failing.

Researchers, Hafstad said, "collect their apparatus, tune it up, spend a lot of time working on it, take a set of readings, and get a Nobel Prize." Once the device works, "the job is done."

But volume production is a game of percentages. What's more, the rigors of battlefield conditions demanded a higher standard of dependability. Fuses would have to work after being dropped sideways in cargo holds, frozen in Norway, baked in Hawaii, and soaked in the humid airs of the Philippines.

Which may explain why, in August of 1941, when Navy Lieutenant Deak Parsons first saw a radio proximity fuse detonate, he didn't appear to care.

At Dahlgren's aquatic firing range, Parsons, Dick Roberts, and Henry Porter took the observer boat seven thousand yards down the Potomac, and tuned the radio unit to the squeal of the fuses. Ten shells zipped by on a fourteen-thousand-yard arc. Nine of the ten fuses inside the rounds failed, but one, curving down toward the impact area, its own radio signal reflecting back off the brackish water, burst perfectly in the distance, fifty feet in the air, before splashing into the drink.

Parsons hardly even reacted.

It was the first time that a smart fuse in a shell had ever worked. But while Parsons knew how badly the Navy needed the fuse, he also knew military ordnance standards. One success in ten wasn't nearly acceptable. The Navy required at least 50 percent reliability to put the fuse into production.

Section T had to make the fuse five times more dependable, a goal that required testing far more prototypes than they could assemble at DTM.

For help, Tuve turned to the Erwood Sound Equipment Company in Chicago. Erwood produced audio gadgets, including portable sound systems with built-in speakers, radios, microphone inputs, and record players with automatic disc changers, all packaged in "attractive tweed covered" carrying cases.

Brothers Joe and John Erwood had founded the company only two years earlier, in 1939. Joe was the mechanical expert, electrician, and experienced machinist foreman. John, a former Navy man with a boxlike jaw and high, puffy hair, worked the business side with help from their nephew "Buzz" Beck.

Merle had first pitched John Erwood on assembling fuse parts earlier that year, at a meeting in the Cyclotron Building.

"Now, of course," he told Erwood, "I wouldn't want you to jump into this thing and commit your company for several years without taking some time to think this over. Go out on the lawn and think it over for fifteen minutes."

Within weeks, Erwood's small shop in a loft on West Erie Street, just north of the Chicago River, began building fuse parts for Section T.

Now Tuve needed them to scale up production.

The sound company was as new to the particulars of radio fuses as the physicists once were to gunpowder and military firing ranges. Erwood's fuse team would, over the coming months, recruit staff from a wide array of professions. They hired a baker, bartender, venetian blinds supplier, window washer, and gas-station attendant. Erwood enlisted a cobbler, a cop, a farmer, an advertising copywriter, a laundry-truck driver, a carnival barker, and a professional wrestler. Women assembled the most delicate fuse components, work that required poised, exacting fingers, and exceptional focus.

For dependable statistics, Tuve needed the company to deliver over a hundred fuses a day. Every new fix and innovation had to be tested in batches of at least fifty, to make sure that a jump in test scores wasn't just random chance. Not only that, but each batch of fuses—every design variation — had to be utterly identical. Unless each configuration was assembled in precisely the same way, it would be impossible to pinpoint which changes were affecting the test scores.

At the same time, Tuve needed higher-quality components. Since each unit had some five hundred components, the overall performance of a fuse could be dragged down radically by mediocre parts. Assuming, for example, that three hundred components were important enough that if any of them failed, the device failed, then even if each part was 99 percent reliable, only 5 percent of the assembled fuses would work.

The fuse, in other words, wasn't an "invention" measured by binary criteria, like a light bulb that suddenly flicks on, or the atomic bomb. It was slowly developed, painfully, through small changes and minor adjustments.

It was a grueling statistical challenge.

The most delicate parts inside Section T's prototype tubes — the microthin filaments — could now survive a cannon blast 90 percent of the time. It wasn't good enough. The truth was, the shift to mass production was proving to be more troublesome than Tuve could have possibly guessed.

Ramping up was causing all sorts of new headaches.

On September 8, 1941, Section T disclosed, in a letter to Van Bush, that they were violating secrecy protocols. The shortage in qualified personnel had grown so dire, the memo said, that they had been forced to employ new hires immediately, without waiting for their pending security clearances to process.

Otherwise, they said, "we would lose at least 90 per cent of the applicants because of the present demand for skilled professional and technical men."

Of course, Tuve was still keen to emphasize the gravity of keeping secrets. Hafstad, who did most of the interviewing, went so far as to borrow a hiring technique from the Navy's Deak Parsons. Hafstad would bring candidates in and discuss their credentials, describing the rounds-per-bird puzzle and the challenges they faced. Then he would pull out a piece of paper and sketch out the "story" of the fuse, including all the particulars and critical statistics.

At the end of each interview, in front of the new employee, Hafstad would tear up the paper, put the shards in an ashtray, and set them on fire.

Merle adopted another, more basic protective strategy. "If we reach out in the United States and pick people," he reasoned, "the probability of getting a traitor is awfully, awfully low . . . one in ten million or something. I don't worry about that. The only people I worry about is people who come to us. So we'll only pick people that are initiated from here."

They hired people they knew: family and friends, and then friends of those friends, expanding their recruiting pool with each new addition.

Dick Roberts brought in his two brothers, forming a "Roberts triumvirate." Walter was a radio-circuitry expert, and Tom helped with test shoots.

Tuve drafted the Danish physicist Charles Lauritsen and his twenty-six-year-old son, Thomas, who came to America with his father as a small child. He hired R. K. Squire from down the block as an assistant radio engineer, and also signed up R. K. Squire Jr. He hired two young brothers, F. R. Nichols and W. A. Nichols Jr., from Silver Spring, Maryland, as a lab aide and a clerical assistant.

Isabelle Lange, a "computer" at the Department of Terrestrial Magnetism, was roped in by Section T for her skill with technical computation and analysis. Likewise enlisted was Marcella Phillips, a visiting physicist at DTM.

A mustached draftsman, H. W. Bixby, was hired to draw blueprints.

Bright, youthful faces arrived and mingled amidst the gray hair, bald heads, three-piece suits, and austere glances of the older staff. Fresh recruits posed for ID photos with the kind, open expressions of eager adolescents, spines stiff like GIs at attention, some in cardigans, or without ties, or with pens in their pockets and the untamed hair of rabid thinkers. Nervous, ready faces. Older physicists offered slightly aloof poses. Technicians relaxed in suspenders.

Merle's team also filled up with young assistants, typists, and stenographers in pearl necklaces, dresses, and blouses with puffed shoulders: Nancy Warner from Hawaii, Shirley Puffer from Bridgeport, Connecticut, and Margaret Dike from Reading, Massachusetts. Evelyn Wood was a twenty-year-old dual citizen born in Paris, and Elizabeth Small, Section T's telephone operator, came from Haddon Heights, New Jersey. Section T's average age was now around thirty.

New employees flooded the vacant second floor of the Cyclotron Building and, as staff expanded further, filled up the Experimental Building. The Department of Terrestrial Magnetism was too small.

"During the last six weeks," Tuve wrote, "we have been continuously confronted with the need for certain types of expansion in staff and in physical facilities."

He was overwhelmed by technical and organizational problems.

Fuse components that worked individually failed when assembled, and every jump in production quantity seemed to spark a drop in quality, requiring closer liaison with industrial partners. Which meant more staff.

The quality of the vacuum tubes had declined. The filaments kept breaking. Putting other work on hold, Tuve formed "Emergency Group X" and assigned the bashful James Van Allen to work with Ray Mindlin and tube designers at Raytheon.

From digging up test rounds at Stump Neck, Van Allen guessed the problem was not the filaments themselves but how they were anchored. So he sketched a new type of spring that wound over a crossbar. Inside a tube, the spring would sit on the top mica platform. The design looked like a crane on the roof of a building holding up an elevator inside by a cable — if the building were the size of an almond.

Devised in mid-November, Van Allen's spring was a breakthrough. Seventy-two tubes with the new design were shot. All seventy-two filaments survived.

Later patented, it was called the "mousetrap" spring.

Just as soon as the filament problem seemed to be solved, however, microphonics troubles arose. One test of ten units at Dahlgren's range on the Potomac saw all ten fuses fail. While only four of the units were duds, six detonated prematurely, likely because of microphonic noises.

Erwood still wasn't delivering complete fuses from Chicago, and Section T's Secret Assembly Shop was struggling to build identical units.

Section T staff strained under the pressure. There was a "dangerous accident" at a firing range, apparently caused by someone disassembling a fuse. Feeling overwhelmed, Tuve's secretary quit because, he speculated, of "the confusion of working with a whole group of unseasoned personnel."

Warren Weaver, a statistician advising Section T, was so annoyed by the disorganization that he threatened to resign. He penned a nasty letter to Tuve, complaining that he had failed to receive copies of several reports.

"I am frankly shocked that you would ever put your signature to such a document," Tuve replied, and then forwarded the entire correspondence to their boss, Richard Tolman. Weaver griped that Tuve, however "ingenious and lively" his ideas, was behaving like "an ill-tempered adolescent."

According to OSRD's London office, the British were increasingly frustrated by a lack of coordination with the Americans. One British scientist unleashed a "torrent of complaints" about rugged tubes and, as Tuve put it, "the assumed way in which Section T conducts its business." Criticizing American tubes, Tuve wrote, was "most audacious, in the light of the failure of their tubes in our tests here." Reports out of Section T *were* in fact scarce, but that was because of the "desperate shortage in scientific and technical men for the development work itself."

By November, Van Bush had heard enough complaints to take extreme measures. He dispatched a personal spy of sorts, a junior researcher named David Langmuir, who might provide an impartial evaluation of Tuve.

"An absolute master," Langmuir reported back, "totally in charge of the situation and absolutely brilliant. He has so many good ideas his thing just can't fail."

Bush was deeply relieved to hear the news. "Of course he is a wonderful leader," he replied. "[But] damn it, he almost drives me nuts."

To cope with the stress, Merle smoked incessantly.

"For the last 16 months," he explained, in a feisty memo, "I have felt nothing but a heavy load at the top . . . like a cork bumping the top of a filled tank." His team was bursting at the seams, and could no longer make do with whatever offices happened to be free at the Department of Terrestrial Magnetism.

On Saturday, December 6, 1941, he drafted what he called a "rambling memorandum," arguing forcefully that Section T was in urgent need of its own facilities to see the fuse project through to completion.

"TIME IS SHORTER THAN WE THINK," he wrote.

The next morning, two privates at a radar station near Pearl Harbor, Hawaii, picked up an odd formation of planes approaching from the north.

Part II

WAR

This was a secret war, whose battles were lost or won unknown to the public; and only with difficulty is it comprehended, even now, by those outside the small high scientific circles concerned. No such warfare had ever been waged by mortal men.

— WINSTON CHURCHILL

12
READY OR NOT

n Washington, D.C., football fans headed to Griffith Stadium. They could not have imagined that in the Pacific, bombers bearing the red symbol of the rising sun of Japan were streaking over Hawaiian sugarcane and pineapple fields. Nor was there any way for them to hear of the surprise attack once news broke.

The twenty-seven thousand or so spectators attending the Redskins game against the Philadelphia Eagles that Sunday were a dutiful bunch. The scrum was the last of the regular season, and the home team was already out of the playoffs. But if the Redskins won, at least they'd finish the year with a winning record.

"Slingin'" Sammy Baugh, the Redskins' quarterback described as "the passin' Texan," wasn't going down without a fight. It was sunny out but cold enough that fans could see their breath. They stomped their feet to stay warm.

For eight full minutes after kickoff, while the rest of the nation began to catch news of the catastrophic raid, none of the onlookers in Griffith Stadium heard a word of the tragedy. Then, in the press box, the Associated Press's sports reporter Pat O'Brien received an odd note from over the wire.

"Keep it short."

He turned to his neighbor, the *Washington Post*'s Shirley Povich. "For five years I've been covering these Redskins games," O'Brien complained, "and now some jerk is telling me how much to write." He pivoted back to the operator.

"Ask 'em who's giving me these new orders." Minutes later, they heard the horrid reply. "The Japs have just kicked off. Pearl Harbor bombed. War now."

For a short while, O'Brien, Povich, and the telegraph operator grappled alone with the "stupefying" news. Could it really be true? Wasn't it a mistake?

The public address system inside the stadium began to page important attendees. Midway into the first quarter, the chief of the Bureau of Ordnance for the U.S. Navy, William Blandy—the man who had first told Tuve of the Navy's need for a fuse—was called over the loudspeakers and asked to return to work.

Another message soon followed. "The resident commissioner of the Philippines, Mr. Joaquim Elizalde, is urged to report to his office immediately!"

When the Redskins' general manager heard the news, he decided not to use the PA system to announce the attack—out of fear of hysteria, he said later.

"Mr. J. Edgar Hoover is asked to report to his office," loudspeakers blared.

By the second half, all but one of the gaggle of photographers covering the game were gone, summoned by their editors. Newspapers began to recall their reporters. The wife of one newspaper editor dispatched a telegraph messenger into the stadium. "Deliver to Section P, Top Row, Seat 27, opposite 25-yard line, East Side, Griffith Stadium," the note read. "War with Japan Get to office."

In a box on the fifty-yard line, one of President Roosevelt's cabinet members, Secretary of Commerce Jesse Jones, was handed a note and left.

"Joseph Umglumph of the Federal Bureau of Investigation is requested to report to the FBI office at once," said the PA system.

"Capt. H. X. Fenn of the United States Army is asked to report to his offices at once."

Admirals and colonels were called away.

A Redskins lineman, Clyde Shugart, like many in the stands, sensed that something serious must have happened. But very few knew exactly what it was. When Slingin' Sammy Baugh threw the winning touchdown into Joe Aguirre's "big mitts," the crowd was on their feet, cheering like nothing was wrong at all.

"We didn't know what the hell was going on," Baugh said. "I had never heard that many announcements one right after another . . . we just kept playing."

It wasn't until fans left the stadium some three hours later that news of the disaster finally reached them at the gates—like a "thunderclap." Outside, newsboys distributed "extra" editions with damage estimates that turned out to be painfully optimistic. Initial reports of military deaths were in the hundreds.

Americans who happened to be listening to the Redskins on the radio heard about Pearl Harbor before the fans actually attending the game. Across the country, regularly scheduled broadcasts were interrupted by urgent reports.

A news bulletin disrupted the New York Giants football game. CBS broke into a broadcast of the New York Philharmonic Orchestra. NBC cut off its Great Plays series adaptation of Nikolai Gogol's satire *The Inspector General.*

Americans enjoying the simple pleasures of a calm, relaxing Sunday were left to cope with the ramifications of a generation-defining event. Churchgoers were returning from Sunday services. Movie matinees were halted by managers who took the stage before confused audiences and read early news reports of the bombing. On sidewalks, pedestrians clustered around car radios, listening for updates. A peculiar sense of fellowship sprang up between Americans of all backgrounds. Strangers chatted with strangers like they were suddenly close friends.

General "Hap" Arnold, head of the Army Air Forces, was hunting quail in California. A sheriff had planes drop leaflets over his location with the news. In rural Virginia, Deak Parsons was late picking up his seven-year-old daughter, Peggy, from Sunday school. She and a neighbor's boy were left playing in a graveyard until he arrived with news that she did not understand.

Across the United States, private aircraft were grounded.

Soon, Americans gathered at Red Cross stations to donate blood. They congregated on Pennsylvania Avenue outside the White House with children on their shoulders, peering quietly at the illuminated windows.

FDR, Eleanor Roosevelt said, was "deadly calm."

He was "completely calm. His reaction to any event was always to be calm. If it was something bad, he just became almost like an iceberg."

Rumors spread that Japanese troops were parachuting into Hawaii and that saboteurs were wreaking havoc across the islands. Word circulated of an upcoming attack on the mainland, including a strike in D.C. by Japanese sleeper agents. Reports of a Japanese aircraft carrier floating off San Francisco closed schools in Oakland, California, the next day. On Wall Street, stocks plunged.

At Fort Lewis, an Army base outside of Tacoma, Washington, thousands of troops were evacuated in case of another raid. In San Francisco, boats were ordered to remain at anchor. Lights on the Golden Gate Bridge were shut off.

Factories were ordered to step up their protections against sabotage. Around Los Angeles, some four thousand antiaircraft troops were deployed. Guards were stationed at the water aqueduct. Blackouts were ordered for the nearby harbors.

Streetlights were dimmed, roadblocks erected, and borders closed. At government buildings across the capital, armed guards were posted.

In New York City, shopkeepers grabbed clothing marked MADE IN JAPAN and tossed the garments into the streets. Police officers visited Japanese restaurants and, after allowing diners to finish, escorted owners safely to their homes.

The FBI was tasked with arresting any suspicious Japanese nationals. Hundreds were immediately apprehended, joining some twelve hundred Germans and Italians already locked up in detention centers in North Dakota and Montana.

In a bonfire at the Japanese embassy in D.C., attachés burned codebooks and other sensitive documents while reporters ogled the glowing blaze.

Recruiting offices filled with applicants who spilled into the streets. Women arrived at defense training centers asking how they could contribute.

Labor strikes were called off. The government placed ads for blacksmiths and machinists. Defense plants were put on twenty-four-hour production cycles.

That Monday in Springfield, Massachusetts, at Classical High School, students filed up the concrete steps and marched under the arched doors into the grand Renaissance Revival–style building as they normally would. Among them was a sixteen-year-old with dark hair, a broad face, and big quiet eyes named Samuel Edward Hatch, called Ed. He had not yet heard of the 130th Chemical Processing Company of the U.S. Chemical Warfare Service.

Around noon, the students gathered in the assembly hall, which quickly grew overcrowded with bright young lives. Hatch perched on a staircase as a loudspeaker carried President Roosevelt's address to the nation.

Japan hadn't attacked only Pearl Harbor. It also attacked the Philippines, Hong Kong, Malaya, Singapore, Guam, and Shanghai. Churchill had declared war on Japan. Canada had entered the scrum. China, at last, would formally declare war.

In the House chamber in the Capitol Building, a Senate committee that included Senator Carter Glass of Virginia escorted the president to the rostrum. Some watching the speech live were ashamed to be crying until they noticed that others around them were also in tears. "There is no blinking at the fact that our people, our territory, and our interests are in grave danger," Roosevelt said.

A declaration of war was approved 82 to 0 in the Senate, and 388 to 1 in the House.

At Hatch's high school, "word came down" to make gym class tougher. Physical education for teenagers would now be preparation for military service.

For his part, Adolf Hitler was skeptical of American resolve. "The Americans are a bunch of rowdies," Hitler said, after the attack. "They won't fight."

Clenching the bow in his left hand, pulling the string taut, Van Bush gazed through rimless glasses at the bull's-eye, trying not to lose sight of it on release, as he often did, and to quiet his left arm, which tended to "wave all over the landscape."

Bush, the scientist czar of military technology, responsible for ushering in a new wave of modern armaments, spent the summer of 1941 trying to master a weapon pioneered some five thousand years earlier, in ancient Egypt.

Every Sunday, he practiced archery on D.C.'s National Mall.

The Potomac Archers club to which Bush belonged was, in his words, a "congenial group" of Washingtonians from many walks of life. If an expert saw a beginner making an error, they kindly stopped to offer tips. If someone lost an arrow—ruining the set—everyone helped search. To locate arrows burrowed under the grass, women slipped off their shoes and felt around with their bare feet.

The odd hobby—one of many—was part of Bush's strategy to "live a reasonably sane life," and not "crack up" under the mounting pressure.

There was a lot on his plate.

In July, when Bush updated President Roosevelt on his progress, the Office of Scientific Research and Development had already authorized 207 contracts, including 155 with academic researchers. He recruited from America's top universities. Brown, Columbia, Cornell, Harvard, the University of Pennsylvania, Princeton, Yale, the University of Chicago, Johns Hopkins, MIT, and Stanford were all under OSRD contracts. Dartmouth would soon join them.

Over the course of 1940 and 1941, Bush and his advisers disbursed more than six million dollars. Over the next fiscal year, as emphasis shifted after Pearl Harbor from development to production, that figure increased to nearly forty million dollars.

In addition to working on weaponry, Bush and his scientists were now also busy with overseeing research and development into new military medicines. OSRD's Committee on Medical Research would work to counter infectious and tropical diseases, develop novel insecticides and aviation medicines (including treatments for altitude sickness), synthesize blood substitutes, treat traumatic shock, and mass-produce penicillin to beat back soldiers' infections.

Under OSRD's Division A, headed by the physicist Richard Tolman (Tuve's immediate boss), researchers were doing more than developing a smart fuse. Outside of Section T, scientists were investigating jet propulsion, new bombs, bombardment defenses, terminal ballistics, and tougher armor.

Karl Compton, the jock-scientist president of MIT and head of Division D, was supervising groundbreaking work on airborne radar systems that could locate planes through darkness, cloud, or fog. His scientists were working on precision navigation equipment, radar to detect enemy vessels at sea, and a better system for aiming antiaircraft guns, including improved range finders.

Under Frank Jewett, the rancher's son turned businessman, Division C had developed new portable bridges, an infrared telescope, and a compass that could work inside the belly of a metal tank. Jewett's scientists were exploring how to use ultraviolet light to land airplanes at night. They had researched sonic weapons that might cause "permanent impairment to the human auditory mechanism." Certain avenues for such a weapon were promising—frightening enough in their potential that "work on ear defenders" had already been prioritized.

The less savory projects within Van Bush's purview followed a similar, morbid logic. OSRD could not know what German's scientists were cooking up. How can one defend against unknown weapons without exploring which ones are scientifically feasible? To guard against horrors, they hunted nightmares.

With Harvard's president James Conant serving as Bush's deputy, Division B was now supervised by Roger Adams of the University of Illinois, one of the premier organic chemists in the country. Division B scientists had already devised a cheap method of synthesizing the high explosive RDX. They engineered sensitive paints and papers that changed color in the presence of toxic gases. Of particular interest to the Chemical Warfare Service, Division B also synthesized seventeen new compounds for "preventing the penetration through clothing and for protecting the skin against the action of [blistering agents] such as mustard gas."

Adams's scientists were not only analyzing known gases but actively searching for new ones. They had developed over thirty gases that ranged across "all the classes of war gases," including tear gases, sternutators—also known as "vomiting gases"—lung irritants, and blistering agents.

In OSRD's final meeting before Pearl Harbor, on November 28, 1941, Adams suggested that blistering agents be tested on human subjects.

Aside from a discussion about isolating uranium isotopes using Ernest Lawrence's modified cyclotron, OSRD's meeting that day was standard. Bush approved new contracts with major universities and companies like General Electric, Western Electric, Eastman Kodak, and the

Bakelite Corporation. He approved money for the California Institute of Technology to research underwater projectiles, and for the Stevens Institute of Technology to study the maneuverability of high-speed vessels. The University of California was granted three hundred and fifty thousand dollars for submarine studies. The Bureau of Mines received one hundred thousand dollars for retrieving "certain types of powders." Johns Hopkins was given additional cash to study war gases. Harvard and the University of Chicago were approved to research accelerants. Wesleyan scientists would study flash bombs.

Despite foreseeing war, Bush was shocked by Pearl Harbor.

Over twenty-three hundred U.S. servicemen lost their lives in the sneak attack. The Army Air Forces planes on the island of Oahu were destroyed. Eight battleships, three destroyers, a minelayer, and three cruisers were sunk or damaged. The attack neutralized much of the Pacific Fleet of the United States.

Like most Americans, Bush's family was affected by the raid. His youngest son, John, was newly enrolled at Haverford College and was currently in training to join the Air Forces. "The nature of the world in which we live has radically altered," Bush wrote, shortly afterward, "and I have hardly a moment to think."

Then, just three days after Pearl Harbor, more bad news. North of Singapore, the British battleship *Prince of Wales* and the battle cruiser *Repulse,* navigating without air support, were bombed by Japanese aircraft and quickly sunk.

Only three Japanese airplanes were shot down.

The incident was a stark warning. The British were aghast, and the U.S. Navy was horrified. These were the two most powerful British ships in Asia. The "most modern British" warships, as Bush wrote later, "depending on their own antiaircraft guns for protection" were simply no match for airpower. Battleships, a centerpiece of U.S. naval strategy, could not defend themselves. Historians later described it as the day the battleship became obsolete. New technologies were changing the nature of warfare, at the strategic level, in real time.

At Pearl Harbor, the American radar that had warned of approaching planes was ignored, in part because the technology wasn't fully respected. During the attack, additional radar units were left in crates on the naval docks.

Out of the 353 Japanese planes in the assault, just 29 were shot down. Most of those were claimed by American planes late in the raid. U.S. antiaircraft guns sputtered lamely against the bombers. Dated fuses in the antiaircraft shells missed their targets and exploded instead, fatally, on Hawaiian streets.

13
NO ALIBIS

On December 10, 1941, hours after the *Prince of Wales* and the *Repulse* plummeted to the Pacific seabed, Tuve received a call from the Navy.

Lieutenant Victor Hicks, of ordnance research and development, told him, in Tuve's words, that "on the basis of secret reports from the Pacific, the Office of the Chief of Naval Operations and the Bureau of Ordnance jointly considered it <u>most</u> urgent to concentrate <u>every possible effort</u>" toward the smart fuse.

The Navy's vulnerability to airpower and the consequences of the rounds-per-bird problem were suddenly, shockingly obvious. Within days, Van Bush and the Navy infused Section T with six hundred thousand dollars in emergency funds.

Money alone wasn't enough. Racing to produce a fuse that worked 50 percent of the time—the Navy's threshold to accept the gadget—was straining other resources.

To deliver the device even faster, Tuve needed more data. He had to increase the number of test shoots. But for that, he needed more fuse parts, extra field crew, and an additional proving ground farther south, where it was warmer. Stump Neck, their main test site, was freezing.

Dahlgren's aquatic range was ideal for complete fuses, but Stump Neck was better for "analytical shoots" of partial devices to isolate sources of failure. At Dahlgren, shells sunk in the Potomac; at Stump Neck, rounds fired vertically were dug up for "postmortems." As winter set in, fuse parts were breaking when they crashed into the hardened earth, making statistical analysis far more difficult.

The U.S. Weather Bureau reported that the dirt was frozen to a depth of up to fifteen inches. Simply gouging the rounds out of the packed ground for examination was becoming "almost impossible," Tuve complained in a weekly report.

Nor would finding another test site be easy. Shoots required at least a mile radius of clear, safe ground. The landing area had to be free of vegetation. The earth couldn't contain water or rocks to a depth of seven feet. The site had to be near D.C., secluded, easily patrolled for secrecy, and, of course, "frost-free."

Faced with their new "overwhelming shooting schedule," Merle's gun crews were hampered by the ebbs and flows of the cold, and by the blistering winds, which blew the test rounds off course and made them toilsome to find.

Tuve's goal was to fire at least two hundred shots a week. But he didn't have that many staff members with field experience. "Urgent requests for acceleration of our tests can only be met by using 'green men,'" he grumbled, on January 7. "The results show clearly in the scores at Dahlgren and in the duds and breakages at Stump Neck."

December's field tests had been rather discouraging. Twenty-nine percent of the fuses shot on December 10 worked. Five days later, only 17 percent did. The week after that, 18 percent succeeded. And on December 31, a 13 percent score.

Merle had to double his field crew, and train the fresh recruits.

The miniature-tube manufacturers, pressed to step up deliveries to meet the new test regimen, also had to add inexperienced workers. The problem was serious enough that Section T had to institute design changes intended, Tuve wrote, to make the tubes "less vulnerable to new hands added to the production lines." The extremely delicate mountings for the tungsten filaments were simplified.

Section T's Secret Assembly Shop, completely overwhelmed, struggled to produce the large batches of exact replicas that were necessary for proper statistics. Neither its machinists nor the Erwood sound company in Chicago, now helping with the task, seemed able to deliver the gadgets exactly as ordered.

Tuve "couldn't get 25 units that were identical."

The smallest, most innocent changes in assembling fuses could radically hurt test scores. With the best of intentions, Erwood took shortcuts that did just that. For instance, at Section T, the technicians had learned to put a miniature kink in the external tube wires. The loops provided slack to absorb the firing shock. Erwood ignored that step, and without the tiny kinks, the vacuum seals broke.

Even the Crosley Corporation, a heavyweight manufacturer, struggled to grasp just how delicate and finicky a problem producing the fuse would be.

Based in Cincinnati, Crosley built refrigerators, cheap cars, and high-quality radios. Just like Erwood, the company was run by two entrepreneurial brothers, Powel and Lewis. Lewis was a practical industrialist. Powel was an inventor and pioneer in radio advertising once dubbed "the Henry Ford of Radio." But unlike Erwood, the company had massive facilities and resources. Over the first half of 1941, Crosley's sales totaled over fourteen million dollars. (The brothers also owned and ran the Cincinnati Reds, who were fresh off a World Series win in 1940.)

Crosley's vice president of research and development, Lewis Clement, whose defining feature was a thin shock of gray hair in the middle of his dark locks, had first learned about Section T's project in October of 1941: Larry Hafstad, Lieutenant Victor Hicks, and an inspector of naval materiel simply arrived at Crosley's headquarters, a nondescript building with big factory windows off the Baltimore and Ohio Railroad. They asked to speak to the vice president of engineering. The only prior warning, Clement said, was a letter stating that "we would be contacted that month on a very important, top secret, top priority job."

In Clement's sixth-floor office, with the door locked and shades drawn, Lieutenant Hicks laid his gun on the desk and asked Clement to sign an acknowledgment of the Espionage Act. Then Hafstad uncovered a prototype of the fuse he wanted replicated.

Crosley dedicated six hundred square feet and fifteen engineers to the task.

But, just like Erwood, Crosley assumed they could assemble fuses better than Section T. They kept "making what are considered 'minor mechanical changes.'" For one batch, Crosley used heavy brass noses instead of aluminum ones, causing the "projectiles to tumble in flight . . . the noses themselves flew off." On another shoot, "all the noses except one were blown off."

"They have achieved at best a poor performance in the field," Tuve wrote, "showing a lack of any basic notion of how such a problem can be systematically approached."

Merle quickly grew allergic to so-called "obvious improvements." He scribbled a new decree for design changes for prototypes in his aqua notebook: "<u>Absolutely</u> only <u>one</u> change at a time and test <u>every</u> change." And he had Hafstad personally pay a visit to Erwood's shop in Chicago. Hafstad preached "to the assembled group that even if they changed the color of the paint" on the prototypes, Section T would need to shoot fifty samples of the new fuses at Dahlgren.

The entire smart fuse project was beginning to feel, as Hafstad later described it, "like an endless series of high hurdles with one crisis after another."

They tinkered, generally, with hundreds of minute adjustments, changing the core design of the fuse. They switched the type of radio transmitter circuit they were using. To reduce microphonics, they swapped a two-filament tube for a single-filament one. They modified the circuitry of the amplifier tubes to make them less noisy. They cushioned the glass tubes in soft rubber envelopes. Through "analytical shoots" at Stump Neck, by testing individual parts, they pinpointed why fuses tended to blow up too early or not at all.

A month after Pearl Harbor, these tests yielded the first design based on statistically targeted changes. They dubbed the new fuse model "Prescription A."

On January 29, at Dahlgren, Deak Parsons had Prescription A fuses fitted in fifty-some-pound shells. Then the prototypes were fired over a seven-mile trajectory at an initial velocity of 2,600 feet per second and an angle of 29 degrees.

Aboard the frigid observer boat, Merle penciled down the results, noting each success, dud, and premature explosion. He carefully recorded the height of each burst, after about thirty seconds of flight time, as the fuses detonated over the Potomac.

Of the fifty rounds shot, twenty-six worked.

Fifty-two percent.

Parsons phoned in the results to the Bureau of Ordnance, and Captain

Samuel Shumaker, director of Research and Development, summoned Larry Hafstad to his office. Bundled in winter gear, the physicist was swiftly ushered in.

"This proves you know what you're doing," Shumaker said. "From now on, the sky's the limit. We're going into production as fast as we can. Good luck."

The Navy initiated production contracts worth eighty million dollars.

On February 24, an ordnance worker at a Navy ammunition depot in D.C. was arrested on a routine traffic violation. He had two aluminum fuse caps in his possession and was charged with larceny of federal property. Tuve stuck a clipping of the incident to the Cyclotron Building bulletin board along with a note. MORAL: DO NOT CARRY SECRET PARTS IN YOUR POCKETS.

The atmosphere at Section T had changed. The scientists toted guns now. After Pearl Harbor, the Navy insisted that all employees tasked with protecting fuses, design specs, and reports be armed, and loaned out revolvers until Section T could order its own pistols.

To carry the guns across state lines for test shoots, key personnel were formally appointed as deputy sheriffs. (According to an unwritten code among peace officers, Merle was informed, sheriffs could carry concealed weapons anywhere, as long as they didn't get into brawls or brandish their sidearms.)

James Van Allen, the physicist so shy that he used to avoid eye contact, was commissioned a deputy sheriff of Montgomery County, Maryland.

He had never fired a gun in his life. At the nearby Marine Corps base in Quantico, Virginia, he was trained by "a crusty old drill sergeant."

Van Allen was scared of his own pistol.

The Navy sent the Department of Terrestrial Magnetism instructions on minimizing "damage that might arise as a result of an air raid." The memo advised adequate fire equipment, including dry sand and hand pumps for incendiary bombs.

Section T's stenographers were tasked with locking up all confidential materials in the Cyclotron Building's safes when the air-raid warnings sounded. DTM participated in blackout drills.

Security everywhere was tightened.

At Crosley, fuses were escorted by armed guard from the assembly room to the loading platform, where sentries kept their "revolvers in their hands." Truck drivers shepherding the units were armed, as were the escorts of the "money shipments" of the valuables on the Baltimore and Ohio Railroad to Washington.

Section T's rugged tubes were deemed so valuable that Tuve had to refuse other U.S. military requests for information about their unique design.

Pressure to deliver the fuse mounted with every Axis victory. Japan expanded its Pacific empire by thrashing the Allies in the Philippines, in the Dutch East Indies, and in Burma. The British endured a humiliating defeat at the Battle of Singapore, losing that city and suffering the largest British surrender in history, with eighty thousand troops captured as prisoners of war. And the U.S. Navy suffered another devastating rout at the Battle of the Java Sea.

Deak Parsons estimated that delays in delivering the fuse and ushering the device into battle in the Pacific was costing the Navy the equivalent of a battleship every three months, a cruiser every month, and a hundred and fifty sailors every day.

In less than a year and a half, Merle's "unseasoned" ragtag crew had achieved the unimaginable, and seemed on the brink of success. They had run through the full gamut of different types of fuses, narrowed their options, redesigned subminiature tubes to withstand twenty thousand times the force of gravity, and reduced the microphonics inside them to over ten times less than other models.

The highest-quality rugged tubes were now being produced by Raytheon and a relative newcomer to Section T's project: Hygrade Sylvania.

Employing some six thousand workers, Hygrade Sylvania built lamps and lighting fixtures. A division known simply as Sylvania made radio tubes. By 1937, Sylvania's workers were assembling—using magnifying glasses and fine tweezers—321 different tube types. The company had a vast red-brick factory in Emporium, Pennsylvania. It even had its own publication, *Sylvania News,* with articles on tube breakthroughs and features like "A Chat with Roger Wise," chief tube engineer.

Like Crosley, Sylvania had the capacity to produce in volume. Production of rugged tubes had gone from one or two "expert hand-operators" to a dozen "girls" on a pilot production line to large numbers of young women working at long factory benches in the usual mode of mass production. As volume rose, the quality fluctuated. But by February of 1942, Section T contractors were producing roughly three thousand tubes a day, with reliability at over 98 percent.

Tuve's success caught the attention of other military services. Now that the fuse worked, the U.S. Army and the British Royal Navy were eager to have it adapted for their antiaircraft guns. But while U.S. Navy shells were five inches wide, Section T's new clients used rounds as small as three and a half inches in width. Shrinking rugged electronics for the Navy had been agonizing enough. But the "Army-British" fuse needed to be almost impossibly small: one and a half inches in diameter instead of two. It was not at all clear the demand was realistic.

Merle assigned the task to James Van Allen and Edward Salant.

Ed Salant was in ways the opposite of the shy, rural Van Allen. An associate professor of physics from New York University and an old friend of Tuve's, he was an extrovert, a "city boy," and a hard-living workaholic. Van Allen weighed one hundred and thirty-seven pounds and was five foot six and a half. Salant was four inches taller and twenty-eight pounds heavier, not to mention fourteen years older. Van Allen had a full head of hair. Salant was impressively bald, with thinning brown wisps over his ears. Van Allen had a handsome, ruddy complexion and an immediately obvious gentleness about him. Salant was curt and pale, with a mole on his left cheek and a scar on his right ankle.

Compared to Van Allen's Section T reports, which often filled many pages with rich, detailed descriptions, Salant's weekly summaries to his supervisor were comically terse. "Worked with you around lab. Collected some equipment and considerable misinformation from colleagues," a report read in full. "Collected and studied diagrams and equipment for tests" was an entire report. "Constructed oscillator test shell. It works," read yet another.

Tuve also appointed Salant to replace the late Dr. Mark Benjamin as Section T's Anglo-American liaison. If Van Allen and Salant succeeded in building a smaller fuse for the British, Salant would represent Section T in England.

As things stood, a fuse for the British looked highly unlikely.

Salant and Van Allen were not alone in the task but they didn't have much help. Tuve could spare only eight men total for the job, and six were crammed inside room 105 of the Cyclotron Building down the hall from his office. The room was meant for two; Salant and Van Allen barely had enough space to stretch their arms.

The project needed four times the space and ten more men. The pressure from the Navy, and the additional fuse requests, pushed Merle to the brink. He pleaded with the Navy to help him find an additional test field, to little avail. He pleaded with Director Fleming, who wouldn't help him rent space outside of DTM or expand staff further inside it. He pleaded with Division A's chairman Richard Tolman, who he believed had an "unconscious dislike for unexpected additional responsibilities, particularly if they involve differences of opinion."

In February, Merle took his case directly to James Conant.

In a letter, he explained to Van Bush's second in command that Section T was "unable to carry out the most obvious and unquestionably necessary additions to our facilities." Neither Director Fleming nor Richard Tolman had an active part in the work, and yet they controlled it. He wanted independence.

Bush, meanwhile, was concerned about appearances. Expanding Section T even further at DTM looked like self-dealing. After all, Bush was the Carnegie Institution president, and the Department of Terrestrial Magnetism was under his supervision. Channeling massive taxpayer funds to his own organization opened him up to unsavory speculations. He weighed the matter and decided Tuve was right.

On March 31, Bush removed Section T from Division A and transferred its supervision out of Richard Tolman's hands and into the Navy's Deak Parsons's. As Parsons was already a close ally of the project, Merle would essentially have full authority and no room for excuses—or "alibis," as he called them.

Section T would also be leaving DTM. Contract negotiations were begun with Johns Hopkins University, where Tuve would have free rein. The plan was to secure up to thirty thousand square feet of real estate for a new headquarters.

Tuve decided to call it the "Applied Physics Laboratory."

The move couldn't come soon enough.

14
DR. JONES'S RAID

On a cold, still morning in late February, 1942, a thick frost reflected glints of the bright sun. The men of C Company, of Britain's Second Parachute Battalion, packed their containers and checked and cleaned their weapons. At Thruxton air base, outside Andover, the parachutists nibbled sandwiches, sipped tea laced with rum, and chatted nervously until nightfall. In groups of ten, they boarded twelve Whitley bombers. The aircraft bathrooms had been removed to save space, and the soldiers were advised to "relieve themselves last thing before emplaning." C Company was formed in part from Scottish units, and as the paratroopers waited to take off, a bagpiper moved among the aircraft playing their regimental Scottish marches.

Twelve Rolls-Royce engines revved to life. On ribbed aluminum floors, the men donned silk gloves for warmth and settled into sleeping bags.

On the English Channel, the *Prinz Albert,* a Combined Operations support ship, lowered six assault landing craft into the water and flashed a message of "Good speed." Each boat carried four black-faced Welsh commandos prepared to give supporting fire to the paratroopers during their risky escape.

The Whitley bombers headed for the coast of occupied France. When the order came to "prepare for action," the troops wriggled from their sleeping bags and fixed their parachute lines. Trapdoors were removed from the floors of the bombers, and the loud piercing air rushed through the cabins, and they jumped in a carefully planned sequence through the dark circles into the moonlit night.

Black and green parachutes bloomed over a snowscape.

The Nazis would describe the mission as "a violent technical reconnaissance."

To the British, it was Operating Biting. And to R. V. Jones, it capped off the greatest advance yet in building a complete scientific intelligence service.

While still viewed with skepticism, the Oslo Report—which had warned of a research station at Peenemünde and of pilotless aircraft—had so far proven right. Jones often returned to it for clues to the next Nazi threat "in the pipeline."

One of those threats was German radar. As Jones knew, British night defenses against Luftwaffe attacks now relied on cutting-edge radar technology. German night defenses in France and Belgium might also rely on some kind of radar. One year prior, Allied intelligence circles had debated whether German radar existed at all.

Jones and his deputy Charles Frank had settled the debate for good. A pair of photographs from aerial reconnaissance, snapped in sequence over a farm on the French coast of the English Channel, had revealed a subtle clue. Two blurred circles perhaps twenty feet in diameter caught their attention. The rings looked a bit like cow troughs. From one photograph to the next, the width of a shadow extending from one of the circles broadened almost imperceptibly. Some object inside the circle had turned ever so slightly between the first and second snapshots.

Turned like a radar antenna.

Reconnaissance flights confirmed their suspicions and more. The Nazis had installed an extensive chain of defensive radar along the French coast.

Jones then helped mount what he called "a basic intelligence assault upon the German defense system," focusing all available intelligence resources and assets toward the problem of German radar. British intelligence eavesdropped on the Luftwaffe night fighters, and tried to decode their chatter. From the planes' transmitters, they mapped the night fighters' areas of operation—which naturally would be protected by new radar. Knowing what to look for, with the help of spies on the ground, they located more radar installations farther inland.

These units were known as Freya radars. But the Oslo Report pointed to a second type of device known as a Würzburg radar, running on another frequency, that seemed poised to cause serious trouble for the Royal Air Force in its offensive bombing campaign.

Würzburg radars, Jones guessed, might be installed on the same sites as the Freya units they knew about. He and Frank scoured their file photos.

What they found was "a small speck." The image was so minute and faint that at first R.V. wasn't sure whether the dot was a smudge on the negative. In a snapshot of a coastal plot near Bruneval, France, on the edge of a majestic cliff below two Freya units, a worn track seemed to run southward from the Freya radars to a villa. But on closer inspection, he saw the path actually ended slightly away from the house—at a mysterious speck above the cliff edge. Why did the path stop short?

A pilot named Tony Hill, later a dependable mainstay for Jones and his growing team, flew out unofficially to get a closer look at the Bruneval site. The "speck" was a small parabolic dish. It turned out to be, in fact, a Würzburg radar. Hill's photos plainly showed the dish atop the four-hundred-foot cliff, not far from a gently sloping gully, leading down to a beach several hundred yards down the coast.

Jones saw it all clearly: The beach, the sloping path, the Würzburg.

"Look, Charles," he told his deputy. "We could get in there!"

It was not in character for Jones to request a raid that risked British lives. But the target was too valuable, and with support from the Telecommunications Research Establishment, which was in charge of radar research, his idea was passed up from the air staff to Combined Operations Headquarters. At the very least, the raid would harass the Germans and might distract their attention.

They were going to steal Hitler's radar.

To plan the Bruneval raid, they needed help from French agents.

The task fell to a Resistance spy named Gilbert Renault, known as Rémy, a former banking executive and movie financier who would build a network of thousands of spies under the guise of a religious organization, the "Brotherhood of Notre Dame." He received instructions by radio in late January, 1942: the Allies needed details of a radar installation near Bruneval, including troop numbers, the lay of barbed wire, and whether machine guns protected the cliff road.

Rémy called on Roger Dumont, code-named Pol, and Charles Chauveau, aka Charlemagne, owner of a garage in the port of Le Havre, eighteen miles south of Bruneval. Pol and Charlemagne stayed in a seedy Le Havre hotel, in an unheated room with damp sheets. Pol spent the night fully dressed, shivering on a chair.

They fitted a car with snow chains and drove north on back roads until they reached a hotel on the edge of Bruneval hamlet.

Monsieur and Madame Vennier, proprietors of L'Hôtel Beau-Minet, were friendly to the Resistance. They confirmed that there were Luftwaffe personnel stationed in the nearby farm buildings. A guard post in a house on the beach below was manned by ten soldiers. And there were machine guns and barbed-wire emplacements lining the cliff path. Mines dotted the beach. A German infantry platoon led by an energetic sergeant was quartered in the Beau-Minet itself.

The lone château near the Würzburg might be unoccupied.

Pol boldly insisted they have a look at the beach, and the two men trotted through the biting wind until they were stopped by a tall German sentry with a sad expression. "Good morning, Fritz," Charlemagne said to him, in German. "Just taking a stroll with my cousin," he continued. "He's from Paris . . . Feels he must see the sea before he goes on home. Shut up in a dark office all day long, you see. You know how they get, desperate? Lucky you're here. Without you we wouldn't have dared go any further. We've heard there are mines. Imagine that!"

"*Ja, Tellerminen,*" the sentry replied. Antitank mines.

"Not good. I wonder if I dare suggest—would you accompany us down the shingle, just for a second? It would give this dear fellow so much innocent pleasure."

The soldier—from boredom, perhaps—agreed. He opened a small gate in the barbed wire and led them down the path to the pebbly beach.

It was no beach for bathing. But the tide was low, and the spies could not see anything underwater that might prevent a naval landing. Charlemagne noticed two bunkers with machine guns. And Pol watched in surprise as the German sentry marched casually over ground marked "Beware of Mines."

The mines were a bluff.

A raid was viable. The British decided on paratroopers, and C Company was chosen. Their commander was Major John Frost, who later defended the bridge at the Battle of Arnhem (and was immortalized in *A Bridge Too Far*).

For secrecy, Frost was told that his unit was training for a demonstration before the War Cabinet. The true plan called for Royal Air Force planes to drop the paratroopers at Bruneval and distract the Nazis with diversionary raids. While men retrieved the Würzburg, another team would secure the path down the cliff and the beach itself, where naval forces would rendezvous with the troops, load the prize, and direct the escape by sea. All this was to be done under the noses of the German garrison in L'Hôtel Beau-Minet and the Luftwaffe personnel in the farm buildings above the château who, after all, operated a radar installation.

To examine the Würzburg—and, if possible, dismantle it—someone with technical expertise had to join the parachutists. A junior scientist on Jones's team, who had a background in radar, volunteered for the job. So did R.V. himself. But neither Jones nor any of his small staff were approved. Should they be captured and interrogated, the security risk simply was too great.

Instead, on February 1, 1942, a radar mechanic named Charles Cox was summoned from his station in North Devon to the Air Ministry building in London.

"You've volunteered for a dangerous job, Sergeant Cox," he was told on arrival.

"No, sir."

"What do you mean, no, sir?"

"I never volunteered for anything, sir," Cox said.

"But now you're here, Sergeant. *Will* you volunteer?"

"Exactly what would I be letting myself in for, sir?"

"I'm not at liberty to tell you . . . I honestly think the job offers a reasonable chance of survival. It's of great importance to the Royal Air Force."

"I volunteer, sir," Cox replied.

Although he was one of the top radar mechanics in Britain, Cox had never been on a boat or an airplane in his life. Before the war, he had been a cinema projectionist. Now, he found himself rehearsing parachute jumps for a commando raid.

Practice runs went poorly, and Cox, married with a newborn at home, worried about being captured. Heightening the danger was a maddening bureaucratic problem: While C Company would wear army gear, Cox would be in an air force uniform. If caught, the mechanic would stand out like a black sheep. Jones tried to resolve the issue, but the War Office refused to issue Cox an army uniform.

R.V. hoped, at least, to soothe the mechanic's nerves. "Don't be worried too much about physical torture," he told Cox. The real worry was sitting in solitary, with little food, and being tricked into talking by a kind German officer who would appear suddenly and insist on better treatment. "Be on your guard against any German officer who is kind to you!" Jones said.

"I can stand a lot of kindness, sir!" Cox replied.

The raid itself was not characterized by compassion.

Bruneval was covered in a foot of snow.

Above the imperial cliff, around 11:55 p.m. on February 27, a Luftwaffe signaler swiveled and adjusted the Würzburg radar dish toward the Allied planes as they puttered over the French coast. He tracked the aircraft, which were moving slowly like bombers with heavy loads, as they turned and headed toward him.

It looked like a bombing raid.

Fourteen Luftwaffe troops rushed for shelter as the alarms rang, while down below, the Nazi officer in charge of defending the rocky beach awoke his men.

Antiaircraft guns fired in vain as the low-flying bombers escaped.

Caught in the slipstream, falling through the night sky, Charles Cox looked up at the long underbelly of the Whitley bomber and its burning engine exhausts. He double-checked his fighting knife and Colt .45 pistol as he drifted to earth.

Landing on the soft snow, Cox and the paratroopers gathered themselves. They felt no wind, and could see no German troops. Major John Frost, his bladder full from too much tea at Thruxton air base, relieved himself—bad protocol but "a small initial gesture of defiance" and a necessary one. The calm hush was a stark change from the biting gusts inside the noisy bombers just minutes earlier.

Cox picked up the trolley dropped alongside him—he would need it to transport the stolen Würzburg—and joined the assembled men by a line of trees. The raid had, Frost learned, already suffered an ominous mishap. Twenty paratroopers, tasked with securing the beach and the path down to it, were missing.

Frost would have to make do. Two sections of his men positioned themselves to ward off the Luftwaffe troops in the farm buildings and the infantry platoon quartered in the Beau-Minet. Cox, other troops, and a small group of combat engineers headed for the Würzburg. Frost himself led men to the château above it.

Through a wide-open door, Frost entered a gloomy, unfurnished hall. He could hear shots from the floor above, directed at Cox and the assault team encircling the Würzburg. Frost and his men moved swiftly upstairs, found a single German soldier firing down at the commotion outside, and shot and killed him.

The fight for the Würzburg itself was one-sided. A lone German sentry fired on the paratroopers and was quickly silenced. Then, a British lieutenant recalled, "We hunted them out of cellars, trenches, and rooms with hand grenades, automatic weapons, revolvers and knives." Hiding on the cliff edge, a Luftwaffe radar operator clung desperately to the rocks, his silhouette betrayed by the moonlight. He was taken prisoner, and the swastika badge was torn off his uniform.

At the radar dish, Cox navigated through the barbed wire as bullets from the Luftwaffe barracks flew past his head. He tore the black rubber curtain from the base of the Würzburg, its instruments still hot from tracking their arrival.

Using a hooded flashlight, he took notes in the dark.

"Like a searchlight on a rotatable platform mounted on a flat four-wheeled truck," he scribbled. "At rear of paraboloid is container three feet wide, two feet deep, five feet high. This appears to hold all the works with the exception of display." He marveled at the "very clean" design.

A British paratrooper snapped flashbulb photographs of the unit, drawing more gunfire to their position, until Frost ordered him to stop.

As the combat engineers sawed off the aerial in the middle of the dish, Cox focused on the gear in the container. He removed the pulse unit and the amplifier, but even with his longest screwdriver he could not reach the transmitter. Instead, he used a crowbar. As he loaded components on the trolley, the gunfire felt closer and more accurate. One of Frost's men lay dead near the door of the château. Gunfire seemed to be coming from all directions, including from the beach.

A peculiar white flare illuminated the night sky.

Carrying most of the Würzburg, the soldiers headed down the cliff path. A machine gun caught the paratroopers against the snow, wounding a man in the stomach. Cox was skidding and sliding down the frozen trail as Frost dashed ahead of the trolley. "Don't come down," they heard a man say. "The beach is not taken yet." Then Frost heard another, familiar call echoing through the darkness.

"Cabar Feidh!" It was the Scottish war cry of his missing troops. Dropped off their target, they had humped two miles back to the cliff path, and they now quieted the German guns guarding the beach and sent their operators running.

Two Allied men had lost their lives, six were missing, and six lay wounded on the craggy beach. In ten minutes, Cox had managed to retrieve all but one of the key radar components. As German troops closed in from above, at around 2:35 a.m., the naval landing craft spotted C Company's green signal flares.

"The boats are coming in," a man told Frost. "God bless the ruddy Navy!"

Cox, the wounded, and the Würzburg were on the first boats out.

On March 2, Jones visited Air Ministry headquarters in London to inspect the Würzburg bounty. The radar, he saw, was far better engineered than British equipment. But the Allies didn't only have the equipment to study. They also had the Luftwaffe radar operator who had been captured on the cliff face. At a British prisoner-of-war camp in north London, Jones and the German spent an afternoon together on a floor, fitting the radar components together.

When Hitler heard of the Bruneval raid, he was furious.

The operation was a complete success. It netted an intimate look at Germany's radar defenses—crucial for any future assault into France.

MI6 now knew the range of wavelengths the Würzburg could be tuned to. And they discovered that the radar had no built-in defense against jamming or other interference.

Scientific intelligence was becoming a recognized branch of espionage work. Among those in Jones's circle who helped cement its reputation were the cryptographers from Hut 6 in Bletchley Park, a prisoner interrogator and German document specialist, radar scientists, a photographic interpretation expert, and a radio amateur with the Royal Air Force wireless listening service.

Jones had learned to focus the "senses" at his disposal—the eyes of photographic reconnaissance, and the ears of the radio listening services—in a concentrated attack on a target. And he had cultivated entirely new sources. But he could take no credit for his most extraordinary source.

That would belong to Jeannie Rousseau herself.

The linguist whose exploits in Dinard landed her in prison, who was now working as a liaison for the Germans in Paris, had reestablished contact with the Resistance. She had been traveling by train at night from Paris to unoccupied Vichy, where she hoped to learn more about the war, when she recognized Georges Lamarque, a young mathematician she knew from her student days. There was no place for them to sit on the train, so they spoke in the corridor. She told him of her work with the French industrialists, and her contacts at the old Hôtel Majestic.

Under a flickering, dim blue light, Lamarque confessed, in so many words, that he was already engaged as a Resistance spy. He was setting up a network of agents that would become known as the Druids—a "little outfit," he said. "Would you like to work for me?"

Once again, she didn't hesitate.

He gave her the code name "Amniarix."

15
THE GARAGE

Rumors spread around Silver Spring, Maryland, of what might be truly going on inside the shaded two-story Chevrolet garage, which kept its old USED CARS sign out front but was now protected by armed guards and barbed wire. Groups of young, smartly dressed women arrived each morning at the Wolfe Motor Company garage, at 8621 Georgia Avenue, and left at odd hours—late at night, sometimes past midnight. Neighbors suspected a high-class call-girl ring. But the security guards addressed many who entered as "Doctor," so maybe it was not a house of ill-repute after all. It could be a secret experimental medical facility.

Several doors down, an Italian-American barber gossiped about the black-draped boxes—filled with tubes for quality testing—that were hoisted up two dozen feet by a pulley and then crashed down loudly onto an armor plate.

"They push-a them up," he murmured, "and they push-a them down."

Section T's new headquarters, known on paper as the Applied Physics Laboratory, was nominally organized under the wing of Johns Hopkins University. Hopkins's board of trustees was initially reluctant to "oversee" work they were told nothing about. (FDR had to personally intervene with Hopkins's president.) Nor did the garage owner, Garland Wolfe, enjoy renting his property and land for an unknown purpose. He upped the insurance cost on their lease accordingly.

The area was tranquil and lightly industrial—the sort of place for bottling factories, printing plants, and machine shops. On Georgia Avenue, a central thoroughfare, banks, car dealerships, and tire shops were interspersed with small family businesses like hardware stores, jewelers, and diners.

While Silver Spring was conveniently located just north of Washington, the facility itself was a work in progress. The building had no ventilation system, and Tuve had authorized renovations to the tune of over a hundred thousand dollars.

"There was rarely a day," a new hire remembered, "that somewhere in the building the banging of sledgehammers was not heard, creating piles of rubble and clouds of dust as walls were removed and rooms reshaped."

Supplies poured into the new lab. Among the band saws, drill presses, putty knives, and protractors were ninety-five workbenches, seven water coolers, and four smoking stands. Tuve ordered a dinette set, a living room set, and four beds and mattresses so that Section T workers could eat, decompress, and sleep there.

No wonder Merle's home and work life were blurring together.

Domestic concerns crept into his turquoise notebook in between jottings about Section T's most urgent problems. Next to details of fuse-assembly plans, he wrote down a tip on cheap tailored suits. Production quotas and troubles with Crosley were interrupted by a grocery list of caraway, eggs, and ginger.

He worried about his son's health. "Tryg seems to have just gone through another allergic explosion," he wrote, next to the name of a company that might supply Section T with parts. "We don't think he has (or has had) a cold."

Tryg was eight, and Lucy was nearly four.

By June of 1942, Navy officers had begun invading the Tuves' home on Hesketh Street, packing the living room at odd hours to discuss Section T's vital work. Almost every night, according to Merle's wife, Winnie, the house seemed to be full of "Admirals." The Navy's Deak Parsons and Samuel Shumaker were regulars. Army officers would join the crowd as their interest in the fuse grew.

Some days, Winnie said, the men would drink "more alcohol than anybody I ever knew" and yet somehow, she marveled, "still stay with it."

Liquor flowed so freely she was afraid to light a match.

Merle struck plenty of matches, an "absolute chimney" as he chatted up his new Navy bosses. Parsons and Shumaker, he knew, could cut through the red tape faster than his old Section T supervisor, Richard Tolman.

The houseguests also seemed to have encouraged another bad habit and military custom: cursing. "The military is full of four-letter words," said a Section T technician. "You fast learn that unless you learn that language, they don't even understand you." Merle didn't curse *at* anyone, but seemed now to enjoy sprinkling his everyday descriptions with colorful jargon.

Winnie did her best to break him of the practice, installing on their dining room table, next to the salt and pepper, a little wooden "swear box."

According to Lucy, her father was "charged a nickel for ordinary swear words and a dime for those that were a little new, a little worse. Sometimes up to a quarter." (These probably included *bastard, goddamn, son of a bitch,* and *fuck.*) Change donated to the box was intended as a "treat kitty" for the children.

Lucy's earliest memories included the strange nightly ritual. "We would sometimes make up to several dollars off of him at a course of an evening's meal!" she recalled. One imagines the steady clink of change over pot roast as Merle let off steam from work, while his two children watched with pleasure their growing pot of coins for purchasing sweets and candies.

Merle was gone often, but he was not an absent father. Lucy described him as "one of those individuals who is ever present when he was in a room or with you." He did not sit and read "silly little stories," but he did teach his children about the natural world. When a thunderstorm scared Lucy—she was barely four—he took her out to their backyard. They sat in the wind, his arms wrapped around her, as the storm neared and lightning flashed, and he told her about St. Elmo's fire, the weather phenomenon where ionized air turns to blue or violet plasma.

He wanted her "to get out there and enjoy it and see it and feel it."

Once, when walking with Lucy, he told her not to step on the ants. "You don't want to kill these creatures," he said. He lifted a daddy longlegs up and carried it safely to a nearby tree. "It wasn't his nature to kill anything," she said.

He knew the fuse would kill, knew it was necessary, and did not like it.

Amid the racket of construction work, Tuve's foul language echoed in the hallways of the Applied Physics Laboratory. According to staffer James Maddox, Section T's leader was "one of the finest but cussedest men you ever met."

"He didn't give a damn who was there," Maddox recalled, "whether it was his wife or the head of the company. If he wanted to cuss, he'd cuss. And he'd give you hell. But then he'd turn around afterward and shake hands with you."

There was no hierarchy at Section T. "Results took precedence over organization," said Wilbur Goss, a young professor newly recruited out of a lonely, two-man physics department at New Mexico State University. "People who could do, found themselves doing."

Formal training was of little concern. Tuve was as likely to assign an Ivy League physics PhD to a project as an amateur-radio operator from the rural South. In fact, he liked to pair PhDs and hams together to meld different approaches.

One recent hire was an *astronomer*.

Paul Ertsgaard, a technician who would play a prominent role at Section T, had only a high-school education. His previous work experience was listed as "Student Assistant at the Bureau of Standards." Another hire, John Doak, was a radio repairman. Joseph Teresi was a telephone mechanic from Pittsburgh.

Section T even recruited a spate of oil men out of Houston.

"The important thing was to get the job done, and he imbued everyone with that spirit," Goss said. "Merle was gifted in making people who are five feet ten inches tall walk eight feet tall. Very ordinary people became rather extraordinary people under the challenges that he posed . . . and with the degree of backing that he provided."

Typical workdays didn't exist either. "Time meant nothing," Maddox said. "After eight hours, you'd do eight hours more." Ralph Robinson, a radio ham from Texas, once took a break after three weeks because he hadn't had a day off and was getting groggy. "You were working at a pace," he said, "that was all that you could physically and mentally put out."

Ralph Baldwin, the astronomer, was so busy—and under such great psychological stress—that he complained his rotary telephone was taking too long to dial out numbers. A repairman fixed it that same afternoon.

As his staff grew, Merle developed a set of "Running Orders" for Section T — guiding principles to help govern the daily organized chaos:

- I don't want any damn fool in this laboratory to save money. I only want him to save <u>time</u>.
- There are no private wires from God Almighty in the lab that I know about — certainly none in my office.
- Run your bets in <u>parallel</u> not in <u>series</u>.
- Our moral responsibility goes all the way to the final battle use of this unit; its failure there is our failure. . . . The <u>final</u> result is the only thing that counts, and the only criteria is does it work <u>then</u>.
- Shoot at an 80% job, we can't afford perfection. . . . Don't forget that the best job in the world is a total <u>failure</u> if it is too late.

There was no resource more precious than time.

To ensure fresh perspectives, Merle shifted personnel constantly. Workers on one project would find notes on the bulletin board and discover they were now assigned to an entirely new one. "This is it! — M. A. Tuve," the notes read. Staff called them Tuve's Monthlies. According to one technician recruited out of a radio station in east Texas, "You never knew what you'd be doing next or where."

The laboratory space at the garage, along with the direct backing of the Navy's Bureau of Ordnance, allowed Tuve to go on a hiring spree. At the end of May, 1942, his staff numbered 115. A month later, the figure had jumped to 196.

Section T was expanding at a frantic pace.

The sudden growth made for some awkward encounters. Tuve, for obvious security reasons, felt it was his duty to "challenge strangers in the laboratory" and demand who they were. "It is uncomfortable for me to challenge the men whose applications I have signed," he noted. So much hiring was now going on that it grew difficult for Tuve to keep track of who actually worked for him.

Amid the boom, security protocols were established. A sophisticated ADT alarm system was put in, with sensors outside to alert the watchmen and, inside, six more quadrants of sensors that were wired to notify the local police.

New hires arrived to hours of photographing, fingerprinting, and lectures about the Espionage Act. At Section T's weekly meetings, held just across the street at the Silver Theater, a movie house on Colesville Road, staff were diligently reminded to tell no one, including their spouses, of their classified work. "Assume always an enemy agent is listening on the line," one circular advised. Instead of saying *fuse* and *Dahlgren*, employees were instructed to instead "refer to the 'gadget' or 'device' or the 'place down the river.'"

Fuse radio frequencies were described in code, and any discussion of fuse countermeasures would soon carry a top secret classification.

They faced security scares and other worrying incidents.

Ed Salant, after a visit to the British Central Scientific Office in D.C. and a meal with its liaison officer, couldn't find his copies of a confidential invoice.

In late May, a subcontracting electrician with a foreign accent, evidently drunk, entered the telephone room and uttered "some very unfortunate remarks disparaging to the Navy." He was arrested and reported to the FBI.

On June 16, a man taking pictures was spotted across the street from Section T's new headquarters. Questioned by the guard, he flashed a white pass that had no photo and read U.S. NAVY DEPARTMENT. The Navy sent him, the man claimed, and then rushed off in a red Packard sedan with the words *Press Car* over the license plate. The Navy press office had no record of ordering photos.

Even Deak Parsons's wife could tell the fuse had improved.

Deak and Martha actually lived at Dahlgren proving ground, within earshot of its aquatic firing range, in the senior officers' quarters in a fine house with a lawn, a view of a creek, and access to a tennis court. On days her husband didn't commute to the new "garage" in Silver Spring, Martha would work in her vegetable garden and listen to the pattern of the gunshots echoing over the base.

She guessed that his secret project had to do with "some kind of shell," and admitted as much one evening when Deak walked in the door.

"Gee, that really works better!" she said.

"What are you talking about?"

"Well, whatever you're working on," she answered. "I could hear it. It used to go bang . . . bang . . . bang . . . . . . bang . . . bang . . . .

bang. And now it goes bang, bang, bang, bang, bang, bang. It sounds better to me."

"You're not supposed to know," he said.

She may not have been permitted to learn what exactly her husband was working on, but she was right. The smart fuse was nearly complete.

Parsons deserved a fair deal of the credit.

With the Navy lieutenant now overseeing the project, bottlenecks suddenly seemed to disappear. An additional testing ground with four separate fields was finally leased at Newtown Neck, near Leonardtown, Virginia. By firing multiple guns vertically at the same time, Section T field crews could shoot seven hundred test rounds every day there. Farmers complained that the racket disturbed their hens.

It was Parsons who had pushed through the contract for the Silver Spring garage, after inspecting it himself. He attended the weekly officers' meetings with Tuve and Larry Hafstad, and was closely involved in the day-to-day operations. When Section T put in a request for additional sidearms, possibly revolvers (for field crew to protect the fuses during transport), Parsons replied that he would see to it that day. He also promised to look into procuring some submachine guns.

Only a few hurdles remained before the fuse could be mass-produced. For one, it wasn't obvious how far away from an aircraft a shell should blow up. Twenty feet? Two hundred? To adjust that span, the fuse could be made more or less sensitive to its own radio signal bouncing back. What distance gave the shell's shrapnel the best chance to bring down an Axis airplane?

Merle asked two physicists at the University of Michigan, Richard Crane and David Dennison, to examine the problem. He commissioned an urgent report on the "Vulnerability of Modern Planes to Shell Fragments." Then he sent the brash physicist Dick Roberts to a Marine Corps base in South Carolina.

At Parris Island, north of Hilton Head, Roberts and seven other Section T men shot fuses against an airplane for the first time. An engineless Taylor Cub, covered in aluminum gauze, was suspended nine hundred feet in the air by a Navy kite balloon six hundred feet above it, and anchored to a buoy in the Carolina waters. The plane swayed a hundred feet in the wind as the men readied the five-inch shells.

The Taylor Cub was eleven thousand feet from the gun. Using two stations closer to the target, observers marked where the shells burst near the plane, filmed them exploding, then plotted the coordinates of the test shots. A safety plane kept the skies clear overhead, while a guard boat below patrolled the inlet.

The shells exploded within thirty feet of the plane, but often detonated behind it. The radio receiver in the fuse, it seemed, needed to be more sensitive. Otherwise the shrapnel, which fanned out in three narrow bands, would often miss.

The proper distance was within seventy feet.

With the goal so close at hand, Merle was even more irritable than usual. Section T had progressed, tiny change by change, from the Prescription A fuse all the way to Prescription F. But he was impatient about the smaller fuse for the British, which was proving hard to adapt to size. And he was anxious about the production lines, and the minute problems sinking test scores. "Foreign matter" found its way into some fuse parts and ruined them. Tube sealant left a damaging vapor residue. Crosley fuses fluctuated for no clear reason, until careful study revealed that the trouble was the summer heat inside the Cincinnati assembly room. Crosley was behind schedule.

"It now appears impractical," Merle wrote to Parsons, on May 20, furious at the delays, "to count on Crosley in any way for supplying fuses for drone tests, fleet tests, British tests or any other important parts of the program."

As backup, he asked Sylvania to build fuses as well as tubes. Sylvania's president, Walter Poor, promised that the company would not make any design changes—even small ones—on its own. If Section T had a fly inside the sample fuse provided to the company, Poor said, then Sylvania would "catch flies and put one in each fuse that we make."

Tuve hated depending on one company or even one team of scientists, engineers, and technicians. Just as he hired multiple tube manufacturers, he often assigned more than one group the same problem without telling them. His approach to the final obstacle to a battle-ready fuse—a safety delay "clock" to prevent shells from exploding too close to the ship —was no different. One engineer assigned to the project learned a week later that there were two other teams already working on it separately. At one point, Tuve had no less than five groups working on safety devices. "If you need thirty clocks in a hurry," as he put it, "get thirty clockmakers to do the job." Hoover, the vacuum company, was working on one approach; Jack Workman, in New Mexico, on another.

The existing safety device was so unreliable that during the Parris Island tests, the gunners took cover before firing in case the shells blew up early.

In the Pacific, meanwhile, time was running short.

16

THREE RUNS, THREE HITS

From the deck of the USS *Lexington*, on May 8, 1942, a barrage of deafening shots flung dozens of five-inch rounds into the clear sky. They pockmarked the air with black plumes, peppering the vista with inkblots of exploding shrapnel.

In the Coral Sea, off Australia's northeastern coast, the 888-foot Navy aircraft carrier was under heavy assault by eighteen fighters and thirty-six torpedo and dive-bombers. The *Lexington* and a second carrier, USS *Yorktown*, formed the heart of an Allied mission to stop the Japanese from invading and occupying Port Moresby, New Guinea, a strategic foothold right off Australia's doorstep.

Frederick Sherman, the *Lexington's* captain, steered sharply at high speed to dodge the falling bombs, jolting his men under the deck and causing the huge vessel to roll, sway, and moan. Bullets from Japanese gunners in the planes struck a violent staccato beat on the metal hull and echoed through the ship's belly. Axis bombs narrowly missed the carrier deck and exploded under the water, unleashing pressurized tremors that popped the American sailors' ears.

The ship had never faced a raid of such intensity.

On the port side, a set of three five-inch antiaircraft guns blazed away stubbornly at the bombers. Jesse Rutherford Jr., a nineteen-year-old from Kansas, hoisted the fifty-four-pound rounds from the ammunition locker at his feet. Like a link in a bucket brigade, he handed them to the "primary loader," a fellow Marine standing at the breech of the twelve-foot, two-ton monster called gun no. 10.

Rutherford was among a small contingent of Marines manning the guns. Since six a.m., the captain had them at the ready wearing "flash gear"—heavy, fire-resistant clothing that included a protective hood and gloves. The Marines had waited five hours in the baking heat before the Japanese attack was spotted.

Most of the *Lexington's* planes were far away, executing their own raid on Japanese carriers. With only a handful of Allied planes remaining, the Japanese easily reached the *Lexington,* prompting ack-ack fire from its gunners. But the small-caliber machine guns and five-inch cannons did not deter the pilots, who flew without hesitation, largely untouched, through the porous flak.

After the first minute of the assault, it became difficult for the men to discern the exact order of events or where the bombs and bullets were coming from. Usually, four Mk 19 "gun directors" with telescopic lenses would have tracked incoming aircraft, determined the height, range, and bearing of enemy planes, and fed coordinates to the gun mounts. But the attack was so chaotic that gunners were given "local control" over where to shoot and had to select their own ammunition.

Inside the shells, time fuses adjusted by twisting a metal ring were all *preset.* The scheme saved men from having to calibrate them during the heat of battle, but it was wildly inflexible. As bombs fell, gunners tried to determine the flight paths of approaching raiders, and then figure out where those flight paths might meet in midair with a shell that blew up at 2.2, 3, or 5.2 seconds after being fired.

The *Lexington's* antiaircraft guns could not protect the ship.

Within minutes, around eleven twenty a.m., it was hit by a series of torpedoes, producing explosions so violent that they froze the elevators and fractured the aviation-fuel storage tanks, which began to leak gas and poisonous vapors.

Beneath the deck, repair teams quickly dispatched crews to plug the holes in the hull, and starboard compartments were "counter-flooded."

On the bridge, Captain Sherman craved a cigarette, but the fumes made smoking too dangerous. In the distance, he saw that the faster, more agile carrier *Yorktown* was also being ambushed. Naval tactics dictated that the ships in the Allied battle group (which included cruisers and destroyers) should form a strategic ring to maximize their antiaircraft guns. But the formation had broken.

A bomb pierced the hull and exploded in the admiral's and chief of staff's living quarters, enflaming furniture and distorting the lip of the deck.

The Marines at guns 2, 4, and 6 suffered a direct hit. Marine Corps captain Ralph Houser, their commanding officer, discovered the gruesome scene. Like victims at Pompeii, the charred bodies were frozen at their gun positions. Wounded men moaned and bled on the gnarled deck. Medics applied battle dressings and tannic jelly to their burns, and administered morphine.

A jagged hole punctured the deck beside gun no. 2.

The explosion splintered a storage locker of five-inch shells, scattering them. Swelling with heat, rounds slipped from their brass cases and spilled firing powder, which ignited in tails of flame and let out angry hisses.

Two Japanese planes sprayed the deck with machine-gun fire, wounding three men working gun no. 10 and ending the life of another. Rutherford was shot several times but refused to stop lifting the heavy shells, one by one, for loading. Bombs hit the water and threw up towering walls of ocean, obscuring the ship's profile and soaking the gunners still desperately trying to save the ship.

The attack lasted only twenty-three minutes. When it was over, the *Lexington*'s gunners had shot down only six of the fifty-four Japanese aircraft in the assault group.

It was now just a matter of time. At 12:47 p.m., the leaking aviation fuel blew up, taking out the damage control station. Two hours later, an explosion knocked an elevator through the flight deck. At 3:25, another blast took out the water pressure in the hangar. At 5:07, Captain Sherman gave the order to abandon ship.

Floating helplessly, the crew were unable to get far away from the sinking hull as the vortex of churning currents pulled them closer, like a magnet. That night, over twenty-seven hundred men of the almost three thousand aboard were safely rescued by Allied ships.

The loss of the *Lexington* reinforced the lesson of the *Prince of Wales* and the *Repulse*. Ships could not defend themselves without air cover. "Air offense is definitely superior to the defense," the incident report dryly concluded.

The Battle of the Coral Sea marked the first time in history that enemy aircraft carriers waged a fight against each other. It was the first battle in history in which neither side's ships ever saw or fired directly on the others.

Naval airpower had come of age.

It was a tactical loss but a strategic victory. Japan failed to take Port Moresby, and a Japanese carrier was sunk. The contest also marked a turn for the Navy, which was preparing to go on the offensive and, in 1943, hoped to claw back Japanese gains in the Pacific and take the war to the nation's island citadel itself.

Japanese resistance would be deadly and savage.

After the *Lexington* sank to the ocean floor, the USS *Yorktown* limped to Pearl Harbor, where the vessel underwent a frantic repair job to return it to action.

Weeks later, the *Yorktown* sunk too.

Sailors met the scientists at the waterfront.

Dick Roberts was impressed by the Navy work party, which swiftly loaded the radios, binoculars, and batches of secret fuses.

August 10, 1942, was less than two years since Roberts initiated the fuse project, with some swagger, on Merle's request, by firing a pistol at a vacuum tube in a bunker underneath a particle accelerator. He could not have guessed where that journey would lead. Now in front of the physicist, on a pier in Norfolk, Virginia, was a six-hundred-ten-foot Navy cruiser known as the USS *Cleveland*.

An imposing metal giant—a freshly commissioned ship—the *Cleveland* displaced some eleven thousand tons of water and carried a thousand men. At the stern was a crane used for retrieving four scouting seaplanes. In its center were stacks of circular towers, curved platforms, and boxy compartments. The ship's core gave the impression of a small mountaintop favela made of iron, where generations of inhabitants added their own ferrous modules as space allowed.

The vessel was heavily armed. Four turrets and twelve guns used for land bombardment dotted the bow and stern. Behind them, encircling a rounded bridge, slanted masts, and dual smokestacks, were twelve five-inch guns in six turrets. The Navy was well aware by now that the guns weren't enough, and had been busy cluttering the decks of cruisers like the *Cleveland* with dozens of smaller twenty- and forty-millimeter guns. The *Cleveland* itself had thirty-two of them. The ship was not designed to handle the weight of the extra guns and their aiming devices, and the boat—like others in its class—had grown increasingly unstable as it overflowed with more and more guns that were fitted like porcupine quills to the deck.

Roberts climbed the gangplank onto the massive carrier. With him was Section T's "Mac" McAlister, from the Smithsonian Institution, and Herb Trotter Jr., a square-jawed physicist from Washington and Lee who looked more like an amateur boxer than a scientist. Lieutenant Deak Parsons was overseeing things.

As the *Cleveland* set off into Chesapeake Bay, the steam turbine engines propelled the sailors, researchers, film cameras, and precious fuses past the York and Rappahannock Rivers to Tangier Island, seventy miles north. The cruiser stopped at the widest stretch of the estuary, and made anchor for the night.

The ship's insides were as alien to Roberts as its cluttered skin. Below deck, he encountered a maze of control rooms, berths, narrow passages, repair shops, ammunition rooms, supply rooms for spare parts, diving gear, and "chemical defense material." The *Cleveland* was a tiny city with a post office, bakery, metalworking shop, optical shop, and even a room for "potato stowage."

Roberts would not be sleeping in the "guest cabin" with its matching bath. He was bunked along a corridor and would have a more plebian naval experience. The ship was on a "shakedown" cruise to test its performance and ready the crew, and the sailors were kept busy with unexpected drills. The boatswain would blow a high-pitched pipe, and sailors would rush to their battle stations, prepare to abandon ship, or respond to "fires," "collisions," and "damage reports." The physicist was sound asleep the next day, at five a.m., when he was suddenly jolted awake while "half the crew ran over" his bunk for a surprise drill.

Tangier Island warmed slowly in the August heat, and as the sun climbed in the sky, Roberts, Parsons, and the other Section T men gathered on deck.

Today's test was against moving targets.

Small drones—remote-controlled planes about the size of an albatross, used for gunnery practice—were notoriously difficult to shoot down. The tiny aircraft were so tough to knock from the sky that even though Parsons had requested six target planes for the trial, the Navy drone technicians opted to bring only four. In their experience, ambitious gunnery officers usually asked for more target planes than needed. Their drones were rarely damaged beyond repair.

The Navy photographic crew assigned to document the trials told Roberts that they had never once seen a drone shot down.

The waterway was cleared. The remote-control drone pilot steadied his hands. A radar, range finder, and mechanical "predictor" would help to aim the guns. Section T fuses, fitted into five-inch shells, were duly loaded. As the first drone left the deck, the *Cleveland*'s gunnery crew was primed and ready for action.

Each pair of five-inch guns on the ship protruded from an enclosed mount that resembled a squat tank with no treads. A standard gun crew consisted of twenty-six men, but twenty-seven were required for firing practice. The mount needed "powder men" to handle the powder casings, two "projectile and rammer men" to prepare rounds for firing, and two "hot case" men to catch ejected casings. "Trainers," "sight setters," and "pointers" were normally at the ready to aim the guns manually using optical lenses. And there would usually be a fuse setter, who was not needed that day and whose job, should Section T succeed, would no longer exist.

Under the guns, in ammunition handling rooms, thirteen of the men operated hoists and supplied powder cases and projectiles to the guns. Both rooms in this miniature, two-story arrangement had managers prepared to supervise the frenetic symphony of churning metal belts, valves, shells, and deafening explosions.

The first drone promptly crashed into the water, defective.

Roberts peered through binoculars at the second drone as it began a run toward the ship from three thousand yards away. The five-inch guns unleashed eighty rounds, and within seconds three shells detonated and struck the drone on the right side. It burst into flames and then spiraled into the drink. The third drone, launched off the starboard side, fell after four rounds. Over forty-five hundred feet away, a shell with a smart fuse sliced it with shrapnel and knocked it into Chesapeake Bay.

Parsons requested another target plane. But the drone operators didn't have the last one ready. According to Roberts, Parsons was irate. He'd asked them for six drones, and they had refused. Why wasn't the fourth drone ready, at least?

"You've wrecked two of my drones," a handler said. "That's very expensive."

When the final plane was ready, over an hour later, its pilot simulated a low-altitude bombing run. The lower height didn't make a difference. Eight shots and it was gone.

Eighty rounds for a single target? Eight? Four? By any measure, the results of the drone trials were spectacular. The *Cleveland*'s captain came down to congratulate Parsons and the Section T men. As the physicists boarded a small launch to return to shore, he ordered life preservers brought for them. To prevent the hundreds of sailors aboard from spreading news of the test—of the wondrous accuracy of some new secret weapon—the Navy canceled their shore leave.

Tuve's boss was elated. "Three runs, three hits, and no errors," Bush wrote Conant, in a telegram. The fuse did exactly what it was supposed to.

Now they just had to put it to war.

The aftermath of a V-1 strike in central London, June 1944.
U.S. Army Signal Corps, National Archives

The view two hundred feet from a V-1 crater in southeast London.
U.S. Army Signal Corps, National Archives

A "Light Rescue Squad" in London searches for survivors of a V-1. *U.S. Army Signal Corps, National Archives*

The damage one hundred feet from a V-1 explosion in Forest Hill, London. *U.S. Army Signal Corps, National Archives*

Sloane Court East shortly after the V-1 strike. *Courtesy of Alex Schneider and the Hatch family*

The wreckage at Sloane Court East. *Courtesy of Alex Schneider and the Hatch family*

The truck Ed Hatch tried to board, shortly after the incident. *U.S. Army Signal Corps, National Archives*

A block away from Sloane Court East, where the V-1 struck. *U.S. Army Signal Corps, National Archives*

The 130th Chemical Processing Company, circa 1944. *Courtesy of Alex Schneider and the Hatch family*

Samuel "Ed" Hatch. *Courtesy of Alex Schneider and the Hatch family*

The silhouette of a V-1 in flight. *New York Times Paris Bureau, National Archives*

A V-1 falls silently over London, June 22, 1944. *U.S. Army Signal Corps, National Archives*

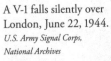

At a Nazi launching site, Colonel Max Wachtel's men fit the wings onto a V-1. © *Imperial War Museum (CL 3431)*

A V-1 rests on a launching ramp at Zempin, near Peenemünde, October 26, 1943. *From V-Missiles of the Third Reich (Monogram Aviation Publications, 1994)*

Engineers at Peenemünde prepare a V-1 for testing. *U.S. Air Force, National Archives*

Colonel Max Wachtel (facing the photographer) readies for a V-1 test at Zempin.
From V-Missiles of the Third Reich *(Monogram Aviation Publications, 1994)*

Colonel Max
Wachtel, 1944.
From V-Missiles
of the Third Reich
*(Monogram Aviation
Publications, 1994)*

A 1942 photo of an Auxiliary Territorial Service spotter (an "ack-ack girl") beside a British 3.7-inch antiaircraft gun. © *Imperial War Museum (TR 453)*

The author holds Merle Tuve's souvenir fuse, of the type used against the V-1s. The label reads "Tuve personal."
Courtesy of the author

Smart fuses are fitted into 105 mm howitzer shells.

Gunners of the U.S. 134th, equipped with the fuse, fire on a V-1.
U.S. Army Signal Corps, National Archives

A V-1 is met by a firework display of antiaircraft fire. The time exposure reveals the bursting exhaust of the pulsejet. © *Imperial War Museum*

Private Margaret Hicks tallies another V-1 (or doodlebug) "kill," August 6, 1944.
© *Imperial War Museum (H 39811)*

A full-scale model of a Japanese fighter hangs between 248-foot towers at Section T's New Mexico Proving Ground.

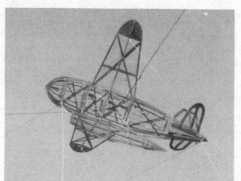

Section T's mockup of the Nazi V-1 is readied for firing tests, March 1944.

Vertical firing tests at Section T's Newtown Neck Proving Ground. *© 1944 The Johns Hopkins University Applied Physics Laboratory, LLC. All rights reserved.*

Section T's 37 mm gun in front of the unfinished cyclotron, 1943. *Carnegie Institution for Science, DTM Archives*

Merle Tuve at the Department of Terrestrial Magnetism, 1936. *Carnegie Institution for Science, DTM Archives*

Merle Tuve (second from right) at a Section T proving ground, 1943. The dirt piles are from digging up test shells. *© 1943 The Johns Hopkins University Applied Physics Laboratory, LLC. All rights reserved.*

Lucy and Tryg Tuve, circa 1940.
Courtesy of Lucy Tuve Comly

Merle Tuve (left), Larry Hafstad, and Odd Dahl huddle around a particle detector at DTM, 1931. *Carnegie Institution for Science, DTM Archives*

Merle Tuve with Winifred, Lucy, and Tryg, circa 1945.
Courtesy of Lucy Tuve Comly

Section T hires,
with Merle Tuve
at center.
*Carnegie Institution for
Science, DTM Archives*

Ed Salant, after the war.
Courtesy of Patsy Asch

James Van
Allen in his
Navy uniform,
in 1945. *Courtesy of
Henry County Heritage
Trust, Mount Pleasant,
Iowa*

Section T's self-named "Wench Bench." Left to right, standing: Lulu Trundle, Doris Poole, Clarice Fillner, J. J. Hopkins, Anna Vantine, Shirley Evison, and Lillian Hines; seated, Annabele Sherbert, Cicily Jones, and Irma Popp. Their coat of arms reads "E Pluribus Oscillator."

Van Bush behind his desk at the Carnegie Institution of Washington, shortly after President Roosevelt approved the creation of the Office of Scientific Research and Development.

Van Bush in his basement machine shop.
A photo from this series first appeared in the June 1948 issue of Popular Science *magazine.*

Jeannie Rousseau during the war, in Paris.
Courtesy of Pascal de Clarens

R. V. Jones. *Courtesy of Rosemary Forsyth*

A wartime aerial
photograph of
Peenemünde.
UK National Archives

Section T's converted garage.

Construction at Section T headquarters, 1944. The original garage building is on the far right.

Section T's Applied Physics Laboratory, January 1945.

At the Erwood Company in Chicago, a worker assembles fuse parts. *© 1945 The Johns Hopkins University Applied Physics Laboratory, LLC. All rights reserved.*

Workers test fuse parts at the Erwood Company. Batteries line the desk in the foreground. *© 1945 The Johns Hopkins University Applied Physics Laboratory, LLC. All rights reserved.*

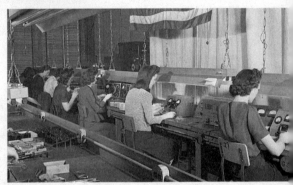

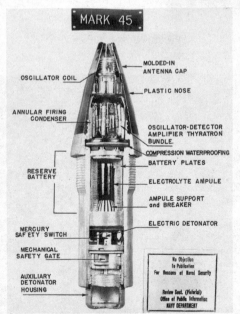

MARK 45

OSCILLATOR COIL

MOLDED-IN ANTENNA CAP

PLASTIC NOSE

ANNULAR FIRING CONDENSER

OSCILLATOR-DETECTOR AMPLIFIER THYRATRON BUNDLE.

COMPRESSION WATERPROOFING

BATTERY PLATES

RESERVE BATTERY

ELECTROLYTE AMPULE

AMPULE SUPPORT and BREAKER

ELECTRIC DETONATOR

MERCURY SAFETY SWITCH

MECHANICAL SAFETY GATE

AUXILIARY DETONATOR HOUSING

No Objection to Publication For Reasons of Naval Security

Review Sect. (Pictorial) Office of Public Information NAVY DEPARTMENT

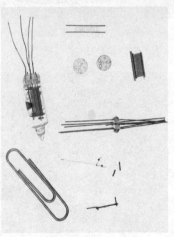

A "rugged" tube and its parts. *© 1942 The Johns Hopkins University Applied Physics Laboratory, LLC. All rights reserved.*

Cutaway of a Mark 45 fuse, the type used against the V-1s. *© 1948 The Johns Hopkins University Applied Physics Laboratory, LLC. All rights reserved.*

17
INTO THE FLEET

Geoorge Hussey Jr., a forty-eight-year-old rear admiral whose scornful face seemed designed by God for belittling sailors, was the new director of production for Navy ordnance. He was just getting settled in when Sam Shumaker, head of research and development, told him about the mysterious project.

"It'll take an afternoon to see it," Shumaker said.

When the pair reached their destination in Maryland—a used-car garage—Hussey was shocked by the secrecy protocols. He'd been at Pearl Harbor two days after the attack, and hadn't seen security this tight since Hawaii.

Through two sets of gates, past two sets of guards, Hussey met the physicist chairman of the clandestine operation. Merle Tuve, very quietly and deliberately, explained to him their work and the details of the device they were working on. One by one, a series of scientists and engineers came in and described their roles.

The utter strangeness of the project began to dawn on Hussey. He owned radios. He had, like almost everyone, accidentally knocked a radio onto the floor before. The electronics shattered. He'd tossed the "sorry mess" in the trash.

Had these scientists really crammed a tiny radio inside a shell, into the nose of a projectile that, they claimed, could now somehow think for itself?

His eyes were "sticking out" of his head, and that was before Tuve showed him the movie of the recent Chesapeake test. Hussey watched the film with amazement. The images being projected were, in his words, a true stunner: "One drone, one shot, no drone. Second drone, one shot, no drone. Third drone, one shot."

"Exercises complete," he marveled, "no more drones available."

When Shumaker took him to Crosley's factory in Cincinnati days later, to see how the fuses were made, Hussey's impression was just as glowing. Inside a modest unmarked building, a pilot production line was "beautifully laid out." None of the operators where chatting, and supervisors circled like hawks.

Three pilot productions lines in all, each isolated from the others, guaranteed that the workers knew relatively little about what they were making. Nowhere along the lines was there a reference to a fuse, and only at the end of the third line, located in a second building, were the complete, seven-pound fuses finally assembled.

Hussey was struck by the workers' extraordinary focus.

Shumaker had taken him on the Section T tour for a reason: If the smart weapon was going to be manufactured in volume—if it was truly going into battle on Navy ships—Hussey's department would be supervising the job.

"George," Shumaker said, "I think that project is ready for production."

"I agree," Hussey said. "We will take it."

"Ok, kid. You're out on a limb for eighty-five million."

Hussey didn't know how shaky the limb really was.

The Navy's new director of production didn't seem to realize that the footage of the Chesapeake trial was edited to be even more appealing. The film was a highlight reel. None of the drones had been downed by a single shot, and not all of the different fuses that were tested had performed to the same standards.

Devices made at Crosley, it turned out, were the ones that took eighty rounds to destroy the drone: ten times the number achieved by Sylvania's factory in Ipswich, Massachusetts, and twenty times that of Erwood's shop in Chicago.

Yet only Crosley was ready for mass production.

The Cincinnati fuses weren't sensitive enough to the radio signal bouncing off the drone. A lower sensitivity made "premature" misfires less common. The higher the signal threshold to trigger a shell, the less likely it was that the microphonic vibrations would set them off early. The downside was that Crosley's fuses "heard" the signal too late. During the Chesapeake trial, their shells often exploded behind the target. Much of the shrapnel missed, just like at Parris Island.

Crosley fuses had to be three times more sensitive.

Merle formed an emergency "Steering Committee" for expediting production. He sent Dick Roberts to Cincinnati to convince John Reid, a top Crosley engineer, to fix their design. But when the hotshot physicist arrived, Reed staged a formal presentation, replete with graphs, arguing that the real trouble with their fuses was that the sensitivity was too *high*. He insisted on lowering it further.

Design changes, in any case, Crosley told Roberts, took at least two weeks to go into effect. If Section T wanted the devices to be more sensitive, the company would need new blueprints. And those would require proper approvals.

Roberts wandered over to the assembly line. He discovered that, actually, simply switching out one basket of existing components did the trick. Over the next month, Section T stuck close to Crosley, finalizing the fuse design and smoothing out what they diplomatically called "friction in human transactions."

By September 19, Crosley's production line was up and humming, and by September 26, it was delivering service fuses ready for Navy ammunition.

The question of safety delay devices had also been resolved. The answer had been right in front of the scientists all along. The Navy's mechanical fuses, as Section T knew, had timers that could be set up to forty-five seconds. But instead of using the timers to explode fuses, one of Tuve's teams modified them to *prevent* detonation. Now the shells couldn't explode until they were safely a half second out of the gun. Spearheaded by physicist Lewis Mott-Smith, it was a shrewd solution. Rather than invent a new device, he had incorporated parts already in production.

Tuve hoped to ship one hundred thousand fuses to the Navy. Yet the sheer size of the supply chain feeding small parts to Cincinnati threatened to delay that goal. At Crosley, production slowed because Hoover, which was assembling the Mott-Smith safety devices, couldn't keep pace with them. When Hoover finally upped its production capacity, its own supplier of parts for the device, the Bendix Corporation, hit a snag. When Bendix got back on track, Hoover was delayed. And when Hoover was running smoothly, Crosley seemed to be slowing down. Seemingly insignificant delays in shipping parts that cost only six cents each could halt the entire show.

Materials weren't always easy to procure. For security reasons, Section T wanted the plastic nose cone, which encased the radio transmitter and antenna, to be made of black Lucite. But the Lucite ran out, and had to be replaced with black lacquer. Beryllium copper, used for condensers that held electric charges for detonation, was scarce. Copper and beryllium—along with aluminum, tin, bronze, brass, tungsten, and nickel—were among items listed by OSRD as "highly essential to the war effort" but "whose supplies are inadequate for existing demands." Substitute materials had to be "proven in" with shooting tests. More delays.

Finally, fuses had to be certified for battlefield conditions. The Navy required that weapons withstand "drop, jolt, and jumble" tests. Fuses had to survive being dropped from five feet. Safety devices were "jolted" thousands of times in a special machine, and rudely "jumbled" at thirty-five rotations per minute.

Section T needed to know—and fast—if a torpedo strike could ruin or detonate their devices. How did the explosive squibs react to low air pressure? How would the constant sway of ships affect the rugged tubes? How would muggy air affect the wiring? How would the gadgets react to winter cold and tropical heat? Fuses had to be frozen and baked, and waterproofed to withstand humidity.

None of the test results posed a serious threat to the project.

Except one.

By the fall of 1942, Tuve was less concerned with the technical and scientific aspects of the fuse than he was with organizational delays.

But the batteries were a special case.

Like any portable radio, Section T's gadget needed batteries. But before the war, no battery existed that was small and tough enough for the job. Just a few years prior, the "energizer" business as a whole had been in serious trouble. As Thomas Edison's vision of an electrified United

States became the new reality, the cumbersome portable radios of the early 1920s went out of style. Americans figured there was no need to bother with ungainly portable sets when they could plug their radios in. Batteries were too big, heavy, and expensive, and they leaked. The National Carbon Company, which owned Eveready (today the company is known as Energizer), was thinking of getting out of the battery business altogether.

It wasn't until shortly before the war that a breakthrough from National Carbon's lab in Cleveland brought the industry roaring back. Harry French, a short, exacting man with battery patents stretching all the way back to 1918, had engineered a design that could generate the same capacity as the existing energizer in half the space. It was also cheaper. (French was so fastidious, one colleague said, "It would take him three minutes to get a cigarette ready to light.")

National Carbon pitched the breakthrough to portable radio makers, who were so impressed with its potential that they designed brand-new radio sets around the smaller battery. By 1939, "convenient" portable radios—on the order of eight-by-ten-by-fifteen inches—were beginning to sell in real quantity.

The radio industry wanted Eveready to go smaller still. But miniaturization was limited by the individual cells inside each battery. Thirty connected cells of 1.5 volts, for example, produced 45 volts of energy. In French's innovative batteries, the cells' casing still took up a lot of space. By wrapping them in vinyl plastic—a new material—National Carbon again halved the final size. Now the critical mix inside the cells—the wet-paste electrolytes that converted chemical energy into electrical power—made up a larger proportion of the battery. The company was pushing the limits of materials science.

But Eveready's engineers had a message for their eager sales force: "Don't ever come back for anything smaller than this."

Not long after, Tuve asked Harry French, National Carbon, and Eveready to devise an even smaller battery that could sustain 20,000 g's. To fit the fuse, it needed to be just two inches wide and two and a half inches long. Eveready would have to shrink their tiniest battery cells in half. (For the early firing tests, Section T had frozen larger batteries, sawed them in half, and sealed them in wax.)

To work on the project, National Carbon assigned a liaison. Jean Paul

Teas was a chemical engineer and battery expert with a degree in bio-chemistry from Cornell University. He was also a first lieutenant in the infantry reserve. Thirty-one but baby-faced, Teas had actually been called up to duty after Pearl Harbor. But Merle intervened with the draft board and kept him with Section T.

Teas worked diligently with Harry French, with the British and Mark Benjamin (until his tragic death), and with Tuve himself on successive improvements to the design. But there were warning signs about its durability.

National Carbon's battery worked, but radical miniaturization came at a price. The units aged quickly. And when they were tested at different temperatures, they performed poorly in the cold and aged twenty times faster under heat. Tuve believed that the fuse batteries would quickly wither in the tropics.

Never one without a backup plan, in late 1941, Tuve set in motion a replacement-battery program that he hoped would solve the lifespan problem for good.

The solution, suggested by Harry French, was inspired by batteries of old. Instead of relying on the damp paste of a "dry battery," which aged poorly, French planned to separate the vital chemicals into a glass vial. He proposed an outdated wet battery. Fired inside an antiaircraft gun, the idea was, a glass jar filled with electrolyte would shatter on a sharp breaker. The centrifugal forces of the fast-spinning round would distribute the liquid sulfuric and chromic acids throughout the stacks of individual cells and activate the battery midflight.

Engineering glass that wouldn't break from a five-foot drop but would shatter when fired from a gun—a force comparable to a forty-foot plunge—wasn't simple. But National Carbon's wet battery tested well and, according to French, could withstand a fall of between four and twelve feet before the glass cracked.

But when Section T dropped them from over three feet, they burst.

"It is now September," Merle complained, in a 1942 memo, "and we do not have a [wet] battery that works . . . production machinery (for which $100,000 extra was allotted) has not been made; no estimates, to my knowledge, or plans have been made for a factory layout on the necessary scale of 5,000 to 10,000 per day." Tuve accused his liaison Jean

Teas of bad management, and sat in on meetings with the battery group, pressing them on quotas, tests, and priorities.

Teas, Merle said, seemed "inclined to give various alibis."

Neither Rear Admiral Hussey nor anyone else in Navy production seemed to be helping matters much, and every lapse and delay frayed Tuve's nerves.

But Section T's chairman was now out of time. The existing dry batteries had a shelf life of merely six months. So be it. The fuse was shipping out.

James Van Allen was commissioned as a Navy lieutenant, junior grade, on November 4, 1942. It marked a singular turn for the twenty-eight-year-old physicist, who was already deputized as a sheriff in Maryland and who feared the pistol he was issued. Eleven years had passed since he graduated high school and was rejected by the Navy for bad eyesight, poor swimming skills, and flat feet.

He was one of three unmarried men from Section T whom Tuve sent to the Pacific to go ship by ship and introduce the Navy to their new invention.

Deak Parsons understood that if the men were captured as civilians, the Japanese would execute them as spies. But as naval officers, they were, in theory, subject to the Geneva Conventions and might only be starved to death.

Van Allen was given three days to settle his affairs. He borrowed three hundred dollars from the Section T cash "kitty" and purchased what he could of the recommended gear: blue, white, and khaki uniforms, binoculars, a leather jacket, a sweater, and old uniform caps for staying warm on deck at night. He wrapped up his personal things and found a safe place to stash them at the Applied Physics Lab. He picked up a pamphlet on the "Duties and Responsibilities of Naval Officers."

Next was a four-day cross-country train ride to San Francisco. Northeast of the city, at the Mare Island Navy Shipyard, he supervised the loading of some fifty-nine hundred Section T fuses into the cargo holds of the USS *Republic,* a World War I–era troop transport. Boxes weighing one hundred and twenty pounds apiece held two fuses already fitted inside

Navy shells, while "loose" fuses—the majority of the shipment—were packed inside pallet boxes in bunches of forty. To conceal the nature of the secret detonators, the cargo was officially listed as "VT" fuses, standing for "variable time."

Van Allen looked smart in his cap and double-breasted Navy dress uniform with gilt buttons. If he looked a bit green, it was only temporary. By November 19, he was asea in the Pacific trekking to an undisclosed location. From the night stars, he tracked the latitude of the ship. As the *Republic* journeyed west, he estimated their longitude by marking the times of the sunrise and sunset.

Days earlier, he had been struggling with the smaller smart fuse intended for the British, and his mind was still in engineering mode. In a modest black leather memo book, he penned notes about vacuum tubes and ballistic aerodynamics. He was, he said later, "steeped in the virtues" of Section T fuses and eager to "save American lives in the Pacific where the fleet was under murderous attack."

His imagination also drifted to dark outcomes. As they crawled along without an escort at fourteen knots (sixteen miles an hour), he couldn't help but notice that Japanese bombers might easily overpower the meager armaments on deck: some light weapons, four three-inch cannons, and a single gun capable of firing the secret fuses stowed among the cargo. It would be prudent, he wrote, to have a "condensed set of working formulas & tables for emergency life boat navigation."

The Thanksgiving menu included turkey soup, homemade cranberry sauce, peas, carrots, sweet potatoes, cigarettes and cigars, and mixed pickles.

The ship docked at Nouméa, New Caledonia, a port nine hundred miles east of Australia and the headquarters of the U.S. military in the South Pacific.

Deak Parsons was waiting. He had already spoken about the fuse with Admiral William Halsey, naval commander of the South Pacific, and received his approval. Van Allen would introduce the fuse to Navy destroyers. Neil Dilley, recruited from the Eastman Kodak company, would handle the cruisers. Bob Petersen, a physicist from Santa Ana, California, would take samples to battleships.

But Halsey had only okayed a "hunting license." The men had no mandate from on high, and Van Allen couldn't compel anyone to use the fuse. He had to sell them on it.

Boat by boat, like a door-to-door vendor, the shy physicist loaded up his motor launch with some one hundred and fifty fuses and pitched captains and gunners.

He told them a very simple story.

The bane of a gunner's job was what was called the "range error." To shoot at an aircraft, you had to determine the plane's bearing, altitude, and range. While the direction of a target and its height in the air were relatively easy to calculate, its distance from the gun was notoriously difficult to gauge. The range was the data that fuse setters needed to calculate when the rounds should explode.

This new device, Van Allen promised, wiped out the range error.

For security reasons, he couldn't tell anyone how the fuse worked. He could only explain its "black box characteristics." The gadgets didn't have to be set, he said, like standard mechanical fuses. They could be fired in the usual way with the usual fire-control systems for aiming. Maybe 15 to 20 percent of the time, the fuses might burst prematurely, and perhaps 15 to 20 percent would be duds.

But 60 percent, he promised gunners, would have an "uncanny capability of bursting in the immediate vicinity of an aircraft and of doing it fatal damage."

They didn't believe him.

18

NEW TRICKS

On New Year's Day, 1943, Merle Tuve, Section T vice chairman Larry Hafstad, and representatives from Navy ordnance visited the White House. At two p.m. in the West Wing, off the Oval Office in the Cabinet Room, with a fine view of the Rose Garden, they briefed Franklin Roosevelt on the status of the smart fuse, and the president watched the footage from the drone tests on the Chesapeake.

Little is known about the meeting, Tuve's first encounter with Roosevelt. It may have been more sales pitch than courtesy call. (FDR's stenographer notes only that the president was occupied watching "Navy Ordnance Movies.") What is clear is that Section T was struggling to put the fuse into battle, and that Admiral Ernest King, commander in chief of the Navy fleet, was a notorious Luddite. King (whom we'll meet shortly) distrusted new "gadgetry" and discouraged subordinates like Deak Parsons who had a mind for innovation.

Over the course of three days, President Roosevelt met with Admiral King, with Secretary of the Navy Frank Knox, and with Tuve, in that order.

Four days later, the wait was over.

On the morning of January 5, the Navy cruiser USS *Helena* was patrolling in the South Pacific, eighteen miles off the island of Guadalcanal.

The crew was relaxed. Hours earlier, the *Helena* had partaken in a predawn, coordinated bombardment of the Japanese-held Munda Point airfield, on New Georgia island. Its gunners fired more than a thousand rounds of ammunition. Topside personnel were blinded by the flashes of gunfire in the darkness.

The Japanese, the sailors understood, might mount a "face-saving" counterassault on the cruiser and on the other ships in the formation. But a Navy air patrol of Wildcat fighters had spotted nothing unusual in the cloudy skies.

Deak Parsons, oddly enough, was hoping to be attacked.

Parsons had hitched a ride aboard the *Helena* for a strangely super-stitious reason: the boat always seemed to find trouble. He'd lobbied its commander, an old shipmate of his, to ferry him and Section T's detona-tors into hostile waters.

"Let's go get into a fight," Parsons told him.

Even so, the raid on the *Helena* was a surprise. The crew spotted four aircraft circling at over eleven thousand feet, but it wasn't clear they were Japanese. Before the sailors could react, another four planes appeared, and then all eight aircraft leaned into steep dives. The *Helena* was recov-ering patrol aircraft with its crane when a bomb exploded alongside the nearby USS *Honolulu*. The gunners did not have time to return fire until the bombers finished the first dive.

The *Helena*'s machine guns returned fire, but the Aichi bombers were out of range of the small-caliber weapons. They were not, however, out of range of the ship's five-inch cannons. Section T fuses with the "elec-tric eye," screwed into the tips of the heavy Navy shells, were slid into the gun breeches, and the *Helena*'s starboard gun crew fired eighty rounds at the raiders.

Within fifteen seconds, a single burst of shrapnel exploded "very close" to a bomber, obscuring it from view, ripping the fuselage and sending the plane, billowing with smoke, tumbling from the sky along with its pi-lot and gunner. A second Aichi quickly followed its example. "This beats target practice," Parsons cracked.

Chased off by Navy fighters, the other bombers withdrew.

That afternoon, the *Helena*'s search team found one of the downed planes floating, its gunner dead and its pilot alive but riddled with shell fragments. Rescued and brought to the ship, the wounded Japanese avia-tor pulled out a pistol and killed himself—a display of resolve that im-pressed the sailors and warned of the fight ahead.

In his write-up, Captain Charles Cecil, a decorated Kentuckian, testified that the battle value of Section T's gadget "cannot be too highly stressed."

Before Parsons personally introduced the fuse, and the invention saw its first early success on the *Helena,* Section T's Navy lieutenants had been in the South Pacific for over four weeks without a single documented use of the device.

That picture now began to change.

The report from the *Helena* helped. So did Admiral Willis Lee.

Commander of a battleship group, Lee was himself a gunnery expert. As a teenager, he won the national rifle and pistol championships. In the 1920 Olympics, his sharpshooting with a rifle won him five gold medals. Now he headed an informal group of fleet gunnery officers known as the Acme Gun Club who met sometimes to "talk shop and drink weak beer." The club would even have its own jokey membership cards. Lee grasped the fuse's potential immediately.

Section T's James Van Allen, now Lieutenant Van Allen, met Lee in his cabin on the USS *Washington* to pitch the new detonators. Naturally, the admiral wanted to know how the fuse worked. It was against regulations to provide that information, but Van Allen needed an ally. He hiked down to one of the ship's magazines, unscrewed a fuse, tossed the round over the side of the ship, and brought the gadget back to Lee. After disconnecting the explosive squib, he sawed through the casing and showed Lee the battery, the safety clock, the antenna in the nose, and the tiny glass tubes that sent, received, and amplified the radio signal. He even explained the mousetrap spring that he had designed to buttress the tiny filaments.

Lee ordered fuses sent to all fighting ships in his task force.

With Parsons's triumph on the USS *Helena,* and Lee's backing, the fuses slowly began to disperse throughout the Pacific Fleet. In February, during an attack on a U.S. troop convoy, gunners using the smart weapons shot down five Japanese bombers—out of twelve total—before any of the convoy ships could be damaged.

When Section T's other Navy liaisons, Dilley and Petersen, returned to D.C., Van Allen chose to remain with the fleet. He penned a fuse instruction manual, and in his new position as Admiral Lee's assistant staff gunnery officer, he traveled around distributing the devices to destroyers, cruisers, and battleships.

Van Allen was captivated by the work, and by the "power of the sea and the way you cope with it on a ship," especially on a battleship. He found much to learn from the Navy: self-sufficiency, comporting oneself as "an officer and a gentlemen," and the value and integrity of "meaning what you say [and] doing what you say you'll do." He set up an ammunition depot on Espíritu Santo, the largest island on what is today the archipelago nation of Vanuatu. He ferried secret fuses to ships on a converted tuna-fishing boat. Called a YP or "reefer ship," the craft was "essentially a floating refrigerator that carried fresh meat" and other perishables.

The job brought him into contact with seamen, petty officers, ensigns, commanders, and captains—the most diverse group of men he'd ever met, hailing from all stations and walks of life. If any of his youthful shyness had lingered, it was gone now. Van Allen developed, a colleague recalled, "a way of speaking that was low-key," never lecturing and always "treating everyone as equals."

Some gunners still flatly refused to use the new ammunition. But he persuaded some to use a fifty-fifty mix of mechanical and smart fuses.

Others even agreed to a 75 percent quota.

By March 31, Section T had delivered a half million fuses to the Navy fleet. They could only hope the weapon was put to good use.

Van Bush and his wife Phoebe were no longer living at the Wardman Park Hotel.

The couple traded their stuffy lodgings for a proper Washington, D.C., home on Hillbrook Lane, in an affluent neighborhood lush with vegetation. The Bushes' large house was east of the Dalecarlia Reservoir, and just south of Spring Valley Park. Even today, the surrounding street names—Overlook, Rockwood, Woodway, and Glenbrook—suggest a respite from the bustle of the city. Its rustic feel may have reminded them of their three-hundred-acre plot in New Hampshire, where Bush liked to retreat and tend to their farmhouse and hundreds of turkeys.

Bush planned to build a machine shop in the basement. He relaxed from his brutally stressful days by tinkering with his gadgets and inventions, and straining to coax color, using various chemicals, from pastoral photographs.

He could not help being full of ideas, big and small, reasoned or bizarre, for how his Office of Scientific Research and Development could contribute further to the war effort. He also dreamed up ways to personally contribute more. He devised a process allowing reconnaissance photographs to be developed and viewed within five minutes of the plane landing. He sketched out an innovative design for manufacturing machine guns, and sent plans to the Army Air Forces proposing a better method for mooring aircraft to ships. OSRD's chairman even volunteered his pet to the cause.

"I have a Saint Bernard dog, a year and a half old," he explained in a letter to an outfit called Dogs for Defense Incorporated, which trained patrol and guard dogs. The canine, he wrote, "might be useful" in military service.

"I would be glad to make him available."

Bush even sold Deak Parsons on a plot to throw off Axis attempts to obtain a fuse for themselves. If Section T created a "fake dud" fuse, he suggested, "which would incorporate all possible leads into blind alleys," these spoofs could be shot ashore "casually" to confuse enemy scientists who examined them.

But in early 1943, Bush was less concerned with new gadgets than with making sure that OSRD's existing inventions were used properly. Devising weaponry posed scientific problems; using them, he'd learned, caused political ones.

The Navy's slow rollout of the fuse in the Pacific was a case in point. The device worked. As a December 31, 1942, report put it: "A radio proximity fuse is here, as an effective weapon." The gadget was safe and far more potent than standard fuses. Production schedules had been met, and it was out to the fleet. And yet, there was still no mandate. Van Allen was still running around "selling" it. The fuse would have to succeed by word of mouth.

While Bush's immediate concern was not with the fuse, it also involved the Navy. He was tormented by warning signs in the Atlantic, where Nazi submarines were mounting a series of increasingly deadly attacks on Allied ships.

In February, he updated President Roosevelt on submarine warfare. The Navy's incorporation of new weapons, he wrote FDR, "is coming on well."

Bush knew it was a lie. The Atlantic passage provided a vital supply route to England, and if the Nazis continued to sink convoy ships, London could be starved out. In February, U-boats sank an astounding 108 Allied craft, and over the course of twenty days in March, another 107 Allied boats were sent plummeting to the ocean floor.

The outcome of the war hung in the balance.

Bush had an idea for protecting supply convoys to Britain. But to put the plan into action, he would have to navigate some delicate political sensitivities.

His problem was Admiral Ernest King.

Capable, curt, tough, and humorless, King was a dedicated Navy man and a true patriot. He was also known as a heavy drinker and womanizer with a red-hot temper. FDR was told that King was "the toughest man in the Navy" and that he shaved "every morning with a blowtorch."

King's daughter didn't disagree with the assessment. "He is the most even-tempered man in the Navy," she deadpanned. "He is always in a rage."

Among the Joint Chiefs, none was more resistant to novel weaponry than King. Reviewing a design of a cruiser set to be fitted with new radar technology developed by OSRD, King bristled. "There's too much radar on this ship."

"We want something for this war," he said, "not the next one."

Bush ripped the admiral for his "terrible blind spot for new things—and about as rugged a case of stubbornness as has been cultivated by a human being."

King "made the decisions," Bush said later, "in spite of the fact that he could not possibly understand fully the technical things involved."

King, in turn, told Bush that civilian opinions didn't interest him, and roasted OSRD's chairman for "trying to mess into things in connection with the higher strategy which were not his business, and on which he could not have any sound opinions." Civilians, King felt, had no role in strategic decisions.

At the heart of their dispute was exactly that: Should Bush's scientists have any input into how their inventions was used? Or should they simply deliver their vehicles, gadgets, and weaponry and leave war strategy alone?

The trouble, for Bush, was that innovations like radar fundamentally changed what the military *could* do. New capacities begged for new strategies.

In May of the year prior, Bush had achieved a major political victory. He'd lobbied Roosevelt to create a subcommittee under the Joint Chiefs devoted to educating the military about new weapons. FDR's endorsement made him the first civilian with formal, direct access to the nation's military chiefs. But Bush and King had very different ideas for fighting U-boats.

To protect cargo and passenger ships crossing to England, King favored "convoying," or providing military escorts to groups of ships. The practice dated all the way back to the colonial era, when Spain used convoys to protect treasure ships from marauding pirates. It was also used during World War I, after the emergence of submarines posed a threat to wartime shipping routes. According to King, military escorts were "not just one way of handling the submarine menace" but in fact "the only way that gives any promise of success."

Bush believed otherwise. Radar, he suspected, offered a new defense against the German U-boats prowling off England in "wolf packs." He wanted the military to use airborne radar to pinpoint Nazi submarines from above and hunt them down and sink them.

On March 24, Bush lunched with Roosevelt at the White House and made his views on the matter crystal clear, a breach of channels that infuriated King. Weeks later, Bush made his case directly to the admiral. "Antisubmarine warfare," he wrote King, is "a struggle between rapidly advancing techniques" that involved "military aspects on one hand, and scientific and technical aspects on another."

King relented in May, allowing three of Bush's OSRD scientists to sit in on strategy meetings on antisubmarine efforts. Bush's small advisory group would perform "research-based statistical analysis" for the newly formed U.S. Tenth Fleet, approved by King, which was tasked with rethinking submarine warfare. Using Allied planes equipped with radar to sniff out U-boats, and a more aggressive naval search-and-destroy policy, Bush's strategy was put into effect. He had helped infuse scientific analysis directly into military tactics.

It paid off almost instantly.

Up to that point, over forty-four months, the Allies had sunk about two hundred Nazi submarines. Over the next three months, they sank nearly one hundred. The ratio of lost Allied ships to U-boats destroyed had been forty to one. It now dropped to less than one to one.

Radar wasn't the only key to the Allies' shift in fortunes in the Battle of the Atlantic. Code-breaking, new convoy tactics, and the introduction of other weaponry played significant roles. But airborne radar was the dominant factor.

After Germany withdrew their submarines from the North Atlantic, Hitler's U-boat commander blamed "superiority in the field of science" for ripping the "sole offensive weapon in the war against the Anglo-Saxons from our hands."

Submarines were not the only mortal threat to London.

19
CHERRY STONE

I n April 1943, at a Luftwaffe artillery school in Rerik, on Germany's Baltic coast, Nazi colonel Max Wachtel lectured a class of antiaircraft gunners. A stout, tidy man with flared nostrils and a puckered brow, he was the sort of practical, competent leader who in peacetime might have managed a midsize factory.

"The Colonel is urgently required on the telephone," an orderly interrupted, halting Wachtel's lesson and, it turned out, ending the class outright.

General Walther von Axthelm's office was on the line. Overseer of the school in Rerik, Axthelm was the top Luftwaffe commander of Hitler's antiaircraft artillery corps. The general felt no need to explain why he was ordering Wachtel to report to his headquarters on Fasanenstrasse in west Berlin, by four o'clock.

No explanation was given.

Not ten minutes passed before Wachtel and his chauffeur rushed off in a dark blue six-seat convertible. The hours-long drive to the capital was no pleasure cruise, and they arrived at Axthelm's office with only five minutes to spare.

Wachtel worried he was being reprimanded. "What's wrong?" he asked the receptionist. "No idea," she replied. "The boss is very secretive."

The heavy double doors of Axthelm's inner chamber swung open, revealing a bald, snappish, forty-nine-year-old general. "Come in, Wachtel!" The pair settled down at a large conference table. Outside the window, sparrows were chirping.

Branches bobbed on the linden trees.

"Have you ever heard of Cherry Stone?" Axthelm asked.

"No, Herr General."

Over a glass of cognac, Axthelm laid out his offer. "I am going to tell you something, Wachtel," he said. "What I'm about to explain is not merely secret, but top secret, and very few know about it. In Peenemünde, up on Usedom, we are testing a device intended to bombard London." It was a drone with a one-ton warhead.

The Führer sought "revenge" for the British bombing of German cities like Lübeck. With devastating losses in North Africa, the Luftwaffe depleted, and the invasion of Russia stalled, Hitler was counting on a surprise attack to ruin British morale. The Nazi code name for the covert weapon was Kirshkern, for the "pit" or "stone" of a cherry. Over time, the plane would have many names.

In military cables, the "device" would be referred to as FZG 76. Short for Flakzielgerät, or "antiaircraft target apparatus," the designation was meant to conceal its true purpose from Allied intelligence. But the superweapon's most fearful alias sprung from Hitler's call for retribution: Vergeltungswaffe 1.

Vengeance Weapon 1, or the V-1.

By no later than the end of the year — by December of 1943 — Axthelm needed a command group ready to launch the noxious drones at Londoners.

Wachtel was delighted by the plum assignment.

Within weeks, on May 12, Wachtel took up quarters in the resort town of Zinnowitz, the beach getaway where the private train shuttled scientists and visitors to the Peenemünde compound. In Hotel Preussenhof, a white-plastered spa tavern, he reveled in his luck and newfound power. He had wide authority to select leaders for his command. And his room had a balcony and a fine view of the sea.

He recruited top-notch men with whom he had fought near Leningrad and during the invasion of France. Each was sworn to the strictest secrecy. On Peenemünde West, at the Luftwaffe test field on the tip of Usedom, Wachtel and his leadership team inspected their new weapon as it was test-fired over the Baltic.

The Nazi drone was unlike anything in the history of warfare. Twenty-seven feet long, with stubby airfoils and a pulsejet engine perched on top like a gigantic, mutated cigar tube, the rare aircraft resembled a winged missile. The final design was a collaboration between engineer Fritz Gosslau and a talented airframe specialist and aviator named Robert Lusser. With its intensely clattering engine, the drone had a top projected speed that was dazzling: over four hundred miles per hour.

The autopilot used a compass, pendulums, and gyroscope.

The scientists at Peenemünde and at the three German companies involved in engineering the drone—Argus, Fieseler, and Askania—faced challenges very similar to Section T's, but with far darker ends. Every technical puzzle they solved made a higher death count in London likelier. Early on, they struggled to keep the V-1 flying. The drone used a starter engine, but to draw enough air through the pulsejet to get it up to speed required a huge initial thrust. It couldn't be revved up slowly, like an aircraft engine. The drone needed a hard push to stay in the air.

The initial launching system involved a short ramp and a booster rocket to kick-start the pulsejet. It wasn't quite robust enough to handle the weapon's full weight. But in Kiel, Germany, a fourth company founded by engineer Hellmuth Walter suggested an unlikely solution.

Under Walter's system, Wachtel's regiment would fire the V-1s off of one-hundred-and fifty-foot "catapult" rails rising up at a slight angle. Underneath the rails was a metal cylinder with a fin that pressed against the belly of the drone. Courtesy of the Walter-Werke, a cauldron of hydrogen peroxide would mix with sodium permanganate and ignite in a furious burst of steam, pushing the fin with a force of sixteen g's and propelling the aircraft to 248 miles per hour in a single second.

To hide and store the revenge weapons, the Luftwaffe planned to build bunkers in the Pas-de-Calais, a region that borders the Strait of Dover and faces the English coast. To fire the drones, Wachtel's soldiers would need to man dozens of catapults at secret launch sites strewn across northern France.

The Peenemünde Wachtel encountered dwarfed the comparatively meager resources of Section T and its converted Maryland garage.

By 1943, the Baltic base had expanded radically. It now housed some six thousand workers, scientists, and technicians—nearly twelve times as many as the American group. Fresh recruits were culled directly from Germany's top universities. For security, the graduates were not told of the nature of their secret work until they actually arrived on Usedom. Advertisements for various technical skill sets were posted in major newspapers without specifying who was hiring, or where the job was. Interviewers provided scant clues to prospective hires.

In the midst of total war, the base remained an untouched "paradise." Romance was easy to find for new hires. The scenery was lush and peaceful. The work felt urgent and vital. The dashing rocket scientist Wernher von Braun could often be spotted on campus, as could the daring aviatrix favored by Hitler, Hanna Reitsch.

Rationing now limited rare foodstuffs, but the Baltic provided fresh fish and eel. Wine could be difficult to come by, but the chemists at Peenemünde learned to distill 75 percent ethyl alcohol and concoct a type of moonshine that they added different flavors to and enjoyed at their many social gatherings.

"We had parties," a scientist said. "Parties with rocket fuel."

In Zinnowitz, top-tier restaurants still remained open where waiters wore tuxedos with tails and white ties, and patrons could dine on white tablecloths. Even as Hitler reduced much of Europe to ruins, the scientific city on the Baltic was somehow spared the air raids, bombs, and bullets.

When Wachtel arrived, a stunning eleven thousand foreigners were employed at the high-security base. Trainloads of unsuspecting Russians, Ukrainians, and Poles had arrived in transports at the remote Baltic island. Polish "workers" were shipped to their distant vocations after being arrested for misdemeanors or simply being kidnapped off the streets of Warsaw. Some were as young as sixteen.

Previously, all foreign workers had been lodged in a camp near the housing estate and in the army barracks. But as more transports arrived to meet the needs of Hitler's advanced weapons program, those facilities quickly filled up. A new labor camp had to be constructed farther south, directly before the guarded entrance to the compound, between the electric railway and the beach.

Here, near the tiny village of Trassenheide, at least eight thousand laborers lived in dozens of barracks. Chain-link fences and watchtowers provided a stark contrast to Peenemünde's bright white houses, school, and shopping center.

"Entertainment" at the labor camp included a dance band and a brothel. For two reichsmarks, less than a dollar, a worker could procure a vodka, two cigarettes, and companionship. Occasionally, laborers were briefly allowed into town. In Zinnowitz, a sign plainly warned: POLES AND DOGS NOT ALLOWED ON THE BEACH.

Toiling underneath the main production hall of von Braun's rockets, in one of the largest freestanding structures in all of Germany, were twelve hundred more laborers. They had arrived from concentration camps and were guarded by the SS.

By the summer of 1943 the base was quarters for over seventeen thousand people. Security was so tight, a local said, that "not even a mouse came in and out."

But the reckless breadth of its endeavors was impossible to keep quiet. Bunkers and launchers were already being built. The V-1 was slated for production.

Peenemünde's secrets were spilling out.

Hidden microphones adorned the rooms of Trent Park, an English country house in a northern London suburb now host to Nazi prisoners of war. German generals were treated well at the lavish facility, where they were provided rations of whiskey and allowed regular walks on the estate. One "tenant" was General Ludwig Crüwell, a panzer commander credited with taking Belgrade in 1941, who was captured in 1942 in North Africa. Joining him was General Wilhelm von Thoma, commander of the Afrika Korps and an armored warfare expert who had been apprehended during the Second Battle of El Alamein. Both men were made to feel quite comfortable and had little reason to suspect that their conversations were being recorded.

Crüwell and von Thoma knew they were being held near central London. They were so close, in fact, that had Peenemünde's rocket program proven successful, they should have been able to hear the explosions. On March 22, 1943, von Thoma speculated to Crüwell that, since there had been no great assault, "no progress whatsoever can have been made in this rocket business." He claimed to have seen a special testing ground at Kummersdorf, the original site of Peenemünde's rocket team. "The major there was full of hope," von Thoma told Crüwell.

"He said, 'Wait until next year and the fun will start!'"

Five days later, the transcript of their conversation reached the London office of R. V. Jones and his deputy Charles Frank, at 54 Broadway street.

"It looks," Frank said, "as though we'll have to take those rockets seriously!"

Jones did not know about Gosslau and Lusser's pilotless aircraft and its powerful pulsejet engine, or of Colonel Wachtel's assignment, or the ingenious steam-catapult system that would soon torment London. But MI6 had heard hints and fragments about Peenemünde and the other superweapon being developed there. Jones's team knew about von Braun's rockets.

In January, a tipster from Sweden reported that the Nazis had built "a new factory at Peenemünde near Borhoft where new weapons are constructed," and that foreign laborers there were not allowed to work long at the same site. A month earlier, at a Berlin restaurant, a professor of the Technische Hochschule was heard gossiping loudly with a German engineer about weaponized "rockets."

Within days of von Thoma and Crüwell's recorded talk in Trent Park, an interrogation of a German tank expert yielded wild claims that the Nazis were building a one-hundred-and-twenty-ton rocket. It had a sixty-ton warhead that would destroy anything within eighteen miles, the tank expert said. "Projectors" for the weapon, which was guided by radio beams and gyroscopes, were already in position, he said.

While reconnaissance flights over Peenemünde tried to uncover details of its "new weapons," R.V. turned to the radio listening service for help.

He acted on an extraordinary hunch.

If long-range rockets were being built at Peenemünde, he guessed, then the test range would span northeast, over the water along the coast. He also knew that in World War I, the Germans had trouble developing the greatest siege weapon of that era, the Paris Gun, because they couldn't track the shells. They didn't know how far the weapon was firing. But now that radar existed, Jones believed that in testing long-range weapons, the Nazis would enlist radar technicians for the job.

Because of the rumored supersonic speed of the rockets, monitoring their path would be very difficult. It would require Germany's best radar units.

Jones knew who the Wehrmacht's best radar plotters were: the Fourteenth and Fifteenth Companies of the German Air Signals Experimental Regiment. If he could find them—and if they were stationed near Peenemünde—he might garner clues about the weapons. He asked his contact at Bletchley Park, Bimbo Norman, and the wireless listeners of Y Service to find the radar companies.

Reactions to the Peenemünde intelligence, meanwhile, were mixed. Photographs were analyzed by Jones, MI6, and the British military. But the winter snow provided a natural camouflage that stymied any clear conclusions. R.V.'s own mentor and Churchill's scientific adviser, Frederick Lindemann, did not believe "rumors" of German rockets. But the Chiefs of Staff were alarmed enough to suggest appointing a special adviser, Duncan Sandys (Churchill's son-in-law), to tackle the rocket question. Jones's group, already in place, was sidelined.

When Sandys had his Photographic Interpretation Unit study snapshots of launch ramps for Wachtel's drones, his top analyst dismissed them as irrigation "mud pumps," an error likely influenced by his background in hydraulic engineering.

It was Jones, not Sandys, who first spotted a Nazi rocket in a spyplane photo: a thirty-five-foot cylinder some five feet wide with fins and a blunt nose.

On June 29, Sandys warned Churchill that a single rocket would cause, in the estimate of the Ministry of Home Security, "up to four thousand casualties killed or injured." Lindemann argued that the Nazis might be using wooden dummies as a hoax. Jones told Churchill he believed the rockets were real.

Peenemünde had to be bombed.

The Royal Air Force planned to target three areas on Usedom: the experimental works, where research on rockets was presumably ongoing, the production facilities, and the "living and sleeping quarters, with the object of killing or incapacitating as many of the scientific and technical personnel as possible."

The Luftwaffe's test site for Wachtel's pilotless aircraft, on the northern end of Usedom, remained unknown to the Allies.

It was not targeted.

20
A LONDON FUSE

Theodore Wolfe, cousin to Garland Wolfe—whose former garage was now occupied by strangers—simply had to know what the new tenants were doing. He parked in the alleyway behind Section T headquarters, in the quiet of night, to eavesdrop on the conversations drifting through the open windows.

The Navy let him off with a warning.

Wolfe's curiosity may have been piqued by the young women who reported him. Late into the evenings, a squad of women (including the vibrantly named Irma Popp, Cicily Jones, and Lulu Trundle) moonlighted at the lab assembling experimental fuses. They had their own coat of arms and motto, "E Pluribus Oscillator," and jokingly called themselves the "Wench Bench."

Rush jobs could take until two a.m.

By May of 1943, Section T numbered 480 scientists, typists, stock clerks, field observers, laborers, guards, and other staff. Tuve hired a legal adviser, a contract specialist, a messenger, an accountant, a cook and a maid, and a houseman.

The Navy Yard complained that the volume of confidential "junk" the lab regularly delivered for incineration was becoming unmanageable, and asked them to drop the hush-hush trash directly at the nearby municipal incinerator.

Field operations had grown so extensive that Section T was compelled to lease its very own gas tank and pump and purchase gasoline in bulk.

Tuve's domain was still expanding. Even with forty-one thousand square feet, the rebuilt garage was overcrowded, and construction had begun on a new building next door.

Security incidents had increased yet again.

The wife of James Day, a radio engineer doing classified research for Tuve, was called anonymously and pressed for her husband's exact whereabouts. Employees were overheard speaking carelessly on the phone. A former staffer bragged about the lab to his brother, who was then caught gossiping about it.

But the most significant breaches occurred at factories.

Twice, Crosley representatives betrayed sensitive details about the fuse to subcontractors. An engineer leaked information to a colleague about a secret project with the Carnegie Institution of Washington and a device with electrical parts that needed to work only once. In a similar breach, a sales representative spread rumors of a gadget in which his components were used just a single time.

In Ipswich, Massachusetts, at the Sylvania offices, a marketer from the Central Laboratory Company of Milwaukee showed up demanding to know more about a highly secret "radio device to be used in a projectile." The salesman's boss, apparently, had somehow learned that Crosley, RCA, and Sylvania were assembling such gadgets and imagined that his company could get in on the deal.

The more Section T grew, the larger the risk became.

By the summer of 1943, Crosley and Sylvania had produced and shipped seven hundred and twenty thousand fuses. Sylvania was assembling eighty-eight thousand rugged tubes every day.

Vast cavernous halls lined with rows of benches were packed with hundreds of young women assembling tiny vacuum tubes according to Section T's strict specifications. Equipped with pliers and magnifying glasses, they toiled to popular songs piped into the assembly areas to help ease the monotony. "Working so fast that observation is difficult," one visitor said, "young women put invisible wires into invisible holes and weld them in place."

In one assembly room, another gawker recalled, the workers joined together in unison when the hit song "Deep in the Heart of Texas" played through the speakers: "Every girl would rap on the bench with her pliers the four booms that ended every line in the song. . . . Before each rapping, the girls would slow up their work in anticipation and gather their forces so as to come in on the right beat."

Sylvania ads placed in the *Ipswich News and Chronicle* called for "More Women War Workers . . . single or married — from 16 years of age and up," and promised "light clean benchwork . . . considered a highly desirable type of work."

In addition to its factories in Ipswich and Emporium, Pennsylvania, Sylvania had "feeder plants" doing subassembly work for the tubes. One such plant was nothing more than an upstairs floor above a small boutique. A manager from Emporium was sent out to "hire a number of girls to start working the following day." Their job was to separate the thin mica discs punched with miniature holes that would gird the delicate electronics in the tubes. Inside a single suitcase was reportedly enough work to keep thirty people busy for a month.

Subcontractors couldn't help but be a bit curious.

Bizarre calls went out to manufacturers like the Sprague Electrical Company in North Adams, Massachusetts, which fielded a request from Lieutenant Sally White of the Navy's Bureau of Ordnance. White asked the founder, Robert Sprague, how much it would cost to make thirty million "toothpick" capacitors — the tiny devices that held electrical charges and looked like Tootsie Rolls. To withstand the higher temperatures inside an antiaircraft round, the capacitors had to be soaked with vinyl carbazole, a chemical not being made in America. To Sprague's knowledge, no company had ever used the chemical that way.

Capacitors, rugged tubes, batteries, hollow fuse cases resembling tiny metal tennis cans, safety devices, Lucite noses and nose caps, antennas, oscillator coils, and detonators arrived from all over the Midwest and the East Coast. Undrafted men joined legions of housewives, schoolgirls, and other patriots in factories in Massachusetts, Vermont, New York, Pennsylvania, Ohio, Oklahoma, Illinois, and Delaware. By the end of the war, 112 companies would help produce parts for the fuse, bringing the cost down from seventy-five dollars per unit to about eighteen dollars. Many had never performed such intricate work.

Nor could any know of the true end product.

Progress on the new liquid battery and work on fuses for other military branches required new manufacturing contracts — including with

RCA in New Jersey and Eastman Kodak in New York—heightening the security risk.

Tuve's laboratory was a victim of its own success. Demand to adapt its device to various caliber guns was higher than ever. The U.S. Army had already placed an order for antiaircraft fuses, and the British Royal Navy was eagerly awaiting the smaller version for its guns. Only the British Army—whose fuse would need to be both thinner in diameter and significantly shorter—seemed uninterested. Its liaison in D.C. told Section T that Churchill's army saw little value in the smart gadget.

A weapon to defend London was not even under development.

But a similar fuse was.

On April 5, 1943, Ralph Baldwin, Section T's thirty-year-old liaison to the U.S. Army, drove down for a routine meeting to the Army War College on Greenleaf Point, in southwest D.C., where the Anacostia River meets the Potomac.

Baldwin was the sort of man who, in later years, would take great pride in tracing his genealogy. In his former and quite recent life, he had been a young instructor of astronomy at Northwestern University with a passion for stellar novae.

He now had more terrestrial, gruesome concerns.

That morning, Baldwin had been brainstorming. Could the smart fuse, he wondered, be adapted for another use totally unrelated to shooting down aircraft? If the gadgets detonated at Dahlgren over the water, from radio signals bouncing off the Potomac, wouldn't they also be triggered by reflections off the ground?

At the War College, Colonel Bjarne Furuholmen beat him to the punch. "Ralph," Furuholmen began, "will this fuse operate above ground?"

"I think it will," Baldwin replied, surprised that the colonel's train of thought seemed to mirror his own. "Were you thinking of antipersonnel use?"

Up to that point, Section T's fuse was truly a weapon of self-defense, a bulwark against air raids. Now, Furuholmen proposed that the invention be used offensively, by American soldiers on the ground, directly against Axis troops.

According to the colonel, explosive shells detonating in "air bursts" over enemy troops were far more devastating than rounds that blew up on impact. The reason was grimly intuitive. When a shrapnel bomb burst against the dirt, the earth absorbed much of its lethal power. But when a bomb erupted fifteen to seventy-five feet above the terrain, troops underneath were exposed to the full blast radius.

Air bursts, the Army estimated, were twenty-five times deadlier.

"We often try to produce air bursts," Furuholmen said, "but it is difficult because there are only two ways of doing it. If the enemy is in [the] woods, some contact fuses will hit twigs and branches and burst the shell above ground. The other way is to use mechanical time fuses. Theoretically, we can set them to burst a fraction of a second before impact . . . but practically we have the same problem as with [the] antiaircraft time fuses." Setting the timers just right was even harder when the enemy was out of sight, or the terrain was hilly, or the weather was bad.

"Can you run a test in some Army howitzers?" the colonel asked.

Section T's astronomer had never even seen a howitzer gun. "What are the shell dimensions?" he asked. "What is the size of the fuse well in the shell?" The "fuse well," the cavity in the shell where the device fit, was a critical variable. For standardization, howitzer shells had the same fuse well as Army 90 mm rounds. And a fuse for 90 mm shells, which weighed only twenty-three pounds, had to be both skinny and stubby. If the fuse cavity was too long, there wouldn't be enough room for the explosives that made the rounds effective in the first place.

To answer the colonel's question, Baldwin asked Section T's research and development team to modify some of their smaller units. On April 29, at Aberdeen Proving Ground, he successfully fired fuses from a 90 mm gun over land. While digging out two of the spent rounds for mementos, he managed to scratch up his arms on oozing poison ivy roots. The shells made for unique desk lamps.

At Fort Bragg, in North Carolina, Baldwin ran further tests in a howitzer on targets shaped like enemy soldiers. Wooden planks, each six feet long and a foot wide—standing erect, or laid over the earth, or buried in the dirt like coffins—were attacked with fused shells. Baldwin's team studied the shrapnel patterns on the planks. Ione Berkeley, a Section T statistician, calculated the optimal height for an air burst over a standing, prone, or entrenched soldier. She devised the most lethal height for a round to explode no matter how a man defended himself.

For Tuve, perhaps because of his Lutheran faith, the antipersonnel program was uncomfortable and "harder to swallow." But new technologies find new uses, and Baldwin's test data was impressive. On Chesapeake Bay in July, as Baldwin watched from a beautiful mahogany yacht, the *Ricochet,* the fuses achieved a 90 percent success rate. (The gunners also almost blew up the *Ricochet.*)

An antipersonnel fuse would work at night, or through clouds, or over rugged, uneven landscape and could be lethal over far longer distances than time fuses. Usually time fuses were set at fifteen seconds, maybe twenty-five at the maximum. But with smart fuses, you could be precise even with sixty seconds of flight time.

More firing tests at Aberdeen and Fort Bragg—including a night test —convinced any remaining skeptics. The fuses, detonating in bright flashes in the darkness, seemed somehow louder to Baldwin in the stillness of night.

The Army decided that, in addition to the antiaircraft fuses on order, they wanted Section T to deliver one million antipersonnel fuses.

"My God," Tuve exclaimed, when he heard the magnitude of the Army's request. "I meant to pull the string on a small toilet and I got Boulder Dam."

Inside Hereford House, a spacious brick building at 117 Park Street in London, across from Hyde Park and Kensington Gardens, the Office of Scientific Research and Development's liaison office was doing a lot more than liaising.

For the better part of two years, OSRD's satellite office had focused almost exclusively on exchanging information about new weapons. The London Mission traded test data with the British on ballistics and bomb damage, new chemicals, submarine warfare, radar, and military medicine. Bush's scientists visited England to share their expertise in optics, rockets, and atomic physics.

During 1943, as OSRD's focus shifted from sharing research to consulting on using weaponry, Bush had twenty-six scientists stationed in

London. Sixty-nine specialists were assigned there not merely for liaison but for fieldwork. (Soon enough, they would also be deeply involved in gathering scientific intelligence.) Two field stations were set up, transforming the small diplomatic outpost into one of the most important scientific agencies in the United Kingdom.

Section T's experts arrived on May 20.

William Parkinson, twenty-five, was a recently naturalized Canadian engineer and a graduate of the University of Michigan. John Doak, a stylish twenty-three-year-old radio technician from Missouri, was recruited out of Texas. He'd with been with Tuve since June of 1941. The senior man was the terse, energetic Ed Salant.

They brought fuses for the Royal Navy.

Salant had been laboring to finish the thinner fuse and to hone the new liquid battery, which remained Section T's main concern in early 1943. The scientists had struggled to discover the exact right mix of chemicals to put in their glass jar and to manufacture the perfect glass, something that broke at high g-forces but not if the fuses were dropped accidentally. Salant was well equipped to grasp the challenge of the wet batteries and to manage his liaison work with the British. On top of a physics degree, he held a PhD in chemistry from the University of London.

Tuve's old friend might have needed the distraction of a trip more than ever. Salant's wife of fifteen years, Thelma Adamson, had recently fallen ill. She was a pioneer of ethnographic fieldwork in the Pacific Northwest, and they had married in secret to preserve her academic prospects. But she had suffered a nervous breakdown, and the year prior, in 1942, was committed to a psychiatric hospital.

Work at least kept Salant's mind occupied.

In England, he and his twenty-something associates helped perform tests at Shoeburyness, a military proving ground at Pig's Bay, east of London. They shot fuses in British guns and oversaw sensitivity tests against "aircraft" made of wire netting, to confirm that the fuses exploded at the correct distance.

Salant was especially anxious about the "rough usage" trials. He worried that the glass jars of chemicals in Section T's wet batteries—which were not in use in U.S. Navy fuses—would break if a British ship took a torpedo hit.

If an ampule broke, it could trigger a cascade of explosions.

At Section T's suggestion, the British and American researchers placed fused shells on a vibrating machine inside a chamber used to blow up large bombs in controlled detonations.

They shook the fuses and held their breath.

The glass ampules of the batteries did not break. Even when the British dropped shells from a height of twenty feet, the little jars held. The final bar had been cleared, and the Royal Navy accepted the smart fuse into service.

Before Salant returned to Section T, in August, he met with Professor John Cockcroft, the physicist who had traveled to Washington in 1940 to share scientific developments with the United States. Despite what Section T had been told, Cockcroft insisted, the British Army *did* want the fuse.

Salant also met with General Frederick Pile, the man in charge of Britain's Anti-Aircraft Command. It was Pile's gunners who had tried to protect London during the Blitz. And it was Pile's guns that would have to protect the capital from the next attack. Known as Tim, he was fifty-eight and had a beaked nose and a modest mouth that, when closed, gave him an expression of mild satisfaction, as if he had just consumed a small snack. He had a reputation for being "imaginative and inventive."

He needed to be. Pile's gunners, at the start of the war, were what he later called "the leavings of the Army intake after every other branch of the Services had their pick." Many recruits didn't belong in the army at all, and were clearly unsuited for the technical challenges of antiaircraft work. To fill the gaps, he lobbied to recruit over seventy thousand women volunteers from the Auxiliary Territorial Service to help work his gun sites. They would become known as "ack-ack girls."

Pile was not averse to change or new technology.

He asked Salant for fuses for 3.7-inch guns, whose shells weighed only twenty-eight pounds. They were roughly half the size of the Royal Navy shells.

The only model small enough, Salant knew, was the stumpy antipersonnel fuse being developed for the U.S. Army. It wasn't yet in production. But not long after Salant's meeting with Pile, Section T set in motion plans to produce two hundred and fifty thousand fuses for the British Army. Their target delivery date was June, 1944.

According to MI6, the Nazi drone would arrive first.

21
AMNIARIX

The Americans were in the skies above Paris.

On April 4, 1943, eighty-five B-17 bombers of the Eighth Air Force soared over the French capital, through a deep blue sky, toward the outskirts of the city. Their mission was to bomb industrial and strategic sites and help destroy Hitler's capacity to wage war. Targeting the Renault car factory, now producing tanks and armored vehicles for the Nazis, the aviators took in a bird's-eye view of the Eiffel Tower and Notre-Dame. The antiaircraft flak was so light that the pilots, stacked in tight formation, did not even bother to take evasive maneuvers.

According to Allied intelligence, one-tenth of Germany's motor transport was being manufactured in the Renault works along the Seine, and on an islet on the river itself, in the suburb of Boulogne-Billancourt. As American bombs hatched from the planes over the island, citizens watched from Parisian rooftops.

On the return trip, four B-17s fell to Luftwaffe fighters.

Allied pilots who lost dogfights in French airspace were by now parachuting onto French soil in greater and greater numbers. Americans, Australians, British, Canadians, and New Zealanders abandoned their blazing, punctured, twisting, plunging aircraft only to find themselves behind enemy lines.

In Paris, at great risk, friends of the Resistance hid soldiers in their apartments while the bombardiers, navigators, pilots, and radio engineers waited for forged documents to complete the perilous journey back to England.

Parisian women posed as wives to safely escort the soldiers.

On July 14, the Eighth Air Force bombed the airport at Le Bourget, in the northeastern suburbs of Paris, again tangling with the Luftwaffe. Nazi propagandists spread lies that the American air raids were actually targeting French industry so that the "Yankees" would have no economic competition after the war.

As the Allies prepared to invade Sicily, Parisian newspapers under tight supervision by collaborators were not read for news of Axis defeats but instead for information about scarcely available foodstuffs. Counterfeit ration cards were commonplace as the city strained under the stresses of wartime occupation. Weight loss and sickness were rampant. Tuberculosis cases were up 30 percent since 1939.

Five hundred pieces of "degenerate art" were burned outside Paris's Jeu de Paume gallery, in the heart of the city. The bonfire consumed works by Pablo Picasso, Salvador Dalí, Paul Klee, and Joan Miró. The museum itself was now a storage site for art plundered from French Jews. There, Nazi Party leader Hermann Göring would view "exhibitions" and select looted valuables for his private collection. Three warehouses across Paris acted as sorting houses for processing valuables stolen from more than forty thousand apartments.

French Jews were hunted and put onto trains.

Paris had begun to resemble a bleak prison, as gardens were converted into parking lots, guardhouses were erected across the city, and massive concrete bunkers and blockhouses were built. Metal grilles and bars were secured to the windows of famous hotels now repurposed as various Nazi headquarters.

The city's grand hotels had transformed into forts of German power. Hôtel Lutetia, a sensual art nouveau building in the sixth arrondissement, was now home to the Abwehr, Hitler's counterintelligence service. (Ironically, its new guests did not sniff out the hotel's best wines, which clever sommeliers had stashed behind false walls.) Hôtel Meurice, overlooking the Tuileries Garden, hosted the military commander of Paris. Hôtel Ritz was the Luftwaffe's.

And the former Hôtel Majestic—often visited by Jeannie Rousseau, aka Amniarix—remained occupied by German military high command. Over the course of her duties coordinating with French industry, Rousseau had encountered, quite unexpectedly, several officers from her time in Dinard. These were the same Germans who had protested that Rousseau, their young translator in Brittany, a vivacious "girl" who spoke such elegant German, could not possibly be a spy. The men who had saved her from Gestapo justice at Jacques Cartier Prison were now involved with a new military program that required various construction materials from French industry—and therefore her help as liaison.

In the evenings, the tight-knit group of Germans relaxed at 56 Avenue Hoche, a nearby apartment building. The abbreviated boulevard, radiating like a sunbeam from the Arc de Triomphe and the Place de l'Étoile, was located only blocks from the old Hôtel Majestic, and made for a convenient meeting place.

Many evenings, Rousseau joined them.

As she had been in Dinard, the crafty linguist was unusually disarming. The German officers were no doubt charmed by the youthful spy, and began to regard her as part of the group, as unthreatening as "a piece of furniture."

The spring of 1943 turned to summer, and she began to hear fantastic details of the officers' new project. "Seduced" by their translator and perhaps empowered by their own privilege, the Germans were bragging about secret weapons. She teased them, gazed at them with doe eyes, claimed disbelief at their masculine boasts, and pried out information critical to the war.

For the first time, she heard about rockets.

Raketen was not a German word she knew, but she didn't know the term in English or French either. Rockets weren't yet widely known.

Her handler, the mathematician Georges Lamarque, urged Rousseau to dig deeper, and learn as much as possible about the secret project.

"Pursue it, go into it!"

Armed with a near-photographic memory, she began to assemble one of the most important intelligence reports of the Second World War, a "masterpiece in the history of intelligence gathering" that would save countless lives.

The technical details that Amniarix recorded were not gleaned from a group of German officers gossiping loosely on Avenue Hoche, but from a single source who has never been identified. What is known is that Rousseau's source was an active-duty captain attached to the experimental facility at Peenemünde. He made frequent trips, she learned, to northern France and northern Germany.

The project was undoubtedly of high priority: it was clear that the captain could easily obtain whatever supplies he wanted from the various German services.

Rousseau needled and flattered the captain, who didn't understand why the French did not love the Germans more, and why so few French citizens seemed to grasp that collaborating was in their own best interests.

"You really think so?" she said. "Who is so sure that you're going to win the war?"

"But we are going to win the war!" he replied.

She made him prove it.

At Lamarque's safe house at 26 rue Fabert, an inconspicuous beige building facing the Esplanade des Invalides, she copied down what she had learned. "Concentrated on the island of Usedon [Usedom]," she wrote, "are the laboratories and scientific research services for the improvement of existing weapons and the perfecting of new weapons. The island itself is very closely guarded." She described three special permits, required to enter the base and printed on orange, white, and watermarked paper, and noted that the administrative services were located both at Peenemünde and at nearby "Zemfin" (Zempin).

The captain told her of experiments with weaponized bacteria, of a rocket as loud as an American B-17 that could reach the stratosphere and was in the final stages of development, and of Colonel Max Wachtel's "antiaircraft" regiment. She learned that Hitler was referring to these advances when he'd spoken recently of "new weapons that will change the face of war when the Germans use them."

She also discovered that test trials had already been made by Wachtel, and that his regiment consisted of sixteen two-hundred-and-twenty-man batteries, which would soon be stationed across northern France between the cities of Amiens, Abbeville, and Dunkirk. To fire its weapons, she reported, Wachtel's regiment would use 108 "catapults" capable of launching one bomb every twenty minutes.

She determined that a man named "Sommerfeld" — Lieutenant Colonel Heinrich Sommerfeld, a professor and ballistics expert — was Wachtel's technical adviser, and that Sommerfeld believed the weapons could destroy London.

When Rousseau showed the report to Lamarque, the mathematician turned Druids spymaster found the claims preposterous. But he had utter faith in Rousseau, who in turn had confidence in her own source. Both knew at once that what she had uncovered was of the highest, most urgent value.

Lamarque, whose own code name was Petrel—a type of seabird—added a note to Rousseau's statement remarking that while the information it contained did seem bizarre, he was certain there must be truth in its outlandish details. Then he passed the dispatch along to Marie-Madeleine Fourcade, code name Hedgehog, who was his own superior and who oversaw the Alliance spy network later known as Noah's Ark. Fourcade, also, immediately recognized the write-up as "completely out of the ordinary run" of standard intelligence reports.

"Who was Petrel's source?" she asked a fellow spy.

"Amniarix," the man said, "the young Druidess whose real name [Petrel] has refused to divulge. An extraordinary girl who speaks five languages."

Fourcade sent the information to London, to MI6.

At the time, Rousseau was twenty-four years old.

The moon over Peenemünde was nearly full.

Operation Hydra, as the bombing raid was known, depended on a cloudless, well-lit night, and clear weather for landing back in England.

On August 12, 14, and 15, the British staged diversionary raids. Mosquito airplanes—wooden, twin-engine fighter-bombers—charted courses near Peenemünde and then veered south to bomb Berlin instead. The idea was to trigger the air-raid warnings at the Nazi base on Usedom and convince their military command that any further raids were also headed for the German capital.

The mission choice of August 17, a Tuesday, increased the chances that the German engineers and technicians they hoped to target would actually be at Peenemünde, and not on leave. The plan called for special flares to be dropped from targeting planes to illuminate the site's production buildings, experimental works, and housing estate for the bombers.

On the evening of the raid, some six hundred Royal Air Force planes carrying nearly two thousand tons of bombs swept over Denmark and the Baltic. Raids on cities normally called for at least 50 percent incendiary bombs, but 85 percent of the tonnage headed toward Peenemünde were high explosives. Densely packed cities were burned; the scattered sites on Usedom would be blasted into fragments.

As yet another diversion, eight Mosquitoes did attack Berlin, dropping along their route, as part of an ingenious hoax, bundles of aluminum shards. German radar confused the thousands of metallized strips, fluttering in the night and descending very slowly, for aircraft. The ruse attracted some two hundred Luftwaffe night fighters and caused such confusion that German planes accidentally attacked each other, and ack-ack guns fired on their own aircraft.

The view over Peenemünde, a British squadron leader said, was "clear as a bell." Railway lines reflected the moonlight, and Usedom's eastern shoreline shimmered distinctly. As the bombers approached, little streams of white gas—the Germans' smoke-screen defenses—began to billow across the targets.

Flare markers intended to highlight the scientists' housing estate were accidentally dropped two miles south, an error made worse by a horrid coincidence. Two miles south of the scientists' homes was where Peenemünde's foreign "workers" lived. For some three minutes, the main British bombing force mistakenly attacked the barracks of the forced laborers. They wrought a ghastly inferno. The night lit up with the colorful targeting flares. The bombs' fireballs mushroomed in rapid succession; flames burned in red, orange, blue, and even green. From the air, the wooden barracks resembled hot coal embers.

Pine trees blazed like torches.

Inside Peenemünde's guarded gates, scientists huddled with their families in makeshift shelters. The early targeting error was soon corrected, and the Royal Air Force began to systematically bombard the brick and wood homes. One German family, their house crumbling and the shelter filling with smoke, dipped their hair in a baby's bath full of water and ran through the fires to the sea.

Some 75 percent of Peenemünde's homes were destroyed.

Twenty-five buildings at the experimental works, "the scientific heart" of the rocket program, were also demolished. One of the two vast production halls suffered several direct hits. The other was slightly damaged by bomb blasts.

Yet no vital targets were fully destroyed. All but one of the top scientific minds behind Hitler's superweapons survived, as did the experimental workshop, critical equipment and documents, and the state-of-the-art wind tunnel.

The Luftwaffe airfield for Colonel Wachtel's pilotless aircraft, untargeted, was untouched, as were his headquarters and a second launching site at nearby Zempin. During the raid, staff heard only the distant buzz of the bombers.

Hitler's test pilot Hanna Reitsch slept through the night.

The camp, Wachtel said, awoke to a normal day. Cleaners were busy at work, polishing the windowpanes of the mess hall to a spotless shine.

R. V. Jones's work at MI6 was taking its fair toll. So far, 1943 had been a long, hard grind. He strained to keep a constant vigil, navigated bureaucratic infighting and Duncan Sandys's petty attempts to corner the intelligence on German rockets, and endured the stress of the lead-up to the Peenemünde bombing raid.

He and his wife Vera needed a vacation.

One member of Jones's scientific intelligence team, Hugh Smith, kindly offered his country farmhouse in Gloucestershire, in southwest England. So R.V. and his family retreated to the cottage, to the green valley of the river Severn. Here he could relax, practice his sharpshooting skills on the darting hares, and unwind at the local pub, founded in 1474, which served beer in pewter mugs.

While he was on gone, his deputy Charles Frank compiled three intelligence reports that seemed to confirm the threat of German pilotless aircraft.

Frank briefed Jones on August 31, when R.V. was urgently recalled to London for yet another meeting with Prime Minister Churchill about the rocket threat. After-action reports of the Peenemünde strike suggested that the Allies had delayed the deployment of Hitler's rockets by at least two months. It would now be up to Jones and his team to focus on the vague evidence of Nazi drones.

The first report Jones reviewed came out of Denmark.

Nine days prior, on August 22, an alien object—a strange, thirteen-foot pilotless aircraft—had crashed in a tulip field near a church on the Baltic island of Bornholm, some eighty miles northwest of Peenemünde. Before German officers could reclaim the secret armament, a Danish naval officer named Hasager Christiansen reached the crash site. Christiansen made a quick sketch of the aircraft and snapped several photographs. The warhead, he saw, was a dummy made of concrete, and on the tail section, the experimental weapon bore the marking "V83."

Christiansen delivered his report to Danish naval intelligence, and multiple copies were sent to London. The scoop was costly. After one set of his photos was intercepted, Christiansen was identified, caught, and tortured. (He managed to escape from the hospital, but required two major surgeries to treat his injuries.)

At first, Jones wasn't sure what the Danish officer had discovered. It might have been an experimental version of a German glider bomb. He did not know what to make of the odd cylinder stuck to the underbelly of the aircraft that appeared in Christiansen's careful sketch of the weapon.

A second report originated from a disgruntled German officer in the Nazis' Army Weapons Office; he claimed that in addition to rockets, elsewhere on Peenemünde, the Luftwaffe was developing its own weapon: a pilotless aircraft officially known as Phi 7. This was not an army project, and the source knew little else.

The third message was from Amniarix.

For the first time, from Rousseau, Jones learned of the existence of Colonel Wachtel and his regiment and of Nazi plans to launch flying bombs using "catapults." He learned how many catapults there would be, and where to find them.

The dispatch became known as the "Wachtel Report."

R.V. also now learned the name of the colonel's scientific adviser,

Sommerfeld, and his prediction of a dire fate for London within the next six months.

"Wachtel's technical advisor," the report read, "estimates that 50–100 of these bombs would suffice to destroy London. The batteries will be so sited that they can methodically destroy most of Britain's largest cities during the winter."

Part III

VICTORY

For years after the war Van Bush would wake screaming in the night. . . . We all suffer scars, you know. I don't know how we'd help it.

— MERLE TUVE

22
SKI SHAPES

After MI6 learned of Amniarix's ominous report, the British government decided to manufacture an extra one hundred thousand Morrison shelters.

Londoners could hide inside iron cages.

Named after Britain's home secretary, Morrison shelters looked like giant rabbit pens buttressed by iron strips and topped with heavy metal sheets. First introduced in 1941, they were designed for citizens without gardens, who couldn't install the buried Anderson shelters, and had become part of many homes' standard furniture. Families ate on them nightly like dining-room tables, and when the bombs arrived, parents, children, and grandparents alike crawled inside the crates—each some six-and-a-half-feet long, four feet wide, and three feet tall—and hoped to sleep. The cages couldn't absorb a direct hit, but their ductile frames could bend to withstand chunks of ceiling.

Fabricating the extra shelters, R. V. Jones learned, would require so much metal that the construction of two British battleships had to be delayed.

Jones said little at the meeting with Prime Minister Churchill. He was still digesting Rousseau's dispatch. Were there in fact not one but two Nazi superweapons in development? Both meant for London? The picture was blurry.

He returned briefly to Gloucestershire but did not shoot hares.

Anxious to learn more about the quality of the information, he inquired about Amniarix. He was told only that the source spoke five languages and was thought to be "the most remarkable young woman of her generation."

Days later, a decoded Nazi message provided Jones with another vital clue to the pilotless-aircraft puzzle. In the communiqué, German anti-aircraft forces were urgently requested to protect one "Flakzielgerät 76," or "antiaircraft target apparatus 76." The message also reported, as if to justify the request, that the British had recently prioritized the search for secret weapons.

Why were secret weapons mentioned? And what did protecting an antiaircraft device have to do with it? If Flakzielgerät 76 really was some kind of antiaircraft target drone, why would the Nazis need AA guns to protect it?

Jones already knew, from Amniarix, that Colonel Max Wachtel's men "were to form the cadres of an anti-aircraft regiment" and use "catapults" to launch bombs at London. Flakzielgerät 76, he realized, could refer to flying bombs.

On September 14, he concluded that the Nazis were "installing, under the cover name of FZG 76, a large and important ground organization in Belgium-N. France which is probably concerned with directing an attack on England by rocket-driven pilotless aircraft." He reported to Churchill that "the German Air Force has been developing a pilotless aircraft for long range bombardment in competition with the rocket, and it is very possible that the aircraft will arrive first."

Around the same time, his inspired hunch began to pay off. It had been six months since Jones had asked his friends at Bletchley Park to help locate Germany's top radar specialists, who might be called on to track the secret weapons during test flights off Peenemünde. MI6 had indeed found the Fourteenth Company of the German Air Signals Experimental Regiment. The unit had moved a Würzburg radar to the Baltic base, and set up a string of detachments eastward along the coast.

It played out exactly as Jones had imagined.

Whatever objects the Fourteenth Company was now plotting were taking off from Peenemünde at some four hundred miles per hour. By decoding radio reports from each detachment, MI6 could follow the course of every test flight.

The Nazis even made reference to FZG 76.

Jones had a front-row seat to Wachtel's flying bomb trials. He watched nervously from London as the flight paths grew more and more accurate.

Luftwaffe agents began to seize plots of land in northern France near the coast, in Normandy and Pas-de-Calais. At each site, French construction crews arrived to erect small, peculiar, identical structures. Local farmers were allowed to keep cultivating, but certain areas of the camps were strictly off-limits.

Rumors of unusual construction reached the ears of a French railway worker, who passed them along to a forty-five-year-old engineer in Paris, Michel Hollard. An expert in machinery, mechanics, and metallurgy, Hollard was a distinguished World War I veteran who had found civilian life anticlimactic. He was a small man with a handsome, chiseled face and a penchant for disguises. During the First World War, to pass the medical test for combatants, he had been so slender that he faked his chest measurements, and so adolescent that he stuck on fake pubic hair. His current vocation, as a traveling salesman hawking gas engines, provided a convenient cover for his true business of running a spy ring for the Resistance.

Commerce wasn't his only smoke screen.

Posing as a Protestant missionary, Hollard traveled to Rouen, the capital of Normandy, and approached a welfare worker in a labor office who would know exactly where French construction crews were being sent. He professed deep concern for the workers, who had been recently uprooted to work on Nazi projects, and who might find comfort in his Bibles and religious literature.

Armed with a list of local German construction sites, he then boarded a train to the village of Auffay, some twenty miles north of Rouen, where he switched identities once again. Donning shabby workman's clothes, Hollard discovered nothing at the village itself. But three miles southwest, near the hamlet of Beuville, he came upon a noisy construction site in the middle of an apple orchard.

Hundreds of French laborers were busy at work. Undeterred, Hollard picked up an unclaimed wheelbarrow from a ditch, and paraded directly past a German sentry. He simply entered the camp. He filled his wheel-barrow with supplies, chatted with a few French workers, and conducted a daring reconnaissance.

Scattered across the site were ten new buildings. Connected by a winding concrete path, they were all single-story and modest in appearance. Most of the structures were about ten feet tall. Their use remained a mystery, but one strange feature of the camp immediately stood out. Marked by a blue rope and studded with wooden posts down the middle, a one-hundred-and-fifty-foot concrete strip had been paved.

It looked like a small runway.

Hollard leaned down as if to tie his bootlaces, pulled out a compass he kept in his pocket, laid it on the ground, and took the bearing. Then he slipped out of camp.

That night, back in Paris, at a hotel near the train station, he fanned out a map of northern France and southern England and plotted the direction of the mysterious concrete "runway." It was pointing directly at London.

Having enlisted a team of reliable French agents, Hollard set out to learn more. He gave the spies bicycles, advised them of the buildings and concrete strip, and assigned each of them a territory of coastal land. Evading detection by staying off radios, his operatives pedaled through farmland along backcountry lanes.

Across two hundred miles, they found sixty identical sites.

To pry out their exact function, Hollard then sought out construction blueprints. He enlisted an engineer named André Comps, who had been hired at one of the camps as a draftsman. From Comps, Hollard got his hands on drawings of every building at a site later called Bois Carré. The name translates to "square woods," reflecting the shape of the forested plot. Comps even managed to duplicate a blueprint for the unusual runway. A German engineer kept it securely in his overcoat pocket, but Comps copied the schematic while the officer was on the toilet.

Hollard crossed into Switzerland and delivered his report to an MI6 agent stationed in Lausanne. Jones received it on October 28, 1943.

Comps's sketch of the installation revealed the concrete platform with its center axis aimed directly at London, and six days later, photo reconnaissance of Bois Carré confirmed the sketch and the "runway." The concrete strip, Jones saw from the blueprint, was intended to support an inclined ramp with two metal rails that rose up at an angle of fifteen degrees, and stretched some one hundred and fifty feet.

Amniarix's "catapults" were real.

Aerial photography confirmed other peculiar features at Bois Carré.

To the east of the site, three narrow buildings perhaps two hundred and fifty feet long, all precisely identical and curved slightly at one end, fed into pathways leading toward the center of camp. They resembled three skis turned on their sides.

For some reason, one of the huts at the camp, according to Hollard's and Comps's dispatch, contained no metal parts. As it turned out, the hut—which was nearly thirty feet high, much taller than the other buildings at Bois Carré—contained no *magnetic* parts. Perhaps using iron in its construction would disturb an aiming system for a weapon that depended upon some kind of magnetic guidance. Inside the tall hut, the blueprint revealed, was an arc inscribed on the floor, suggesting that a weapon could be swung to test its compass. Paths from the "ski" buildings fed into the hut, which had a low, wide arch, perfectly suited for a small winged aircraft.

MI6 called these military stations "ski sites."

They had to be bombed.

While Amniarix and Michel Hollard risked their lives to reveal the looming threat against London, Jones faced yet another petty political battle over scientific intelligence regarding Nazi rockets and flying bombs.

Spies were executed as analysts bickered.

Despite Jones's successes with uncovering radio-bomb targeting, the Bruneval Raid, identifying the first rocket, and his clever strategy to track the flying bomb trials—and despite the trust the prime minister had in his skills and judgment—the British air staff now in effect sought to demote Jones and his team. Scientific intelligence, in the eyes of some, had grown too important to leave to Jones now that London's very survival depended on it. So the air staff brought in an outsider, Air Commodore Claude Pelly, who had no previous experience in intelligence, to coordinate analysis and plan countermeasures. R.V. was asked to merely supply raw information to Pelly's new committee.

Instead, Jones penned a scathing letter to the chairman of the Joint

Intelligence Committee. "It has been my duty since the beginning of the war," he wrote, "to anticipate new applications of science to warfare by the enemy." Unless he had failed in that duty, he argued, there was no plausible reason for a new committee. Its formation, he wrote, "indicates a lack of confidence in my methods which—if I shared—would lead me to resign." Pelly's committee was not only unnecessary but would, he believed, waste valuable time.

His own team, he added, would not be stepping down. "My section will continue its work regardless of any parallel Committees which may arise and will be mindful only of the safety of the country."

"I trust that we shall not be hindered."

He did not receive a reply to his fiery letter. But he did soon afterward come down with tonsillitis and the flu, running a temperature of 108. While he was sick in bed, his deputy Charles Frank brought him reconnaissance photos from above Zempin, where Amniarix claimed Max Wachtel and his men were stationed.

The snaps revealed the same iconic concrete strip discovered at sites across northern France. They also showed, startlingly, an installed catapult, a breakthrough finding that allowed Jones to photograph identical launching ramps at Peenemünde proper, where they had been mistaken for irrigation pumps.

One photo revealed a flying bomb perched on a ramp.

Jones was forbidden from returning to the office, but he disobeyed the doctor, remarking that he only needed to last another six to eight months. The clock was ticking, he now knew, on both D-Day and the flying bombs.

From the German radar unit's tracking of the trials, and the photos of Peenemünde and the ski sites, MI6 now had a robust picture of Wachtel's secret weapon. The V-1 warhead probably weighed between one thousand and three thousand pounds. Its top speed might be four hundred miles per hour, and it likely flew at an altitude of sixty-five hundred feet. The pilotless aircraft should have a maximum range of one hundred and fifty-five miles, a wingspan of twenty feet, and a length of perhaps nineteen feet. Its guidance system was magnetic.

R.V. uncovered yet another odd construction detail among the blueprints delivered by Michel Hollard. One building at Bois Carré—the "Stofflager"—was divided in two by a blast wall. Each side had its own entrance, and there was no connecting door between them. Whoever designed that building was very serious about keeping the contents of the two compartments separate.

Stofflager was a maddeningly ambiguous term. The word *Stoff* means "material," and *Lager* is "storage." *Stofflager*: "a place for stuff." But *Stoff* is also fuel and a common military code name for chemicals. In World War I, A-Stoff, Bn-Stoff, and T-Stoff were tear gases. Was the building for fuel for the V-1s?

Would they be armed with explosives? Something worse?

In the fall of 1943, Adolf Hitler offered hints to his supporters, and to the world, of the Nazi plan to wreak havoc on the British. On September 10, only two days after Italy's surrender to the Allies was made public, in a strange nighttime broadcast that seemed prerecorded—and, for some reason, speeded up—the Führer promised that a "reply" to the destruction of German cities was close at hand.

"The technical and organizational conditions are being created," Hitler said, "not only to stop [the Allies'] terror for good, but also to pay him back."

Joseph Goebbels, the Nazi propaganda minister, assured citizens that "the theme of 'revenge,' discussed by the German people with such hot passion," was more than "a mere rhetorical or propagandist slogan" meant to boost morale.

In early November, Hitler informed his senior commanders of his plan to strengthen the Atlantic defenses, "particularly in the region from which we shall be opening our long-range bombardment of England." Four days later, in a broadcast from a Munich beer hall, he directed an explicit threat against England.

"Our hour of revenge is near!" he said. "Even if at present we cannot reach America . . . at least one country is near enough for us to tackle."

The V-1 was meant to be ready by winter. But the flying-bomb trials at Peenemünde weren't complete, and production of the drone at the Volkswagen factory at Fallersleben, in the heart of Germany, was hampered by design changes. And the Fieseler plant in Kassel, the second main production line, was hit by the Royal Air Force as part of the Allies' routine strategic bombing campaign.

Colonel Wachtel trained his unit, now designated as Flak Regiment 155(W), while the Nazis raced to finalize and mass-produce the pilotless aircraft.

The Allies did what they could to delay them. Operation Crossbow, the Anglo-American effort to neutralize Hitler's revenge weapons, targeted the ski sites in early December. Several large bunkers designed to launch the armaments—which were far more conspicuous from the air—had already been hit.

The ski sites would prove harder targets.

On December 5, under poor weather at night, the American Ninth Air Force bombed several ski sites. But the strikes were so ineffective that Wachtel did not even mention them in the war diary of his regiment. The colonel did take note of the ski site air raids nine days later, on December 14, when three more V-1 launching camps were attacked but did not sustain any significant damage.

The targets were too small and spread out. The launching ramps were long and narrow. The ski-shaped buildings were only fourteen feet wide.

The British Royal Air Force decided the tactics weren't working, and planned instead, with help from America, on large-scale daylight raids.

Two days before Christmas, American bombardiers, pilots, and navigators were briefed on the ski sites, the pilotless aircraft, and the dire threat they posed. "If these missile launch sites are not destroyed within the next three months," one briefing officer warned, "the city of London will be totally destroyed."

On December 24, over five hundred fighters and seven hundred bombers of the Eighth Air Force targeted twenty-three ski sites in northern France, delivering 1,745 tons of bombs on Wachtel's catapult camps. The Ninth Air Force hit additional sites. The raids represented the largest single mission in the European air war to date.

Only two ski sites were destroyed. Only three were damaged in the massive operation, according to German records. Craters pockmarked the farmland around the ski shapes like a deranged game of tic-tac-toe. Brute-force bombing, the British would admit, felt a bit like using "sledgehammers for tintacks."

Wachtel, meanwhile, was stationed safely outside of Beauvais, north of Paris. At Château de Merlemont (a small "schloss," as he put it), he enjoyed Christmas Eve with a party, presents, fine French wine, chocolate, and cigarettes.

The raids were scattershot, but they were also relentless. Wachtel's launching camps were exposed, and his regiment did not have air support for protection. "At the slightest suspicion of the sound of aircraft," he noted, in January of 1944, the French construction workers would "throw up work and leave." Even some of the German foremen would run off at the hint of bombers above.

The ski sites were practically unusable.

Wachtel's identity had also been revealed somehow. From an intercepted radio message, he learned that Allied intelligence had put a price on his head. His regiment was already using a new code name; he now needed an alias. He received new identity papers in the name of Martin Wolf, and grew a beard.

The Nazis had a plan B.

23
RANCH COUNTRY

Van Bush knew too much to be fearless.

By the time Tuve's boss began poring through intelligence reports on Nazi secret weapons, the tinkering engineer was neck-deep in the world of espionage.

He was impressed by the wild creativity of clandestine warfare. Under Division 19, Bush's scientists helped the OSS (the precursor to the CIA) make deceptive gadgets and props. "If they want to make a fountain pen that does things no self-respecting fountain pen would ever do," Bush had said, "we will make one." Division 19 products included incendiary briefcases, notebooks, and pencils; counterfeit Japanese yen and fake Swiss passports; bombs disguised as crustaceans; a silent flashless pistol; and hallucinogens that could, Bush said, "give a man the symptoms of schizophrenia for some seven or eight hours." Another weapon, code-named Aunt Jemima, was an explosive powder for smuggling into China that looked like flour and could even be baked into bread or muffins.

Bush—as chairman of the three-person Joint Committee on New Weapons, the novel body that advised the Joint Chiefs—was also privy to the stream of intelligence reports on German and Japanese wartime advances. And he helped set up an operation, called Alsos, to study the scope of Hitler's nuclear program. As Italy fell, it was his researchers who would follow in the Allied troops and interrogate Axis scientists. He had oversight of the Manhattan Project and was thoroughly aware of the research that both sides had done into horrific weaponry. Better than anyone, he knew what was scientifically possible. And it scared him.

On Christmas Eve, 1943—while Colonel Wachtel sipped French wine and the Eighth Air Force bombed the ski sites—the wiry, salt-and-pepper-haired inventor and scientific war czar mulled the sinister unknowns. Peenemünde had been raided, he knew, but how long would that delay the German attack on London?

What was the compressed air and cooling water at Wachtel's ski sites for? These items were of particular concern to the Allies: the equipment could easily be used for refrigerant to store unconventional weapons for the warhead.

Was the payload of the pilotless aircraft bacteriological? Could Hitler attack London with Agent X, or botulism, which caused vomiting, paralysis, and death? Would the Führer dare to unleash perhaps the world's most lethal biological weapon on civilians?

Had the French workers at the ski sites been immunized?

If the payload was chemical, which gases did the British suspect?

Assuming many of the ski sites were camouflaged in the woods and difficult to bomb, had the British thought of burning the woods with incendiary bombs? What could the Americans do to help figure out the list of components needed at the ski sites, where they were manufactured, and their shipping routes?

Bush's questions were sent to the Secretary of War's office with instructions to get them to Field Marshal John Dill, Churchill's man in Washington.

Eleven days later, on January 4, 1944, Bush's Joint Committee on New Weapons delivered a warning to Roosevelt's Joint Chiefs based on the "growing mass of intelligence" regarding Hitler's vengeance weapons. "An attack by new means," the briefing advised, "may now be imminent." At any minute, biological agents could be released "suddenly on a large scale" over "civilian populations and centers with very large consequent casualties." If germ warfare wasn't employed, "a gas attack, using well known gasses, distributed from the air with little warning on a large city," would certainly cause "very heavy casualties in the population."

The effort to counter Peenemünde's bombardment weapons fell under the code name Crossbow. The new British committee, which had undercut R. V. Jones and his team so unexpectedly, was known as the Crossbow Committee. On oral instructions from the U.S. secretary of war, an American "Crossbow Committee" was now also quickly set up to study German rockets, pilotless aircraft, and the peculiar ski sites—and to help stop the looming attack on London.

Bush was intimately involved in its formation.

He assigned three scientists to the new committee. One was a bomb damage expert, another an electronics specialist, and—in case of a gas attack—the third was an organic chemist with close ties to the Chemical Warfare Service. Bush also put Major General Stephen Henry, who was overseeing the committee, in touch with a pool of scientists who could consult with his advisers in other key areas: aviation, armor, building construction to advise on the ski sites, new bombing tactics, automatic pilots, magnetic controls, propellants, explosives, rockets, and radar.

The scientists did little to settle Henry's nerves.

Botulism was not the only possible threat. According to George Merck, the pharmaceutical magnate working on biological weapons, anthrax was too. Yet those biological weapons had limitations. Both agents would be difficult to produce in bulk, both were unstable and affected by heat and rain, and both could be filtered out by standard gas masks. It would take thousands of tons of botulinum toxin or anthrax bacteria, Henry's committee was told, to match the effects of chemical weapons.

Mustard gas, on the other hand, especially if delivered in the hot, humid summer, would have grotesque effects. Seeping through cloth, it would cause burns, edema, and blistering of the groin and the sensitive skin on the elbows, neck, and behind the knees. Its lethal effects would be similar to severe fire burns or traumatic shock. According to a report out of Bush's Division 9, mustard gas would produce high casualties even if civilians wore gas masks. At high enough concentrations, the toxin might cause a 50 percent casualty rate from skin burns alone.

While the warhead of the pilotless aircraft was a mystery, the Allies did have valuable clues as to its general design. At Bush's urging, engineers from the National Advisory Committee for Aeronautics analyzed its possible characteristics. From British intelligence, including the report of the Danish naval officer Hasager Christiansen and the images of the experimental aircraft "V83," aeronautics experts sketched out what Hitler's twenty-foot drone might look like.

They called it the "Peenemünde 20."

The P-20 was a Frankenstein airplane combining elements of a small Nazi glider, the HS 293, and the odd craft that Christiansen had found in a tulip field. Resembling a bloated, winged missile, it had a 3,200-pound warhead and a prominent, cigar-shaped engine stuck to its underbelly like a suckerfish on a great white shark.

The V-1 design reached the American Crossbow Committee in February, as Bush's scientists continued to ponder countermeasures. The Allies were already bombing the ski sites, and the committee had authorized a full-scale replica site at an air base in Florida to test bombing techniques. But if airstrikes failed, or if the V-1s were launched another way, they would need to be shot down.

Peenemünde's vengeance weapons could do far more than steal civilian lives and cause panic in London. Depending on the timing of the attack, Henry's committee saw, the pilotless aircraft could threaten the D-Day invasion.

Targeting the drones would not be easy. The V-1s looked to be exceptionally fast. Depending on the engine, they might reach four hundred and fifty miles per hour, making it tricky for antiaircraft guns to get a fix on them. Gunners would find it "extremely difficult" to gauge the V-1s' rapidly changing distance from the gun. Their speed exacerbated the range problem. Standard fuses would blow up late or too early.

Henry's committee asked for smart fuses. So did the British.

Section T's gadgets were not authorized for use over land. While there was no risk when Navy ships fired them over the water, Army use over land was a separate question. If intact duds were recovered by the enemy, German scientists might reverse-engineer the design and use the weapon against the Allies.

Even for use against the V-1, the Combined Chiefs of Staff (the supreme military command for the U.S. and Great Britain) would have to approve the devices "as a special case." The fuses should be allowed, the Combined Chiefs decided, but only when fired "where duds cannot be recovered by the enemy."

In short order, Tuve received an emergency request to deliver fuses to the British and to the American ack-ack crews in England. Scientists trained in the fuse would have to accompany them. Just as James Van Allen was sent to the Pacific, the cantankerous Ed Salant would be dispatched to England.

Salant was looped into the Crossbow reports and, at a meeting at the Army War College in D.C., was shown drawings of the Peenemünde 20.

Section T had a new target.

With its twenty-foot wingspan, the drone was significantly smaller than a standard enemy aircraft. It was a little over half the width of a Japanese Zero. The size of the pilotless aircraft meant a weaker radio signal bouncing back. To ensure that the rounds blew up at the right distance "for the particular target in mind," as Tuve saw at once, their sensitivity to radio waves had to be dialed way up.

Smart fuses had to be tested against the P-20.

Tuve did not need an Army or Navy proving ground to conduct the emergency trials. By 1944 he had his own test site, on a dusty ranch in the desert.

Between the Sandia and Manzano Mountains, patrolling cowboys worked security in the barren foothills. Strewn over twenty-nine thousand desolate acres southeast of Albuquerque, on a tract longer than Manhattan and over twice as wide, Section T's New Mexico Proving Ground was never meant to be this expansive.

Its purpose was initially modest.

Over the first half of 1943, Section T had erected a slew of buildings on the baking, chalky earth, to support a weapons-testing facility. Living quarters made of cinder blocks sprang up for the guards, a guard foreman, two Navy gunners, and a cook. The facility had a 1,700-gallon redwood water tank and a shop building, storage sheds, and a windmill to pump a water well. It had two half-buried storage ammunition magazines with heavy steel doors for black powder, detonators, and primers. They looked like miniature bomb shelters. The ranch had two light magazines with concrete roofs for projectiles and propellants. It had corrals and a barn for the horses.

The original idea was to have a single 57 mm gun on the ranch, but a five-inch Navy gun and an Army 90 mm were quickly added. The first ammunition magazine they built was big enough for five hundred rounds, but they were soon testing more than eleven hundred rounds over the course of a single shooting program. At the outset, the idea had been to construct a single wooden tower some one hundred and eighty feet tall, for suspending targets. But the proving ground now had two timber towers, each rising two hundred and forty-eight feet.

The original budget estimates were just "shots in the dark."

Seven miles of road were paved, eight and a half miles of phone lines laid, and two and a half miles of fence installed. Boundary warnings were posted across a fifteen-mile span.

South of the headquarters and the barn, the five-inch gun and the 90 mm were installed. Beside the guns, for vertical firing, like some modern Stonehenge, timber posts supported a gigantic concrete block—a fine improvement over the days of holding up wooden boards. Both cannons could also face several thousand yards east, toward the soaring timber towers, toward whatever target Section T wanted strung up as a bull's-eye. Along the gun line were concrete-reinforced shelters and observation posts, including one built on a precarious mound of cragged rocks that required a cable line to help haul up observation equipment.

Behind the imposing scaffold towers and the flat, parched dirt and sparse tuffs of vegetation, the gently rolling hills on the edge of the Cibola National Forest offered a scenic and surreal backdrop to the secret tests of the fuse.

Headed by the University of New Mexico's Jack Workman, Section T's satellite team at the ranch would soon number over one hundred, counting twenty-four full-time guards. The isolated range was an ideal site for firing against suspended targets, including a frame model of a Japanese airplane. Test shells weren't loaded with high explosives: the setup was meant to plot out the ideal "burst patterns" of rounds.

Using fir-wood boards, Workman's group also performed tests with high-explosive rounds to study the fragmentation patterns of the shrapnel. Later that year, they would run damage tests with various explosives against sixteen Brewster dive-bombers and two Martin B-26 medium bombers. The aircraft would be "reduced to complete wrecks" from the shrapnel. The Martin bombers came to look like whale carcasses sunk to the ocean floor and picked at by hagfish.

By the time Tuve visited the ranch, in August of 1943, the experimental proving ground was up and humming. For Merle, the visit was an opportunity to get out of D.C., and to take a short break from his frantic pace of work.

For a few months that summer and fall—before they learned about the V-1—there had been a brief lull in activity at Section T. The atmosphere of "gasping urgency," as Merle called it, eased ever so slightly.

Merle's son Tryg was nine now, and Merle decided to take him along for the westward adventure. Over a three-day train ride in Pullman sleeper cars, father and son traversed the country, leaving Washington for Chicago, where they turned southwest to Denver and then finally down to Albuquerque.

Visiting the secret proving ground was a chance to show Tryg another side of America. They explored the Sandia Crest in the mountains east of Albuquerque, which soared over ten thousand feet high. They took an overnight trip to the red rock formations south of Denver, visited the Eagles Nest peak, and slept in a six-dollar-a-night cabin. Tryg was nearing the age that his father had been when he first became a Boy Scout, and Merle still dreamed of moving his family to a small town, to the American West, or at least owning a home in the countryside.

They surveyed dude ranches as potential vacation getaways, and looked at properties for sale. Merle made notes in his teal scratch pad as if calculating the pros and cons of a peaceful, future life. "Nicest kind of place for families." "Lovely valley." "Nothing fancy but a <u>beautiful ranch</u>." "Deer come down to flats in winter."

Tuve wasn't shy about bringing Tryg—and later, his daughter Lucy —onto the New Mexico Proving Ground itself. He told Lucy it was a "School of Mines." On one visit, she remembered crawling around with Tryg inside an old shot-up bomber that was riddled with holes from shrapnel from smart fuses.

Even Merle's vacations merged into his war work.

But Section T's brief respite—if a reprieve from "gasping urgency" can be called that—ended as quickly as Tuve's dreams of southwestern living.

Designs for the Peenemünde 20 reached New Mexico on March 3, 1944. Workman already had skeletal models of a B-25 bomber and a Nakajima 97. Now, by the next day, at the emergency request of Ed Salant, his team had assembled a full-scale model of the V-1 and covered it in chicken wire to reflect radio waves.

Workman's men hung the P-20 some two hundred feet in the air between the wooden towers at various angles. Over March 5 and 6, they shot 296 rounds of highly sensitive fuses from Sylvania at the target to see how the radio beams reflected off the fuselage. From exploding shells on wooden boards, they knew "the main fragmentation zone" of a

90 mm shell, and from the pattern of the tests—assuming that the Nazi secret weapon would travel four hundred miles per hour—they learned that the most effective bursts reflected off the underside of the pilotless aircraft. Rounds that exploded twenty-five feet from the "drone" were most lethal.

An analysis by Bush's scientists, submitted to the American Crossbow Committee, reported their belief that if the antiaircraft guns were accurate enough, the fuse had an 86 percent chance of knocking a V-1 out of the sky.

Within days, the test results were on a plane to London.

On March 16, Section T's Jean Paul Teas boarded a flight to the British capital with the data from the New Mexico shoot. Teas, the baby-faced battery expert who helped develop the liquid chemical battery, was now an Army captain and had been assigned to be the Army ordnance representative in England.

In April, the radio fuses began to arrive in bulk—fifteen thousand were shipped on the twenty-ninth, with another thirty thousand being prepared. As they came in, they were transferred on a special train, under armed guard, to an ordnance depot hidden in Savernack Forest, in Wiltshire, England. They were tested at the Shoeburyness range, and a British colonel "translated" the Army instruction manual for the fuse. At the request of General Pile's Anti-Aircraft Command, twelve gunnery instructors were "indoctrinated" with the basics of Section T's gadgets. So were five American antiaircraft battalions stationed in England.

One remaining trouble was that the fuse took up too much space in the small British shells, squeezing out the explosive that made the rounds' shrapnel deadly. But Bush's scientists had an answer for that, too: a super-explosive called RDX, developed by OSRD and made in Tennessee, that was one and a half times more powerful than TNT. Filling British shells with RDX eliminated the problem. A second issue was more serious. While the smart weapon could now be fired over land in certain cases, it wasn't cleared for firing over cities. Even if there was no chance of the Nazis finding duds, the risk to civilians was too high.

But the Allied guns still encircled London.

In May, the Allies noticed a new type of ski site appearing in northern France. By the twelfth, as the American Crossbow Committee reported, a total of twenty-three "modified" launching sites had popped up. They were far simpler, "very skillfully camouflaged," and "discovered under rapid construction." By the end of the month, thanks to R. V. Jones, the Allies knew that drone trials on the Baltic had proven successful. The V-1 was ready to launch.

A chemical warhead was now "practicable."

24
CHEMICAL BOYS

This past week, we studied about chemical agents," Ed Hatch informed his parents, as if he were checking in from boarding school. The last few days at Camp Sibert, Alabama, the eighteen-year-old reported, were "practically all class work, tests, problems . . . and lectures." His unit, the 130th Chemical Processing Company, was poring over recipes for chemical solutions and suspensions.

"I enjoyed [it] very much," he wrote, "and today I am reading over my notes."

Despite the stakes involved, Hatch conveyed the innocence of a diligent schoolboy. He was of Eastern European, Jewish heritage, raised in a lower-middle-class neighborhood in Springfield, Massachusetts, not far from Connecticut's northern border. His mother had left Lithuania in her youth to flee the wave of pogroms that claimed thousands of Jewish lives. His father was a machinist.

Hatch had graduated from Classical High School only seven months earlier. And yet Camp Sibert was a universe apart from his hometown, and from the steps of his school auditorium where he'd sat and listened with fellow students, the day after Pearl Harbor, as President Roosevelt declared war on Japan.

This was not Chemistry 101.

The 130th was taught the fundamentals of gas warfare. At Sibert, grunts didn't just learn the textbook basics of mustard gas, tear gas, phosgene, and lewisite (a powerful blistering agent that smells like geraniums). The purpose of the thirty-seven-thousand-acre camp of rolling

Alabama farmland was to provide ample space for large-scale "live agent" training. Troops here fired twenty-eight-pound chemical mortars loaded with burning white phosphorus. The camp had a field for testing chemical mines, and a toxic-gas yard of six square miles set aside as a "maneuver area." At its center, Sibert housed a decontamination zone where recruits learned how to "reopen" terrain that had been gassed. Soldiers were trained to decontaminate trucks, airplanes, shell holes, walls, floors, and roads paved with gravel, sand, and crushed-stone macadam. The land was repeatedly gassed, treated, and gassed again.

Airplanes simulated chemical attacks from above.

Depending on the unit, trainees practiced with phosphorus grenades and grenade launchers, portable flame throwers, submachine guns, or bazookas. Some GIs would run field laboratories to test chemicals used in an attack. Others would run smoke generators for troop camouflage, operate decontamination equipment, or handle toxic gas. Hatch's unit, as the name implied, did "processing."

They processed clothing, applying a cutting-edge technique developed by one of Van Bush's scientists, at DuPont's laboratories, in 1942.

In case of a gas attack, the 130th had the job of protecting Allied troops by infusing their clothing with defensive chemicals. The procedure was a bit like doing laundry. Using industrial "predryers," dryers, and "impregnator units" where the clothes were soaked in a secret chemical formula that included chlorinated paraffin, the men practiced plugging the microscopic holes in military uniforms.

Hatch's unit was a hodgepodge group, a cultural grab bag of American miscellanea. Nicknames were handy. Weston Strout, who had worked at a shipyard in Maine before enlisting, was called "Red." Vincent Zanfagna from downtown Boston, one of four brothers serving, was known as "Amigo." Ohio's Lloyd Mullins went by "Moon." New York's Cristino Garcia was simply "Joe."

Their names ranged from dull to diverse to downright poetic. Thomas Parks and John Powers joined Benedetti DeFillipo, Thaddeus Grygo, Tony Jablonski, Orvo Korpi, Zoel Gerard St. Gelais, and, from Georgia, Emory Barefoot.

The soldiers hailed from all sorts of professions: shipbuilding, lithography, shoemaking. They arrived as assistant engineers, sawmill and paper-mill workers, machinists, and ranch hands. Boyce Stone, from South Carolina, made textiles. Stephen Chunco fashioned eyeglasses in New York. Floy Perryman was a schoolteacher in Indiana. William Pickens was a Kentucky farmer.

Robert Cooke, a clerk at a Portland, Maine, hotel was now Sergeant Cooke, the energetic platoon leader who led the 130th in calisthenics, obstacles courses, and marches. To keep them in good spirits, he shouted marching cadences. "In the spring, she wore a yellow ribbon . . . she wore it for her lover who was far, far, away. Hooray! *Hut two three four.*" The tune surely reminded married GIs of their wives back home—Chester Peterson of Jeannette, Donat Patry of Geraldine, James Caruso of Clovia, "Red" Strout of Myrtle, and Philip Conley of Helen.

Hatch was a high-school graduate, a rarity for the unit. Perryman, the schoolteacher, attended college for four years, and Richard Pelton, a clerk from upstate New York, had two years of college under his belt. But most had only grammar-school educations or a few years of high school. Gordon Rust, of New Jersey, was in his senior year of high school when he was called up. Hatch's Army buddy Teddy Booras, from Lynn, Massachusetts, had finished only a year of grammar school.

Recruits as young as eighteen were mixed with Army veterans in their late twenties and thirties to form a disciplined, energetic squad. When the 130th underwent chemical-handling training, camping outside the nearby Huntsville Arsenal, they went at it with such "zeal and industriousness" they received a commendation. When the GIs returned from Huntsville in late March, 1944, they were called up to head oversees. Three weeks later, waiting for their train to take them to the staging area in New Jersey, they sang for hours to the tunes of the Camp Sibert Band.

Sibert's commander had never seen anything like it.

On the convoy over, Hatch was seasick for days, his stomach "zigging" while his head "zagged." Even so, he found the journey "most enjoyable." And even though London had morphed into a war zone, as he wrote his parents on May 10, just three days after arriving, he was exhilarated.

"I never felt better in all my life."

For months, American soldiers had been arriving in England in herds. Two years prior, in 1942, Yankees were a novelty. When the Eighth Bomber Command first landed in Britain, swarms of locals surrounded them "like monkeys in a cage." Spectators peeking through the barbed-wire fence were so persistent, a sergeant quipped, "it was almost impossible for us to change our clothes."

In May of 1943, some one hundred and thirty thousand American troops were posted in England. By the time Hatch arrived, that figure topped one and a half million. The U.S. military footprint in the capital included over one thousand offices, warehouses, and apartments. The British Isles, according to General Dwight Eisenhower, himself stationed in London, were turning into the "greatest operating military base of all time."

Grosvenor Square, home to the American embassy, was now a full-blown American colony. The ballroom of the Grosvenor House Hotel hosted the largest military mess hall in the world. Each day it served six thousand meals.

Living quarters for U.S. troops spread from there across North Audley Street, Park Street, and Upper Grosvenor. Hyde Park, just steps away, was boxed in by Americans to the south, north, and east. GIs were soon billeted farther north near Regent's Park, east in Forest Gate, and south in Chelsea.

Along Chelsea's Sloane Court, where the 130th was living, the respectable brick apartments spanning the residential lane were now practically a barracks. Twenty prime addresses there had been set aside for American soldiers.

Out sightseeing, the 130th donned their spiffy Class A uniforms with polished buttons. Hatch visited 10 Downing Street, and Scotland Yard ("the English FBI"). He beheld Buckingham Palace and the changing of the guard. He soaked in Piccadilly Circus ("the Times Square of London"), and wrote home about how, in peacetime, movies had played there "across the length of the buildings, and at night all the people used to stand around for hours to witness" the spectacle.

He did not describe how Piccadilly Circus had changed.

If Grosvenor Square was the beating heart of the American "occupation," a miniature Washington, D.C., rising up within London, then Piccadilly Circus was the new Las Vegas. Throngs of servicemen roamed its restaurants, music halls, and bars in search of liquor, love, food, and camaraderie. To their British competition, the invasion was unfair. GIs ate better food, sported sharper uniforms, and made more money. As one

British soldier recalled, "The Yanks were the most joyful thing that ever happened to British womanhood. They had *everything*—money in particular, glamour, boldness, cigarettes, chocolate, nylons, Jeeps."

At the American Red Cross's Rainbow Corner, in Piccadilly Circus, grunts basked in an oasis of pool tables, pinball machines, jukeboxes, hot showers, hamburgers, and Cokes. Outside, in the darkness of the blackout, like some surreal mating ritual, prostitutes roamed with flashlights, beacons in the murky night.

Americans driving on the wrong side of the road caused crashes, revelers sunk to drunken brawls, and quickly formed romances bloomed in darkened doorways. The visitors, the British complained, were "oversexed, overpaid, overfed, and over here," while the English, the Yanks replied, were "undersexed, underpaid, underfed and under Eisenhower." GIs joked that British condoms were too petite.

"Let me close with assuring you," Hatch wrote his parents, "that I am attaining a world of knowledge + practical experience each day, that I have an appetite like a chow-hound, that I sleep in an aristocratically dignified mansion, that I enjoy British wit and humor, that I think the English pecuniary (or currency) system is inefficient, that I miss you all a great deal, and that there is no place like home."

Amidst the blackout, Hatch and the other men often met at the Rose and Crown pub, steps away on Turks Row, and shared their Yankee cigarettes with the aged English veterans of the South African Second Boer War, of 1899 to 1902.

In mid-May, the 130th set up their chemical-impregnating units in a warehouse at 4 Crinan Square, in north London off the Regent's Canal. Their "M2 Water Suspension" plants could chemically protect nearly eight thousand pounds of clothing every twenty-four hours, or over twenty-five thousand uniforms a week. They assembled their equipment and waited for orders. In the meantime, they washed laundry for U.S. military police in London—a good backup use of their rigs.

The London edition of the *Stars and Stripes,* the U.S. military newspaper, kept the troops up to date on the progress of the war. The Nazis had lost the Crimean port of Sevastopol to the Russians. The Allies were pushing north against Nazi troops in Italy and closing in on Rome. Ground ops were advancing in New Guinea. Oil fields were bombed in Ploiești, in occupied Romania. Setting the stage for the D-Day invasion, raid after raid of Allied aircraft targeted rail yards, bridges, ammunition dumps, and power plants in France and Belgium. Refresher courses trained glider pilots to use Tommy guns, mines, and booby traps.

"'D-Day' Fever Has All U.S. Steamed Up," one headline read.

"Garlands of colored lights" that once illuminated British piers and beachfronts, where couples danced to swing bands, had been replaced by barbed-wire enclosures where signs forbade civilians from loitering or speaking with troops.

A Nazi spokesman in Vichy warned residents in occupied France: "The last breathing space before invasion has arrived. We cannot give you any last minute instructions . . . as we do not know where the main Allied blow will fall."

On May 27, 1944, in Hendon in north London, R. V. Jones boarded a single-engine Percival Proctor, a tiny aircraft big enough for three or four people. With him were two members of his scientific intelligence unit: James Birtwistle, a former RAF Fighter Command intelligence officer, and the pilot Rupert Cecil, an Oxford-educated biochemist who had earned a Distinguish Flying Cross.

The Percival took off for an air force base on Thorney Island, on England's southern coast. From the base, in preparation for D-Day, a string of bombing missions orchestrated by Jones were being flown against Nazi radar stations in France.

Airborne, R.V. was struck by the hills of bright rhododendrons.

It had been an eventful few months. Weeks earlier, the Luftwaffe had begun yet another assault on London, a "Baby Blitz" that saw the highest level of casualties since May of 1941. At night during the bombings, R.V., his wife, their daughter, and their newborn son slept together under the dining-room table.

All of the resources of the Allies were gearing toward D-Day. And yet, while Jones's group was the de facto hub of scientific intelligence, formal responsibilities for countering Axis technology remained dangerously dispersed. As late as April, as R.V. discovered to his horror, there was no coordinated plan in place to knock out German radar in France. He'd had to personally intervene, through backchannels, to establish a special task force.

His years of carefully mapping Nazi radar installations—including the Freya and Würzburg installations in France and Belgium—now provided the basis for a run of dangerous, low-flying assault missions. The Allied air forces took great care in their targeting. For every attack by British and American planes on radar that covered the D-Day landing zones, two installations were bombed outside them, as a diversion. Reconnaissance flights confirmed the damage.

By the time Jones and his men visited Thorney Island in late May, over six hundred Royal Air Force sorties had been flown against twenty-nine sites, wrecking twenty. By early June, as few as six of the ninety-two targeted German radar stations remained operational.

As D-Day neared, so did the threat of Hitler's V-1s. The drones were almost ready, and the Normandy landings were certain to provoke their release. Given the size of Wachtel's regiment (information provided by Amniarix), Jones estimated that it would likely take just one week for the Nazis to complete launch preparations.

Jeannie Rousseau's reports had proven to be of the highest worth in meeting the V-1 threat. Thanks to her, Jones had ordered reconnaissance photos of Zempin, Wachtel's testing ground near Peenemünde. She was the reason the catapults were identified, which led to the photograph of a V-1 on one of the launching ramps. That was how MI6 knew the approximate size of the drone, data that allowed Section T to calibrate their smart fuses against the mockup in New Mexico.

Rousseau's work had not ended there. That past winter, she had continued to feed MI6 key intelligence about Wachtel as he attempted to evade Allied detection. In one dispatch, she reported that the colonel and his men had been secretly transferred to a château thirty-four miles north of Paris. The castle was well defended by electric wires, ack-ack guns, and patrolmen. His regiment, she wrote, was now being referred to in German army communications by the code named "Flakgruppe Creil." She even reported their new symbol, a "W/8" on a shield.

German for "1/8," an amused Jones realized, is *Achtel.*

Wachtel himself, Amniarix revealed, was shuttling between Paris, Berlin, Belgium, and Zempin to supervise the final touches on his superweapon.

MI6 wondered what else the French spy might know.

Rousseau's insights had proven so valuable that British intelligence hoped to interview her directly. Perhaps, their reasoning went, she knew even more than she realized. Critical, neglected details in the background of her memory might be brought to the fore under a careful, friendly interrogation. And so, days before the Normandy invasion, MI6 arranged to smuggle her to England.

The twenty-five-year-old was provided identity papers with the name Madeleine Chauffeur, and told to proceed to Tréguier, in Brittany. The small town was an hour and a half west of her former "sanctuary" of Dinard, at the intersection of the Guindy and Jaudy Rivers, which flowed into the English Channel.

Two other French agents bringing classified reports, Yves le Bitoux and Raymond Pezet, planned to evacuate with Rousseau. But the Resistance spy tasked with shepherding the three of them through the beach minefields to the boat had been arrested. He had betrayed them. Troops were waiting inside his home, the town was encircled, and Rousseau, Bitoux, and Pezet were surrounded. Amniarix was discovered first. She had brought black-market goods with her — nylons — and pleaded with the Germans that she was "just a poor little French person" hoping to earn cash from selling stockings in Brittany. They arrested her.

When marched out, she warned her fellow spies, who were waiting nearby, by babbling loudly in German. The ruse worked, and Pezet escaped. Bitoux was from Tréguier and was worried about reprisals against locals. He turned himself in.

Rousseau was shipped to Ravensbrück, the Nazis' brutal concentration camp for women. Soon enough she was sent to Königsberg, a punishment camp far to the east. Hard labor here meant hauling rocks and gravel outdoors, in the numbing snow and biting wind, to construct an airstrip. Women limped back to the camp through the glacial dark, desperate for their nightly meal of hot soup. The head guard, a monstrous, rotund woman known to the prisoners as "La Vachère" — the Cowgirl — would mock the starving inmates and kick the vat of soup until its nourishing warmth spilled into the snow. The Cowgirl watched the dying prisoners rummage desperately for bits of food in the icy slop.

25
THINGS CARRIED

On June 6, 1944, along fifty miles of coastline, thousands of warships and landing craft crowded a gloomy dawn horizon for the greatest invasion in military history. Nearly seven thousand vessels, including over four thousand landing craft, ferried more than one hundred and fifty thousand Allied troops across the English Channel toward five landing zones.

American flags with forty-eight stars fluttered across the massive flotilla, rising above the sea like a bed of nails. Kite balloons floated hundreds of feet above the armada, poised to shield the infantry from strafing Luftwaffe gunners. Engines groaned, winches spun, chains clattered, and troops clambered down wet ladders and webbed nets into the beaching transports. Combat engineers carrying blasting caps and explosives dived into the frigid waves to destroy obstacles like defensive cables and underwater mines.

Massive LSTs—tank-landing ships with giant bow doors that could open at sea—carried in their steel bellies hundreds of peculiar boats with wheels. Designed by one of Van Bush's protégés at the Office of Scientific Research and Development, the engineer Palmer Putnam, the odd steel chariots backed off the LST ramps into the drink. They puttered strangely low in the water. The Army called them DUKWs, a tortured acronym indicating that the vehicles were amphibious and equipped with all-wheel drive and dual rear axles. GIs called them "duck" boats.

Allied ships carried medicine, first-aid kits, and hundreds of pints of blood. Medics carried a blood substitute called serum albumin, a mixture of globular proteins derived from whole blood that helped prevent shock and death in cases of massive blood loss. The lifesaving substitute was developed by doctors working for Bush and OSRD's Committee on Medical Research.

Thanks to Bush's recruits, medical officers also brought penicillin as an all-purpose antibiotic. Two billion pills of a new strain of penicillin were available that day, a striking feat considering that when work began in 1941, there wasn't enough penicillin in the United States to treat a single person. A team headed by OSRD's Robert Coghill had cultivated the antibiotic and increased its yield by one hundred times. The medicine was brewed inside huge vats of a liquid byproduct of corn.

D-Day troops carried secret "beach barrage rockets." They were intended to strike Nazi positions on the coast during the vital period after Allied air support operations were over, but before mortars and field artillery could make it ashore. With a range of eleven hundred yards, the rockets produced virtually no recoil and used lightweight launchers. Self-described "rocketboatmen" waited patiently in landing craft to fire the missiles, devised by OSRD's Section L, to soften the shoreline.

Troop carriers, landing craft, cruisers, and battleships carried hundreds of radar jammers and electronic countermeasure devices. The gadgets broadcast clogging, confusing beams toward any German radar installations that might still be working. To install the devices, countermeasure specialists culled from Bush's scientists had canvassed British ports from Exeter, in southern England, all the way to Scapa Flow, north of mainland Scotland. Many of the jammers now among the Normandy fleet had been devised by OSRD recruits.

These scientists were a linchpin of the larger scheme to keep the Nazis ignorant of exactly where and when the invasion would arrive. Troop movements had been faked by large tent cities in open English fields, where no one lived, where dirt roads leading into the woods ended nowhere, and where stoves were kept smoking for Luftwaffe reconnaissance photographers. Car headlights were rigged to look like airplane blinders and driven up and down fake runways at night. Inflatable rubber tanks, trucks, and antiaircraft guns were ballooned to replace the real deal, while "fleets" of fake wood-and-canvas ships floated on old oil drums.

Allied airplanes hauled hundreds of dummy parachutists—three-foot-tall burlap soldiers complete with miniature boots and helmets. Floating in the distance, they were easily mistaken for soldiers. Real paratroopers were guided by British electronic-targeting gadgets. American bombers executed raids on Nazi sites along the Normandy coast using a ground-scanning radar devised, with British help, by OSRD scientists at the Massachusetts Institute of Technology.

Outside of the Normandy landing zones, Allied aircraft carried "Window"—bundles of the same metallic strips dropped in the air by the British before the Peenemünde raid. To Nazi radar, they looked like flocks of bombers. Below the fluttering strips, decoy vessels tugged blimps with special radar reflectors that made them read like imposing warships. A phantom fleet complete with air cover approached the French coast in the Pas-de-Calais region, northeast of Paris, filling German radar screens and confusing the Luftwaffe dispatched to meet it. Thanks to "the most sophisticated faking in the history of man," as one historian put it, a "ghostly procession of nonexistent battleships, cruisers, destroyers, transports, landing craft, and air squadrons swam into the Germans' ken."

The Normandy invasion was not opposed by sea or by air.

Off Utah Beach, as Allied planes saturated the coastline with bombs, the rocket boats neared the shore ahead of the first wave. "Boat, machine guns, rockets, and six man crew, everything was expandable as hell," a soldier recalled. One unit was closer to the beach "than the hitter is to the left field screen at Fenway Park" when it let loose its barrage of missiles against Nazi machine-gun and mortar emplacements.

Gunfire pinged the armored hulls.

The troop transports made land, and the hinged steel ramps slammed into the waves and wet sand as the soldiers rushed from the waterway. The guns and shells on the beach were so loud GIs could not hear themselves talk.

Behind them, Landing Control Craft, or LCCs, carried special navigation equipment invented by Bush's scientists. They chaperoned amphibious tanks from four thousand yards offshore to four hundred. As the violent struggle began, the LCCs assembled, led, and dispatched wave after assault wave from the transport area.

A medical boat sank, spilling packets of blood plasma. The bobbing soldiers made it ashore, stuffed their pockets with the substitute serum albumin, and administered the lifesavers to bleeding GIs until they could be evacuated. Medics applied tourniquets and dressed wounds that would be treated with penicillin.

Out of the sea on Utah and Omaha Beaches, the First, Fifth, and Sixth Engineer Special Brigades piloted hordes of duck boats through underwater passageways still bristling with piercing beams and deadly mines. They risked drowning from punctured hulls, overloaded cargo, bent propellers, and busted rudders. Receiving gunfire, the peculiar ships that reached the beaches rose majestically from the brine to reveal craft far taller than their seaborne profiles suggested. In place of propellers, six wheels gripped the sand and strained to ferry high-priority engineering equipment, artillery, and ammunition onto shore.

Duck boats would carry 40 percent of all supplies up the beaches. The stunning figure contradicted the Nazi skeptics who'd claimed the Allies would not be able to amass enough equipment to muster a fighting force that could challenge them.

The day after the invasion began, D+1, Allied ships carried additional curious contraptions to France. Temporary harbors, one in the British landing zone and another in the Americans' off Omaha Beach, sprang up. Composed of piers attached to metal pylons on the sea floor, they could rise and fall with the help of hydraulic jacks. Steel pontoons, floating roadways, connected the piers to shore while breakwaters beyond them, guarding against waves in a great arc, were formed by deliberately sinking old ships and linking together huge watertight chambers.

To protect the harbors, British trawlers carried troops from the American Chemical Warfare Service along with their M1 mechanical smoke generators. M1s weighed three thousand pounds when empty and fifty-four hundred when filled with fog oil, fuel oil, and water. While the Battle of Normandy erupted, the generators belched white plumes, shrouding the artificial harbors to protect them from attack. The chemical formula was the brainchild of OSRD's David Langmuir.

Allied hulls also ferried Section T fuses designed for Army 90 mm guns. The smart weapons crossed the English Channel in crates of ten, or with single fuses in hermetically sealed metal containers, each with a key and a tear strip for opening, like a tin of sardines. Twelve such containers were packed with a special wrench in padded cardboard cartons enclosed in a steel box. They were headed for gun positions on the beaches to help guard the makeshift harbors.

Transports carried key personnel like Captain Bert Mitchell, who had worked on Section T's fuse at National Carbon. Mitchell trained the antiaircraft battery commanders on the ins and outs of the new gadget in a hayloft of a French barn.

Only senior officers were provided details of the weapon, which the Army called "Pozit." They were told that the device sensed enemy aircraft "electromagnetically" and exploded within sixty feet of a plane. They were warned that the Nazis must not get their hands on it, and that if capture was imminent, the fuses must be destroyed—an incendiary explosive would do it. Enlisted men were taught to remove the plug and gasket from the shells and screw in the fuses like light bulbs.

As the Allies pushed deeper into France, American antiaircraft gunners waited for the Luftwaffe to bomb the harbors. Yet the fuses were not needed.

The Germans had bigger problems. And bigger plans.

Within days, an airplane carried two of Bush's scientists to France. One was Guy Stever, a twenty-seven-year-old radar expert (and later chief science adviser to Presidents Nixon and Ford). The second was Bob Robertson, a forty-one-year-old mathematician and physicist from Princeton and an essential OSRD liaison.

On the morning of June 6, minutes after the invasion of Normandy had begun, Stever and Robertson met with R. V. Jones. Their mission, joint with the British, was to follow in behind the troops as the Allies made headway into France, and to gather intelligence about Hitler's military science.

The V-1 ski sites were a key target of interest.

The pair landed in a C-47 on a small dirt runway above Omaha Beach. The Nazis were still shelling the French town of Carentan when they drove through it, up the Cotentin Peninsula, toward the port of Cherbourg. Stever saw the wreckage of the gliders that had tried to land on D-Day and crashed into trees, walls, and hillsides. The men camped with an engineering unit in a field surrounded by winding country lanes, woodland, pastures, small trees, and high hedges.

They slept in pup tents to the cracks and whistles of shells passing overhead: German rounds from the southwest; American shells retaliating from the northeast. The Norman hedges seemed to offer an illusion of security, as if their tents were merely seats in a cosmic theater, and the countryside were a vast planetarium, and they were really invulnerable to the shells flying under the stars.

26
WACHTEL IS HIDING

Merlemont, the country manor Amniarix discovered and reported to MI6, was bombed by the Allies. The damage was minimal, but there was no way for Colonel Max Wachtel to be sure that his unit wasn't deliberately targeted. He relocated himself and his top men to Château d'Auteuil, five miles to the southwest.

Château d'Auteuil was for civilians.

The elegant brick mansion—a sixteenth-century castle once home to Louis XV—was formally off-limits to the German air force, as were its grassy parks, moat, and the vast pastures of its stud farm. Even Walther von Axthelm, overseer of the pilotless-aircraft program, could not visit in his Luftwaffe uniform. License plates starting with *WL*, for the Wehrmacht Luftwaffe, were switched out before cars entered the estate. Nazi officers here drove private cars with French registrations, and wore badly fitting suits when they prowled the château grounds.

A sign suggested that the German occupants were merely civil engineers, administrators of construction projects for the Reich. Wachtel himself presented as a bearded, forty-six-year-old colonel and functionary, Martin Wolf.

Months earlier, by the Baltic Sea, under the canopy of roadside trees, Wachtel briskly explained his metamorphosis to his regiment. "Men!" the newly transformed officer shouted. "I am Colonel Martin Wolf and have replaced your former commander Wachtel. Colonel Wachtel has taken on another role." The regiment was not amused; the news meant that the job ahead was serious.

As Wolf, he had authority to travel wearing other uniforms than the Luftwaffe's blue-gray attire. He hired a French tailor on the Champs-Élysées to custom-fit him in the field gray of the army. In small groups, his regiment also traveled to Paris to exchange their aviator uniforms for the brown regalia of the German engineering corps that had become their cover story.

Nazi high command was not taking further risks.

The Germans suspected that the thousands of French construction workers had betrayed their ski sites and catapults pointed at London. Replacement sites for launching the V-1s, Wachtel realized, had to be built under the tight supervision of the German military. The solution was convict labor. To divert Allied attention, resources, and bombers as the new network was built up, the Luftwaffe authorized some construction and repair work on the original ski sites.

The new "modified sites" didn't have the telltale ski shaped buildings, so easily spotted from the air. Nor did they have much construction at all. Simple concrete platforms with rails were built for the launching ramps, but they could be camouflaged easily. Instead of installing the catapults beforehand, the Nazis would use prefabricated ramps, which came in six-meter sections. Designed by the Walter-Werke, the same company that engineered the chemical-propellant system, the new launching ramps could be stored offsite and assembled quickly.

Roads around the redesigned camps were improved modestly, and small garages were erected. Concrete floors were laid for the nonmagnetic buildings to set the guidance systems. But that was the extent of it. No specialized equipment would be brought near the improved camps until it was time to launch, and the V-1s themselves would be stored in local caves and tunnels. Instead of requiring eight weeks to construct, like the ski sites, the modified posts took just eight days. They were designed to be ready to launch within ten days of the order.

Detecting them from the air was nearly impossible.

Every effort was aimed toward greater secrecy. On March 5, Hitler himself ordered that new V-1 production facilities be established underground. (Production was ongoing in central Germany, at the Fieseler and Volkswagen plants.)

After the bombing of Peenemünde, the Nazis had established additional factory lines for secret weapons at the Dora concentration camp, a subcamp of Buchenwald. Underground, in the gypsum mines south of the Harz Mountains, amid wooded hills, valleys, and alpine forests, the rockets once constructed at Peenemünde were built on assembly lines in cavernous tunnels under vile conditions.

By early 1944, over eleven thousand prisoners were camped at Dora. Marched through a huge wooden door, slave laborers encountered a world of blackness lit only by sparse acetylene lamps, which threw murky rays over the high walls. Their wooden shoes slipped on the sticky, wet rocks and rough ground. They slept in half-stuffed straw mattresses and fought off lice, pneumonia, and starvation.

V-1s would soon emerge from the dank, sooty depths.

Wachtel's battle headquarters was already underground. Outside the French village of Saleux, not far from Belgium, a series of bunkers and tunnels was carved out sixty feet below the earth. Some one hundred and seventy-five convicts and Russian prisoners of war dug out the secret command post beneath the French cottages, under tight supervision, over the course of two months. When the order to launch the V-1s arrived, Wachtel and his command staff planned to descend underneath the bucolic countryside huts, down one hundred damp, cool steps to the bunkers, and commence the attack.

Wachtel's regiment had already slipped discreetly into positions in France, where they awaited orders a safe distance from the new launching sites.

The drone weapon was ready. In May, Wachtel completed the final stages of flight testing. Over the course of four days, his regiment fired twenty-nine pilotless aircraft near Peenemünde. Twenty-two launched successfully, and only two fell short. Fritz Gosslau, the ambitious engineer who had birthed the V-1, was eager to see his creation finally used against London. He was thrilled that "batteries with especially well-trained personnel achieved results up to 82 percent!"

Hitler ordered Wachtel to begin the bombardment of England with his "revenge" weapons by the middle of June. The code word to commence operations would be *Rumpelkammer*, meaning "junk room." Given the order, Wachtel's men would finish construction at the modified sites. They would lay power and telephone cabling, install the prefabricated ramps now camouflaged in forests and hidden in shelters, ready the V-1s, and prepare the chemical-propulsion equipment. Final supplies were al-

ready being gathered at depots. Wachtel's unit wondered only whether
their secret mission would begin prior to D-Day.

"Every man in the regiment," the unit's war diary read, "is convinced
that the [V-1] will be on the other side before Sammy and Tommy are
over here."

On June 5, Wachtel was only hours away from his forty-seventh birthday. On the sixth, in honor of their colonel, his men hoped to take a few
hours' break from their normal duties and relax a short while. But Wachtel's birthday was quickly forgotten when news of the D-Day invasion
reached Château d'Auteuil. That afternoon, at 5:45, the headquarters received word from command. *Rumpelkammer.*

Instead of the ten-day preparation window the regiment trained for,
the unit was ordered to open fire on London within six days.

"The fight for the flying bomb," Wachtel noted, "has reached its decisive phase." He shaved off his beard and removed his civilian clothes. His
men slipped out of the brown attire of the engineering corps. When they
emerged from the château, they boasted the aviator gray and red collar
tabs of the Luftwaffe. They tossed their cover story aside and marched
down the stairs to their command bunkers. Wachtel's regiment — numbering some sixty-five hundred men — crept out from their farmhouse
billets and began preparing the launching sites.

For six days, across sixty-four camps in northern France, Wachtel's
unit worked without sleep to assemble the catapults. V-1s arrived from
supply depots code-named Nordpol, Leopold, and Richard. Wings and
warheads were carefully attached. "Transport cones" were removed from
the noses, exposing the tiny odometer propellers that would be set to target the British capital.

Still, the sites remained unfinished, and in the early hours of June 12,
Wachtel was nervous and agitated. "Catastrophic supply conditions," he
believed, argued for a delay of forty-eight hours. But Nazi command was
unsympathetic.

Wachtel and his men raced to ready as many catapults as possible, toiling through four o'clock that morning. By seven a.m., they were right
back at work. Phones rang nonstop in the underground command post.
Eager arrivals from across the German military began to descend into the
crowded bunker to get a glimpse of military history and the dawn of a
ruthless new form of warfare.

"Soldiers!" Wachtel addressed his men. "Führer and Fatherland look to us. They expect our crusade to be an overwhelming success. As our attack begins, our thoughts linger fondly and faithfully on our native German soil. Long live our Germany! Long live our Fatherland! Long live our Führer!"

Representatives from the army, Luftwaffe observers, and engineers from Peenemünde poured into the dugout headquarters. At five thirty p.m., Wachtel's commander, General Erich Heinemann, arrived. The officers assembled at a large wooden table and gathered news from the launching batteries.

The news was discouraging. Only seven of the sixty-four sites appeared able to launch. Critical items had still not arrived. Special fuel for the air compressors needed for the pulsejet engines was missing, as were some of the chemicals essential to the steam-propulsion system. At many V-1 sites, there was little or no emergency lighting. The troops were exhausted, and under such extreme pressure from the strict deadline that the final tests were being ignored.

Reports came in by phone of incomplete and untested ramps. From some of the V-1 camps, Wachtel and his headquarters heard no news at all.

The officers stared at a map of London, backlit and translucent, mounted on a wall in the command bunker. The drones did not have to be very accurate to hit their mark. London was a wide target. But the V-1s were still aimed at a bull's-eye, a symbol in the heart of the city: the Tower Bridge.

On the see-through map, V-1 strikes were to be triumphantly plotted by points of light. To help track the weapons' flight path, some had been fitted with radio beacons. General Heinemann stared eagerly at the map, joined by ordnance expert Werner Dahms. The officers sipped glasses of sparkling wine.

At eleven p.m., Wachtel asked for, and was granted, a delay. No launches would take place for several hours, until the early morning of June 13. Across northern France, drones were lifted by cranes to the catapults. Chemical vats were positioned in trolleys behind the sloping rails to propel the clattering pulsejet engines. Around four a.m., the first pilotless aircraft lifted off noisily from its ramp in the French Pas-de-Calais and soared over the darkness of the English Channel. But less than half of the ten Nazi drones launched that morning reached London. On the map in Wachtel's bunker, only four points were illuminated.

Wachtel was disgusted by the meager output. "It did not work," he vented, at six a.m., in his personal diary. "The mission was ordered too soon."

27

A COMET GONE WRONG

The early hours of June 13, 1944, were unseasonably cold. In the night sky, a quarter moon shone dimly through the clouds over darkened homes. The blackout would not be lifted until after five a.m., thirty minutes before sunrise.

In east London, in the district of Woolwich, on a high plateau south of the Thames, British firemen awoke calmly to the sounds of air-raid sirens. They expected to have ten minutes, at least, between the cautionary mechanized wails and the distant hum of Luftwaffe engines approaching the capital. London had not been bombed for some weeks and these were the first sirens since D-Day.

The firemen donned their battle trousers, shirts, socks, and gym shoes. Without panic, they trudged from their watchmen's hut and lumbered across the tarred parade ground and roadway toward the small, brick, concrete-roofed fire station of their army base. But something that morning felt different.

Shooter's Hill, where the antiaircraft guns were positioned a mile to the southeast, was obscured from view by their base's main office building. But the caliber of gunfire echoing in the gloom was unusual. Lighter clatter from 40-millimeter Bofors guns, instead of the normal blasts of 3.7-inch cannons, suggested a target flying much lower than most Luftwaffe bombers.

Searchlight beams, which typically swept the sky excitedly, crossed the low-hanging clouds with eerie purpose. Locked briefly in a glowing ray, an enemy aircraft, one of the firemen recalled, rushed by them "at incredible speed." It was far faster than British Spitfires or even the Luftwaffe's breakneck Messerschmitts.

Crimson flames jetted from the clacking UFO.

Three hundred feet away, the firemen knew, there was a vantage point on the western edge of Woolwich Common that offered a sweeping, panoramic view over central London. Sprinting over for a better look, they caught a full view of the aircraft. It had to be an ultra-fast Luftwaffe bomber executing a sneak attack.

The object soared over St. Paul's Cathedral toward the heart of the city. It raced over the ships floating on the Thames, downstream of the Tower Bridge, where supplies were being loaded for Normandy. Gunners on the armed ships opened fire with all they had, illuminating the early morning with tracer bullets—dashes drawn in space like so many bioluminescent discharges. West of the Isle of Dogs, above Rotherhithe, the aircraft's flame died, and the engine quieted. The firemen waited—it felt like an eternity—for the impact sound of the "crashing" Luftwaffe pilot.

After the distant blast, they returned to sleep.

Witnesses who glimpsed the four Nazi drones that reached English soil that morning came to similar conclusions as the firemen. The burning objects looked like crippled aircraft. A farmer's daughter near the southern coast assumed some pilot was flying with a damaged engine. So did a London guard. In Sussex, a mother decided the plane was a wounded English bomber limping home.

From below, observers saw "nothing but a black shape with sheets of flame spurting out behind it." Dark silhouettes appeared over farms like burning black crosses knifing toward a bull's-eye. American pilots settled inside the cockpits of their Thunderbolt fighters, awaiting takeoff, and then gazed up to glimpse "three lighted objects which resembled balls of fire" three thousand feet overhead.

The sputter of the pulsejet engines' steel shutters, opening and slamming some fifty times a second, reverberating through the night, was unlike anything British ears had ever heard. Bystanders described it as "coughing clattering" like a "diesel truck," only it groaned in cycles like "one long washing machine." It resembled a "steady rattling noise," a "Model T Ford going up a hill," a "two-stroke motorcycle engine," a dentist's grinding drill, or a blacksmith's piping blowtorch held close to the ear.

Three of the four "buzz bombs," as the British would soon call them, detonated on agricultural land. One left a crater in a field of young greens and lettuces, decimating the crops for 240 feet in every direction. Another landed in a wheat field, its blast mowing some eighteen acres to a bizarrely uniform height. The third drone found a strawberry field, wrecking greenhouses and killing chickens.

The fourth V-1 landed in London at 4:25 a.m., in Bethnal Green, on a railway bridge supporting the main line from Essex. The blast ripped up the tracks, demolished a row of homes, and ruptured others within 320 feet. A ten-year-old girl staying with her grandparents, visiting the scene after the "all clear" sounded, beheld nearly half the bridge lying in the road. Next to the gnarled railway stood a lonely wall, and on the second floor of a building with no façade she eyed a fireplace, and above it, "a large glass mirror, not scratched, slightly moving."

Six died, twenty-eight were injured, and two hundred were made homeless.

The full truth of the early-morning incident was kept secret. From scientific intelligence and eyewitnesses, British authorities knew that Hitler's vengeance weapon had finally arrived. But the Ministry of Information persuaded the press to censor its news reports. The public had no urgent need to know, the War Cabinet decided, that a novel form of aerial warfare had just been unleashed on the city. Instead, London's *Evening Standard* informed its readers only that "several people had been killed when bombs fell in a working-class area."

Within two days, the railway bridge was back in service.

Compared to what London had already suffered, the damage was slight. Churchill's science adviser, Lord Cherwell, exclaimed to R. V. Jones that the "mountain hath groaned and given forth a mouse!" But Jones did not share his former mentor's enthusiasm. Wachtel's first effort, he warned, was likely no more than an operational hiccup. The bombardment would begin soon.

Seventy-three drones reached greater London on June 15 and 16.

The British government could no longer hide the nature of the weapon. Too many citizens spotted the "airplanes on fire," heard the pulsejets groan like outboard motors, saw the engines quit, tail flames die, and the objects arc to earth. But the Home Secretary, Herbert Morrison (who gave his name to the steel-cage Morrison shelters), could at least assuage fears and urge general calm.

In the House of Commons, Morrison proclaimed that Hitler's secret weapon had been fully expected, and that "available information does not suggest that exaggerated importance need be attached to this new development."

London newspapers announcing the "Pilotless Warplanes" assured readers that "Our Scientists Will Defeat It." The *Evening Standard* ran a column on "How the Robot Works." An article on "How to Spot Ghost Planes" included the craft's telltale characteristics: its "terrific speed" and straight course, the flames from its exhaust, and its low rhythmic vibrations. "When the engine of the pilotless aircraft stops," the *Evening News* advised, Londoners should take cover, as "it may mean that the explosion will soon follow—perhaps in five to 15 seconds."

On June 18, a Sunday, R. V. Jones was in his office at 54 Broadway Street. He was on the phone with Bletchley Park when, shortly after eleven a.m., he noted the growl of an approaching V-1, and then heard the engine die. He mentioned to Bletchley that a flying bomb was near, and that he needed to hide under his desk. After the explosion, he walked outside and saw with horror where it had detonated: four hundred and fifty feet away, at Guards' Chapel, during Sunday-morning service.

It took forty-eight hours to remove the one hundred and twenty-one bodies. The blast stripped the trees along Birdcage Walk, carpeting the ground with fresh green leaves.

That same day, the 130th's Ed Hatch penned yet another calming letter to his parents back in Massachusetts. "Today marks the 6th week I have been in London," he wrote, "and I visited Kew Gardens, a very lovely park. We have been having a few air raids lately, because of the robot planes" sent by the Germans.

"They are nuisances, and as yet have not caused too much damage."

V-1 assaults were becoming commonplace.

On Sloane Court, the "military street" filled with Americans, the men of the 130th climbed to the top floors of their buildings to gaze at the buzz bombs with binoculars, or rushed outside to meet up with the Women's Army Corps soldiers, who lived behind them, to watch the V-1s from the street.

"It has never awakened me from my slumber," Hatch promised his parents on June 22, speaking of the drone. "It's not really as bad as the paper paints it."

Yet despite press censorship of the death toll, and the brave faces of the Americans and native Londoners, it was becoming clear that far too many flying bombs were getting through. British defenses were three-fold. Over the Channel and coast, the Royal Air Force hunted the drones and hoped to shoot them down. South of London, in the "Kentish Gun Belt," General Frederick Pile's antiaircraft battalions took on the V-1s that slipped past the planes. Giant blimps acting as feeble nets, barrage balloons strung up with cables, represented the last defense.

In the first two weeks of the siege, the Allies estimated that 1,585 drones were launched, and that over 1,100 successfully crossed the Channel. Royal Air Force pilots managed to shoot down only 315 of them. General Pile's ack-ack gunners downed a mere 142. The balloons caught a piddling 33.

Five hundred and fifty-eight struck greater London.

The defense strategy focused overwhelmingly on aircraft. But the V-1's top speed was 408 miles per hour, which strained the capacities of the Royal Air Force. The Gloster Meteor, the first British jet fighter, could match that speed, but it wasn't yet in service. Of the fastest RAF fighters, Spitfires could reach around 365 miles per hour, Tempests hovered around 385 miles per hour, and Mustangs could hit 390. Instead of catching the drones, pilots had to fly above them and swoop down like eagles to gain velocity. One RAF Mosquito was pushed to such limits, Churchill's Crossbow Committee reported, that "'bits fell off' and the aircraft was in danger of breaking up."

Among the experts on the Subcommittee on Flying Bombs, there seemed to be very little debate over using proximity fuses, and that attention focused largely on fitting them to rockets. To the British, the American device remained an unproven technology. Instead, the committee puzzled over how to make airplanes faster. The paint on the Tempests was causing drag, and wing surfaces could be cleaned and polished. Engine manufacturers could try to boost power. Propellers might be replaced. Exhausts could be reshaped more aerodynamically.

Parachutes might be affixed to blimp wires to drag any snagged drones to earth. Paper bits could be dropped over the air intakes of the V-1 pulsejets.

Even if the British had wanted to use Section T fuses, they couldn't. General Pile's gunners were hamstrung, sandwiched between the Royal Air Force fighters and the barrage balloons. Not only were they subject to a byzantine set of rules of engagement that changed with the weather, but the American fuses were restricted for use over land. On June 20, an unnamed member of a Crossbow subcommittee suggested trying smart fuses against the drones. If Pile's guns were moved, relocated en masse from the Kentish Gun Belt to the coast, his gunners could have free rein firing over the water. That would also allow gunners unobstructed views of the V-1s. Section T's devices, the subcommittee was advised, might offer a rounds-per-bird ratio twice as effective as standard fuses.

But no immediate action was taken.

By the end of June, Pile had 376 heavy guns deployed between London and the southeastern coast in a belt stretching from East Grinstead to Maidstone. But his gunners rarely had clear windows to engage their drone targets. And their antiaircraft radar sets, sunk in protective hollows to keep the Germans from jamming them, struggled to pick up the low-flying V-1s. When Pile sought to move the guns to higher ground, the air force decided to claim the sites he had chosen for itself, and to further extend the hapless barrage balloons.

The Royal Air Force, not Pile, had priority.

Even when Pile's gun crews could engage, the V-1s were strangely elusive targets. The ghost planes, which flew straight and could not make sudden maneuvers, should have been easy marks. Yet their speed made them difficult to track, and at two thousand to three thousand feet, they flew too high for light guns and made awkward marks for heavy ones. According to Pile, the resultant shooting "was both wild and inaccurate." His gunners were hitting only 9 percent of the drones.

In London, the ack-ack guns went silent. Shooting the V-1s as they flew over the capital, after all, could only succeed in bringing the pilotless aircraft down on their intended target. Gunners were restricted from firing, and Londoners, as they had in 1940, faced a depressing quiet as flocks of noxious drones moaned and blustered, dove, and wrecked the city anew. After three weeks, Churchill disclosed, Hitler's secret weapon had claimed 2,752 lives and injured some 8,000, devastating figures not seen in the capital since the end of the Blitz three years prior.

Londoners feared the government had no answer. "The lack of guns," one resident complained, "strikes me as an admission of failure in defense."

The V-1s arrived in packs at all hours of the day and night. The unrelenting buzz bombs, according to the British air chief, were "harder to bear than the storm and thunder of the Blitz." One British statistician found herself having to make a dozen trips to an air-raid shelter over a single morning. Thirty alarms might sound in a day. Eric Stern, a U.S. Army translator staying next to the 130th on Sloane Court, later recalled that the drones arrived so "thick and fast," and that "it was 'Alarm' and 'All Clear' so often," that "one didn't know which was which after a day or two."

Pedestrians going about their business eyed doorways or basements to dive into at a moment's notice. Air-raid wardens grew tired from long shifts. Endless cycles of suspense and brief periods of relief exhausted the psyche. The city was again on the frontlines of the war.

"The effect of the new weapons," General Eisenhower said, "was very noticeable upon morale." It felt like London, as the novelist Evelyn Waugh later fictionalized, had been suddenly "infested with enormous venomous insects."

The drones earned their psychological toll. The explosions, a British rescue worker said, made "a bigger mess" than the bombs ever had. Their payloads were dynamic, detonating outward, leaving small, shallow craters at the centers of immense blasts. As the British pilot Herbert Bates put it, a V-1 explosion would be

> heard at a great distance; its effect would be felt over miles. The blast struck houses and large buildings and disintegrated them like a tornado against a house of straw. A single bomb could lay a county parish in ruins. It could blast the houses of a dozen streets. It could make homeless . . . hundreds of people. It could blow the bodies of its victims into small pieces, far afield.

By midsummer, over two hundred and seventy thousand homes and other buildings had been destroyed or damaged. If that rate of destruction continued for two months, it would equal that of the nine months of the Blitz. Total casualties from the V-1s were on track to match those of September, 1940, the deadliest month of the Blitz.

Londoners grew sensitized to sudden noises, sitting up in their beds at night in angst or rushing for shelter at the sounds of trams, trains, mo-torcycles, or even bees buzzing. A sign rattling on the back of a truck could clear the sidewalks. "Kids playing happily, grown ups going about their affairs," a citizen wrote, "next minute, the streets clearing as if by magic. . . . You can feel the uneasiness."

The sequence of noises was particularly hideous. "The stages were so prolonged," a British woman explained. "First the sirens. . . . Second . . . that sinister throbbing drone, and, third, the 'stop' and the tense, terri-ble seconds of suspense, waiting to know whether it was overhead." The pulsejet was so loud, one heard it for at least five minutes before the en-gine cut. The "nonstop clackety-clack-silence-BANG!" seemed to repeat itself ad infinitum, suggesting a malignant assembly line, a morbid con-veyor belt, an inevitable "grisly transport service."

Even worse than the deep roar, the "raucous spluttering," was the "sudden horrifying silence" when the engine cut. The sudden lull, preg-nant with fear and helplessness, felt like an especially cruel form of men-tal torture. One witness called it a "terrifying silence." Others described it as "deafening" or "ominous." A twelve-year-old girl said it felt like "the world stood still and held its breath."

"As it stopped," said another Londoner, "your heart would stop."

Lasting a dozen seconds, the quiet marked the final moments of too many lives, and was the heaviest burden of all. When the pulsejets stopped and the silent dive of a drone began, one felt, a British historian wrote, a "mingled wave of fear, profanity and piety." Prayers and shouts would fill the air begging the drone to continue on to someone else's home, followed by pangs of survivor's guilt if it did.

Twenty thousand Morrison shelters had been delivered, and another four thousand were being distributed weekly. Children who had returned to London after the Blitz were evacuated once more, and families with means began their exodus afresh. The British government discouraged citizens from meeting in large groups.

The 130th was adjusting to London's new morbid rituals. Occasion-ally, Ed Hatch and his roommate, Teddy Booras, sought an air-raid shel-ter until the "all clear" sounded. Subway stations refilled, George Or-well wrote, with "sordid piles of bedding cluttering up the passage-ways and hordes of dirty-faced children playing round the platforms." When Hatch rode the Underground, he was moved by the same sights: families sleeping under blankets on the platforms and stairwells.

Hitler's drones, at least, did not carry chemical payloads. So the 130th spent their days engaged in other useful duties. On July 1 and 2, the unit performed rescue work around the city. They helped the British rescue squads remove debris left by the V-1s and searched for wounded Londoners in bombed-out homes.

Monday, July 3, promised another day of the same. Hatch and Booras awoke early and breakfasted at their usual mess hall. When they returned to Sloane Court, they learned the schedule of their weekly chores. They gathered the trash, cleaned their quarters, and swept dust into the hallway. Outside, men of the 130th gathered on a truck for rescue work. Once the floor was clean, Booras decided to take the trash down to the cellar alone. The pair could alternate days, he said, and Ed could do it tomorrow. Booras walked down the stairs—a fateful, absurdly trivial decision. Hatch went outside to board the pickup. But in yet another pivotal trifle, the truck was packed. Not one extra soldier could fit on the transport.

The sky promised rain.

Hatch did not know who shouted out at 7:47 a.m., but he heard them. "Buzz bomb!" Gazing to the sky, he saw a Nazi drone gliding as if in slow motion, its engine off, utterly silent, approaching from the southeast toward the truck.

He sprinted to the street corner, to the northwest, and dove on the pavement with both arms outstretched. For some reason, he glimpsed his mother's face. The explosion levitated his body and flung it through the air. The next thing he knew, he was enveloped in a blinding cloud of dust particles and debris.

Neighboring troops in the Sloane Court apartments had even less warning than the 130th. Eric Stern, the interpreter, was lingering outside his second-floor room, a raincoat draped over his left arm. He was waiting on an Army buddy, Dick Bierregaard, who was combing his hair in the bathroom upstairs. Stern did not hear or see anything until after the warhead landed in the street. He suddenly found himself, as if for no reason, flat on the floor covered with chunks of ceiling, his lungs filled with thick clouds of suffocating dust. His Army uniform was in shreds, and he was bleeding from his hand, head, neck, back, and legs. He wandered, in shock, back to his room and lay down in bed.

After a few minutes, remembering Bierregaard, Stern got up and climbed the staircase. But when he opened the bathroom door, Bierregaard was gone. The entire room had disappeared, along with the front half of the building.

Jean Castles, of the Women's Army Corps, was living behind Ed Hatch and Teddy Booras's apartment. She recalled the sounds of a motorcycle. The concussive blast knocked her against the door and rocked the building up and down in loops. With her roommate Dot, she hurried outside to help, and then, in a grotesque scene, collected blankets to cover the bodies and detached arms and legs.

In the middle of the road lay a severed head.

The sights would later give her nightmares, but in the moment, Castles did not get sick at the grisly spectacle. It felt, she said, like "living in a play, where you see all that happens, feel it, and are a part of it, yet you are in a dream."

A British bystander went from body to body, rolling soldiers over and checking for signs of life, "but they were all glassy-eyed." Much of Sloane Court East was decimated. Apartments, naked to the street like grim dioramas, revealed desks and broken furniture. Partial stairwells clung to exposed walls. Inside the ruins of brick and timber, piled high and stretching half the street, a fire sparked. Trapped soldiers could do nothing about the inferno and were severely burned.

The nature of the V-1 blast made identification difficult. As many as sixteen bodies were never identified. The men of the 130th on the truck did not have time to get off and were slammed, still inside the vehicle, against the collapsing building. One soldier was trapped in the rubble for four days, a "lucky" accident of physics, geometry, and falling timber. Eric Stern's friend Bierregaard survived the three-story fall from the bathroom with only two broken legs, as if he were carried down, as if on a plate, by a piece of wall. Other men, according to the unit history of the 130th, simply walked from the ruins "amidst smoke and flame."

Teddy Booras lost his life inside the cellar where he'd taken the trash. If Hatch had gone in his place, or if there had been room for one extra soldier on the transport, the Army death notification would have gone to Hatch's parents.

While the British rescue, fire, first-aid, and ambulance services began their difficult work, a roll call was held for the detachments living on Sloane Court East. Name after name was called and lost into silence.

Six, 8, and 10 Sloane Court were gone, as were sixty-two men of the 130th Chemical Processing Company. "Amigo" Zanfagna and "Red" Strout were among the fallen. As were Floy Perryman, the schoolteacher with a college degree, Stephen Chunco, who'd made eyeglasses before the war, Boyce Stone, who'd worked in textiles, William Pickens, the Kentucky farmer, and Robert Cooke, the platoon leader and former hotel clerk. The losses of Philip Conley, James Caruso, Donat Patry, and Chester Peterson widowed Helen, Clovia, Geraldine, and Jeannette. After the tragedy of Guards' Chapel, the incident marked the second deadliest V-1 attack in England, and the greatest loss of any American unit in London during the war.

Wachtel kept launching drones toward the capital.

28
"TURKEY SHOOT"

Merle Tuve imagined horrors and fantasies.

By the summer of 1944, Section T's chairman was a member of a new, high-ranking committee formed to consider the future of science and warfare. How should scientists be used in national defense work after the war?

Tuve studied a column in the *Times-Herald* that lamented the Nazis' "robot bombs," warned that "war seems to inspire the fiendish side of scientific minds," and predicted that the next battle could employ "terrible robot armies."

Mortal man, the writer cautioned, as Nazi drones continued to devastate London, "certainly is turning the world God gave him into a nightmare."

Could science become, instead, as another of Tuve's clippings argued, a "world-unifying force" that could be "systematically and rationally applied"? How could America avoid the lapses in preparedness seen after World War I? Would a postwar "scientific high command" be run by "big business labs"? Could science ensure a lasting peace?

Meanwhile, Rome was liberated. The Russians had ended the siege of Leningrad and were now pushing west. The Allies were making strong inroads into France. It finally felt possible to imagine a new world order, to dream of a postwar life occupied by normal pleasures. In one cartoon in Merle's files, passed along by a green Section T technician, a naughty drawing depicted a young woman sitting in a filing-cabinet drawer, as a matron scolded a very embarrassed gentleman: "Can you explain what this girl is doing in Confidential . . . Post War Plans?"

The Army and Navy would want the top scientific minds for military science, but as Tuve knew, most researchers wanted to resume their academic careers. "Can't be full time, [they're] not interested," he scribbled, in his notebook. "Scientist does not like destruction."

Merle solicited opinions on the postwar picture from military leaders he trusted, like Deak Parsons, and from business leaders at General Electric and Crosley. He even reached out to his childhood friend Ernest Lawrence, who was busy at Los Alamos and whom he addressed as "Dear Dr. Lawrence." For reasons that weren't clear, Ernest replied that while he "had better be on the side lines in the discussion," he "certainly would say that a bold program should prevail."

At Section T, at least, a bold program *was* prevailing. Merle's upstart group had transformed into the nation's third-largest research group, after MIT's radar team and the Manhattan Project. Section T employees now numbered more than a thousand men and three hundred and forty-four women, including key contractors. The Applied Physics Laboratory in Silver Spring hosted over six hundred and fifty people, with an additional sixty-seven military personnel attached for liaison and field training. Any hints of the used-car dealership had completely vanished, and with all the additions—a third story and extensions in nearly every direction—the once modest garage now spanned an entire city block. The old Wolfe building, new Navy buildings, and yet another edifice under construction would soon join to form a complete quadrangle surrounding a central courtyard.

The telephone switchboard was so overloaded in early 1944 that larger equipment, requiring an entirely new setup and room, needed to be purchased. In the meantime, employees were urged to reduce their call volume. A public address system was installed throughout the lab to broadcast news.

Tuve was mayor of a small city, boss to hundreds of physicists and engineers, fifty watchmen, seventy-five stenographers, clerks, and typists, and dozens of janitors, laborers, and field observers. Section T had five drivers, four carpenters, and two photographers on payroll. Thirty staff assistants, and thirty-seven technicians were women, and several departments were either half or majority women.

Starting in July, "Section Tuve" had its own fledgling newsletter, nicknamed "Apple Juice" for the combined initials of the Applied Physics Laboratory and its nominal parent, Johns Hopkins University: APL-JHU-S. Run by editor in chief Dotty Dietz, its staff included associate, art, and sports editors. The inaugural issue included a section for "Gripes," like the lack of an elevator operator. Next to a for-sale notice for a girl's bicycle, it carried a request for employees not to tear up confidential, discarded papers, because that made them harder to burn.

The lab had a library filled with hundreds of scientific magazines, mathematical tables, and textbooks. Topics ranged from aviation, to radio circuitry, alloys, statistics, chemical explosives, materials science, vacuum tubes, and antiaircraft gunnery. Its catalog included three copies of Alan Hynd's *Passport to Treason,* about Bill Sebold and the Nazi spy ring that sought the fuse.

The building at 8621 Georgia Avenue was such a hive of activity that the guards needed a night registrar to track the constant comings and goings. Some four hundred cases of Coca-Cola were being consumed at the lab every month. Section T staff held picnics with beer and music. For morale, movies were screened: of the Chesapeake drone tests, of the experiments at Parris Island in South Carolina, and of the shoots at the New Mexico ranch. The athletically inclined joined a Johns Hopkins league softball team, the Silver Spring Giants. Plans were in place for an astronomy club.

Section T's work had also expanded and changed. The smart gadget for antiaircraft shells was in service. But research continued on dozens of new models and improvements. Between April 1 and June 30, 1944, Tuve's lab produced and released nine separate versions of a single fuse. While those efforts proceeded, Section T took on other projects, like a gun-aiming "predictor" for the Navy.

Tuve did not take his eye off Hitler's drones. On June 17, he arranged for his British liaison, Ed Salant, to visit England in case the fuse was called on.

Against Japan, it was already proving its worth.

"Great rearing clouds," a young lieutenant observed, from the bridge of the battleship USS *Washington*, "marched along the borders of the sea, presenting all sorts of grotesque shapes like a giant parade of animal crackers."

The misty cloud formations, illuminated in a "beautiful peach color," stretched endlessly into the distance. Above the sunset, sailors saw a lone Japanese bomber, fatally wounded by an American B-24, nose-dive into the glowing ocean.

Within hours, a climactic battle would fill the skies.

The U.S. Navy was preparing for a bold assault on the Mariana Islands, at the edge of the Philippine Sea. If the daring incursion was successful, the Navy would secure a precious foothold deep in the Pacific. Only fifteen hundred miles from Tokyo, the islands would afford the Allies with submarine ports and, most critically, with air bases for long-range bombers that could reach Japan.

Japanese high command had been waiting for just such a decisive engagement, a final chance to destroy the American fleet. Days earlier, on June 15, Admiral Toyoda proclaimed that "The fate of the Empire rests on this one battle." The Japanese massed nearly all their serviceable boats: nine carriers, five battleships, thirteen cruisers, and twenty-eight destroyers. The Navy summoned no less than fifteen aircraft carriers, seven battleships, twenty-one cruisers, and sixty-nine destroyers.

Both sides expected that naval airpower would decide the fight. The American battleships formed a protective shield in front of their carriers.

The *Washington* was the flagship of the battle fleet.

James Van Allen was on it. The Section T physicist was now an assistant staff gunnery officer (and also a junior officer of the deck) for Admiral Willis Lee, the gunnery expert who had first vouched for the fuse in early 1943.

Van Allen did not need to be there. The fuse had already developed a well-deserved reputation. Over the course of 1943, 75 percent of rounds fired by Navy five-inch guns used standard ammunition, while 25 percent used Section T fuses. Yet the proximity fuse was credited for 51 percent of downed airplanes, a rounds-per-bird advantage of three to one. Twelve antiaircraft guns armed with smart fuses in effect became thirty-six guns. The gadget took on Japanese Mitsubishis at the battle for Guadalcanal, defended the Seventh Fleet off Makin Island, and helped save a convoy off the Russell Islands. The smart weapon was finally gaining widespread acceptance.

But in early 1944, a devilish problem had emerged. The dry batteries were running out of power. With Navy vessels patrolling the tropics, the scorching cargo holds could reach 120 degrees. The heat had a devastating effect on the batteries' useful lifetime. The liquid, glass-jar energizers were nearly ready, but in the meantime, dying batteries were turning smart fuses into duds.

Van Allen personally devised a technique for replacing them. He asked to return to the Pacific to set up "re-fusing" depots. Then he hopped between New Caledonia, the Solomon Islands, Australia, Eniwetok, the Caroline Islands, Manus, and Espíritu Santo, training soldiers on the battery replacement technique. He removed thousands of old "energizers" himself, disassembling the guts of the devices in a delicate surgical operation. The procedure covered his hands with sticky, fine explosive dust and turned his and the sailors' skin bright yellow.

Now he would get his first taste of battle.

On June 19, Admiral Ozawa, the commander of the Japanese fleet, launched an initial salvo of sixty-nine planes—including forty-five vicious Zero fighter bombers—toward the American carriers. From the bridge of the *Washington,* Van Allen eyed the first wave of Japanese fighters and bombers approaching, and saw three planes escape the net of U.S. Hellcats and begin flying toward the bridge.

Were the pilots steering into the ship? Kamikaze tactics hadn't yet been used. As Van Allen watched, Lee's gunners fired the smart weapon he'd helped invent. Two of the planes were shot down and fell flaming into the sea. The third veered past at such close range that Van Allen could see the fear on the pilot's face.

Forty-four planes slipped through to Lee's ships. The Japanese fighters swarming around the *Washington,* one lieutenant said, "seemed to come in ones and twos from every direction. . . . The attacks seemed to go on and on without end." The Japanese fighters charged over and over. Again and again they were downed by Admiral Lee's gunners. The Hellcat pilots were just as ruthless, shooting down twenty-five planes in a manner reminiscent of "a big Ferris wheel at a country fair." Japanese planes entered at the top, pursued by Hellcats, and exited smoking and burning from the bottom while American pilots climbed the wheel again to the apex.

It was hard to keep track of how many planes were gunned down. The "whole thing became a blur," a sailor said. The blue sky was peppered with exploding shells and lit with burning planes struck by the fuse. For one pilot looking down, the view of the USS *Indiana* in the middle of Lee's formation—a six-mile circle of twenty-four warships rattling off a twenty-thousand-foot umbrella of antiaircraft fire—was his "most memorable scene of World War Two."

Another pilot compared it to "an old-time turkey shoot."

Not one of the forty-four planes made it past Lee's gunners to the carriers. A Navy translator monitoring Japanese communiqués overheard an enemy broadcast advising the next wave of planes to avoid Lee's battleships entirely.

After the fight, Van Allen and the sailors went down to the wardroom. On a table covered with nice linen, they dined on roast beef and strawberries. The reserved physicist never forgot the contrast between the violent battle and the face of the Japanese pilot and the strangely refined banquet hours later.

The Japanese lost 92 percent, or 395, of the attack planes they flew into battle. Their air force never recovered. Hellcats claimed most of the credit, and radar developed by Van Bush's scientists played an outsize role, but the fuse also contributed. Soon enough, Van Allen and the *Washington* were circling the Marianas, living once again, a sailor on deck complained, "a very boring existence."

The *Washington* anchored on Saipan on July 7, 1944.

Even the staunchest critics of the fuse were now convinced. And with the new liquid battery and other fixes, its performance would only improve further.

Soon, the island of Tinian was also taken and occupied by American Marines. Construction began on two airfields suitable for the long-range B-29 bombers that could reach Tokyo. Japanese attack planes, stationed over seven hundred miles away at Iwo Jima and Chichi Jima, were initially kept at bay by Navy carriers, but eventually American planes on Tinian and Saipan had to fend for themselves. During the day, American pilots defended the Marines and controlled the skies, but they did not have the fighters they needed to protect the soldiers at night.

Thousands of American troops, with no caves for protection, camped out in the open, exposed to Japanese raiders who came in the dark. Antiaircraft artillery battalions, equipped with 90-millimeter guns, could do very little to ward off the bombers. U.S. casualties were mounting at a frightening rate. "Almost every night," an American gunner on Tinian recalled, "would see three to four thousand shells fired with only a hit or two in several nights."

Section T smart fuses, designed for Army guns, changed that picture immediately and dramatically. "On the very first night they were used," the gunner reported, "17 [Japanese] planes were shot down, 12 the next night, and 11 the third night." Within days, the enemy bombers were "reduced to a few nuisance raids."

All the Japanese pilots could manage to do was to harass the Marines as they slept, and the vital airfields that could finally reach Japan were completed.

29
ACK-ACK GIRLS

After the bombing of Sloane Court, Ed Hatch was treated at a medical clinic for minor injuries and spent a week sleeping in an air-raid shelter. He was transferred, with the other survivors, to a recuperation camp outside London. But the camp turned out to be just as vulnerable as the center of the capital, and the buzzing Nazi drones tormented the remnants of the 130th. A ghost plane detonated some fifty yards away and sent several of the men to the hospital for "nerves."

Replacements were called up. But the new recruits could not absorb the required technical knowledge. Eventually Hatch and the men were sent to the Midlands of England between Birmingham and Liverpool, out of the reach of German buzz bombs, where their duties were peaceful. They set up a grim laundry service, washing salvaged clothes from the battlefields of France.

Newspapers did not report the tragedy in much detail. Censorship rules forbade publishing the number of dead or the location of the explosion. The *Daily Telegraph* declared only that "a number of bodies were unearthed" when a V-1 "fell on a building" and that American soldiers were among the victims.

The Associated Press was slightly less vague. On July 6, they published a short description of how the soldiers in the truck died "waiting to go on rescue work to help victims of other bomb hits." A Women's Army Corps witness, Corporal Mary Lou Bernick of Sarasota, Florida, was interviewed but declined to describe the carnage. "I don't feel like talking about that part of it." No mention was made of the number of deaths or the 130th Chemical Processing Company.

In news reports of the V-1 bombardment, figures were redacted. The practice did little to calm Allied nerves: "Sections of the world's largest city have been affected by robot raiders which explode with terrific impact . . . (Twenty seven words were censored at the start of this dispatch)." "(Forty words censored) schools, hospitals and churches have been blasted with loss of life." "(Twenty-five words censored) American soldiers were killed and hurt when bombs fell on a club."

No aspect of life in the British capital was untouched by the siege. Concerts, dances, sporting events, circuses, and street fairs were abandoned for fear of mass casualties should a V-1 strike. Twenty-five city boroughs canceled various entertainments, and attendance dropped at those remaining. At the Comedy Theatre on June 12, before the bombardment, 435 of the 700-some seats were filled. After one week, just 222 seats were filled. Days later only 181 were.

Musicians at one venue were interrupted by the obscene clacking of a drone, while spectators were told to "lie down!" as a buzz bomb detonated six hundred feet away. Brave cinephiles ignored air-raid warnings and slumped in their seats until a V-1 explosion sounded. At a screening of *Gone with the Wind,* moviegoers noticed that the battle-scene noises were "having a little help from outside."

Bicyclers, tennis players, golfers, and cricketers had their exercise interrupted. Church services paused while the V-1s rattled overhead.

Hospitals were bedeviled by the droning flocks. The flying bombs, a British historian wrote, almost "seemed to seek out the sick and helpless." Over the course of the siege, one hundred hospitals would be destroyed or damaged by the ghost planes, including seventy-six in the city of London. Nurses and doctors labored, day in and day out, under the vilest, most extraordinary stress. In one particular infirmary, they endured no fewer than 190 air-raid warnings over a single day.

Their work followed a cruel pattern. The less severely wounded arrived first, and were treated for shock. Hours passed, and the badly injured emerged from surgery. "The most harrowing part," a nurse recalled, came later, "when bewildered survivors looking for missing relatives trekked round the wards and hospitals until they found their loved ones or were persuaded it was better to leave."

One hundred and thirty-eight patients lost their lives to V-1 strikes on hospitals, as did twenty-four staff. A doctor severely injured in a drone strike crawled out of the rubble into his own operating theater, where he underwent successful cranial surgery.

One hundred and forty-nine schools would be damaged across southern England. In the village of Crockham Hill, a ghost plane hit a nursery school full of sleeping infants, claiming the lives of twenty-two children under the age of five.

Beer deliverymen had to be paid extra to enter greater London. A pub lost its staff because they thought it unsafe to be working near so much glass. Windows had to be kept open to avoid eye injuries from shards flying "like sheets of rain." A sixteen-year-old salesgirl, given a prize spot at a cosmetics counter, realized with dismay that it "had a lot more glass display cabinets than the other counters."

The blasts blew tiles off roofs, doors off hinges, and leaves off trees. Clothes were shredded and ripped off survivors, who often emerged seminude and covered in brick dust like so many terra-cotta statues. Lumps of brick and concrete, plaster ceilings, and splinters of asbestos piled high in the streets.

Residents grew apprehensive about muffling noises. Running water, vacuum cleaners, or even bacon and eggs frying in a skillet could mask a buzzing drone nearing overhead. Showers shortened. Citizens stayed dressed at bedtime, ready to dash to a shelter. Shy Londoners worried when bathing naked.

Households hired "roof spotters" to alert them of approaching V-1s, but other mammals also proved themselves capable alarm systems. Geese in Hertfordshire squawked at pilotless aircraft. Dogs dove anxiously for cover at the slightest hum. In Sussex, one pet cat would dive under the Morrison shelter long before its human owners could hear the ominous racket of a ghost plane. An elkhound would bark shrilly and run under the stairs. A springer spaniel would "suddenly wake, standing in the middle of the room and 'point,'" giving its owners enough warning to retreat to the glassless hallway.

Prime Minister Churchill could no longer downplay the bloody truth to Londoners. On July 6, he disclosed before the House of Commons that, after less than a month, the death toll from the buzz bombs was already approaching three thousand. "Everything in human power," he promised, was being done to meet the threat, under the counsel of "a great number of scientists and engineers."

More British citizens had already been killed by Peenemünde's super-weapon than had been lost over the first fifteen days of the Battle of Normandy.

By the middle of July, Churchill was holding meetings every other night with General Frederick Pile and Royal Air Force leadership. Pile argued forcefully that the existing three-prong defensive strategy against the flying bombs wasn't working. RAF airplanes, which still had first priority over his antiaircraft guns, were simply not gaining enough of an advantage over the Nazi's terror weapons.

"All right," Churchill replied, "from next Monday for a week General Pile is to have a free hand."

It wasn't that simple. Even as Pile gained access to new American radar, aiming devices, and smart fuses, his gunners still remained in the wrong place. He would have to move all of the guns in the Kentish Belt to the coast. Even then, his batteries would need to be trained in the new devices during combat.

He took the gamble, and formulated a plan to establish a coastal gun belt stretching some fifty miles between Beachy Head, near Eastbourne, and St. Margaret's Bay, just north of Dover. The new sites would restrict the Royal Air Force and allow gunners dominion over the seaside. They could fire the smart fuse.

The airplanes had their chance. Now it was Section T's.

Pile's gun crews included women because antiaircraft guns got little respect. The general was plagued by chronic staffing problems. Already in 1940, he had been short nineteen thousand men. It made sense to propose filling out his ranks with volunteers from the Auxiliary Territorial Service, the women's branch of the British Army. Women were already employed widely as drivers, cooks, and orderlies. But Pile sought female soldiers, receiving equal pay and facing equal risk.

The British undersecretary of state for war called his proposal a "breathtaking and revolutionary" idea. But Churchill approved Pile's plan. Over bizarre objections—that women couldn't be heard over the guns, that they might "smash valuable equipment in a fit of boredom," or that men and women working side by side would somehow result in a "musical-comedy-chorus atmosphere"—mixed batteries were formed in 1941 and were immediately successful.

"The girls cannot be beaten in action," one commander said. "They are quite as steady, if not steadier, than the men." For "political reasons," men still fired the guns. But women spotted enemy planes, employed "predictors" to calculate the right fuse time, and operated searchlights, radar, and range and height finders.

Pile's gunnery crews, then, tasked with defending London from the V-1 bombardment—in what was known as the Battle of the Flying Bombs, and what he referred to as "the first battle of the robots"—were largely female. Some seventy-four thousand women served on gun sites, and by the end of 1944, there were more ack-ack girls than men serving in Pile's Anti-Aircraft Command. Many women joined up as teenagers. Some recruits looked to be as young as sixteen.

A typical heavy mixed battery was staffed by 89 men and 299 women. Under their charge were eight 3.7-inch antiaircraft guns weighing over twenty thousand pounds each, a gun-control room and ammunition depots housed in trench shelters, a command post sited in a tubular, corrugated-steel Nissen hut, radar aerials, predicting computers, tents, thousand-gallon water tanks, and electrical generators. They had cook-houses, latrines, and showering facilities if they were lucky.

Uprooting everything to the coast was a titanic undertaking.

On July 14, 1944, the "great trek southward" was initiated. The transfer relocated twenty-three thousand troops from the Kentish Gun Belt and elsewhere, over three hundred heavy guns, and three thousand miles of communication lines. Sixty million pounds of ammo and another sixty million pounds of supplies were loaded onto eight thousand vehicles. Over a few days, the trucks of one command traveled 2,750,000 miles.

Molly Franklin, of the 571st Heavy Anti-Aircraft Battery, traveled with her unit "atop the furniture in the trucks" and gazed at the English countryside. Within twenty-four hours of receiving Pile's order, they were packed and moving. She arrived at Warden Bay on the Isle of Sheppey—a secondary location sited to defend against V-1s being fired up the Thames. "God forsaken place," a recruit said.

"Not a pub in sight."

"No buses."

Phyllis Ramsden, barely in her twenties when she joined the 478th Battery, also made the journey by truck, motoring from northern England to Hastings on the coast. They drove through towns with no signposts, which had been taken down after the disaster of Dunkirk. If the Nazis ever invaded Britain, at least they'd get lost. When Ramsden and her unit arrived at Hastings, they were billeted in a field under canvas. They "hid under the hedges for 2 days while the guns and instruments were connected," she said, and dug slit trenches so that if the V-1 drones were shot down overhead they could dive for cover.

Vee Robinson, who enlisted in the 536th Heavy Anti-Aircraft Battery at age eighteen, traveled by train with the other women from all the way up in Scotland, a trek that began at four p.m. and lasted until nine thirty the next night. Posted in the coastal marshlands, they were billeted in an abandoned boarding school. They had no blackout shutters or window curtains or any light bulbs, for that matter.

The women "lived like gypsies," slept in their clothes, and used dugout latrines. Where empty buildings were unavailable, sheds were erected. Plotting rooms were set up in Nissen huts or, in one case, in a converted women's bathroom. Shelters were dug for generators, and "tracker" towers built of steel scaffolding.

Within three days, Pile's guns were firing.

Lying in bed, Robinson wrote later, the "sky was alight and the noise deafening. The guns were much nearer to our sleeping quarters than ever before . . . we felt more of the blast when the guns were firing." On one of the first nights, a buzz bomb landed "with an explosion nearby and our windows shattered inwards.

"We were showered with glass." She was steps from a concentrated, "almost touching line of guns" that, she found, "were never silent for long."

By July 19, there were forty-seven batteries and nearly four hundred heavy antiaircraft guns stretching along the southeastern coast, dotting the edge of the land like tilted drilling rigs and arranged in arrowhead formations like so many convex shields. In a nod to a British military nickname for the V-1s — "divers" — Pile's curtain of defensive gun positions became known as the "Diver Belt."

The shift to the coast paid dividends quickly. During the first week, the percentage of V-1s shot down by Pile's gunners rose from 9 to 17 percent.

British Lancaster bombers were flown to Cincinnati, Ohio, to expedite the supply of smart fuses from Crosley and deliver them to the Diver Belt.

The British code name for the fuse was "Bonzo."

One "ack-ack girl" recalled the Sunday the fuses arrived, when all the female recruits had to remove the nose cones on their entire stock of shells, and refit the heavy rounds with special Section T fuses. Their "delicate" hands were deemed more suitable for the job, although as she remembered wryly, that "didn't stop us from having to do manual duties with our dainty sensitive fingers."

Ack-ack women and British gunners weren't the only ones receiving new fuses. American batteries had also now joined the fray. By early August, six hundred and eighty-four heavy guns guarded the Diver Belt and the "Diver Box" at the mouth of the Thames, including eighty American 90-millimeter cannons sited along the southern shoreline.

"On a clear day," said Ralph Griffin, an American gunner stationed near Dover, "we could see the Buzz Bombs almost as soon as they were launched."

The coast of France was so close to one American position that troops could "use binoculars to tell time by the clock in the city hall tower of Calais."

Nearly within sight, an average of 106 times a day that summer, at all hours of the morning, evening, and night, Wachtel flung his drones over England. At the base of the launching catapults, cauldrons of chemicals bubbled together furiously, slinging the ghost planes at 248 miles per hour within an instant and leaving a long trail of chemical smoke along the slanted rails pointed at London. The vats breathed life into the pulse-jet engines, and set the shutters of their intake valves beating fifty times a second. Fuel mixed with air and belched rapid explosions. Odometers in the noses counted down to their target. Guidance systems kept the flying bombs steady as they spent fuel and sped even faster.

At night, a GI recalled, "they'd look like stars."

Under moonless, dark skies, night firing was an otherworldly event. At first, the drones appeared as mere "pin heads" glowing in the black, specks of fire groaning in the distance. At the sight of a V-1 quickening in the darkness, gunners would focus "on the little ball of fire" and a "nervous tension would fill the air."

Radar sniffed out the targets. The motors of the guns spun to face them, and "the stillness of the night," a U.S. soldier recalled, was met with the "roar of the big guns" that "hurled their missiles skyward to meet the incoming target." Teams of men passed rounds forward in a line and rammed them into gun breeches. Nearby cannons and the staccato of various machine guns joined the firing, "filling the sky with a pyrotechnic display more spectacular than any Fourth of July."

The "tenacious fingers" of search beams locked onto V-1s, tips of guns flashed bright cotton-candy explosions, and the "sky became livid with bursting flak," as the multicolored tracer bullets drew curved lines into the sky.

"There was never such a terrific concentration of Ack-Ack," a gunner said.

Finding their marks, direct hits illuminated the night with a "terrific burst of yellow flame" followed by, after several seconds, a concussive blast wave that jarred the men and women, shook the earth, and whipped the tents.

Many had no ear protection. From German guns in France, "shell fragments fell like rain" and no one ventured far without their steel helmet. The U.S. 134th's camp in Hythe was dubbed Shrapnel Heights. The 127th, stationed on Dover's chalky White Cliffs, also endured shelling from Nazi guns. German cannons fired on the area so frequently, the British called it "Hell's Corner." Once, an American recalled, "red hot shrapnel came through the roof and bounced around in the barracks and came to a stop behind [a man's] bed and then laid there and glowed in the dark."

At first, the American gunners found the V-1s to be maddeningly elusive targets. "But after we got proximity fuses," one recalled, "we started to knock them down. We got to where we could get them if they were in range."

Seventeen percent of V-1 "kills" grew to 24 percent.

By the first week of August, practically all the heavy guns on the Diver Belt were equipped with Section T fuses, which were guarded twenty-four hours a day.

Section T was on the coast also, training the gunners.

Paul Teas, the youthful battery engineer and liaison, worked out of

General Pile's headquarters in northwest London. He visited nearly every battery site, British and American, offering guidance about the new Bonzo fuses.

Ed Salant, the blunt workaholic, arrived on July 30. He practically lived among the Diver Belt batteries, motoring in an Army jeep along the narrow country roads in the blackout, scrambling among the gun sites to brief gunners.

When Salant saw a buzz bomb shot down by a fuse for the first time, it left a lasting impression. He knew—far more than any gunner ever could—what he was truly looking at. He knew the long odds of the smart fuse project.

The physicist was inspecting a radar unit when "suddenly a target was detected." The V-1 "came flying along some distance away," he said, "in a region covered only by batteries with time fuses." Without the smart fuses, the forward guns failed to shoot down the drone heading toward London. "It got through."

The V-1's flight path was in line with Salant's position. In the sky overhead, yet another drone passed the forward guns. Then another. "In as many minutes, four came racing low over the Channel between about one and two thousand feet up."

The battery Salant was visiting *was* equipped with the fuse. Four targets flying toward London approached directly above the Section T liaison.

To earsplitting claps, the gunners engaged the drones that had slipped through. Under enormous pressure, some seventeen thousand times the force of gravity, the antiaircraft cannons launched twenty-eight-pound-shells fitted with Bonzo fuses toward the ghost planes, jolting subminiature glass vacuum tubes that were rigged together tightly like packs of dynamite. The firing shocks broke the glass jars of battery acid, and the rotations infused the energizer into the battery plates and activated the smart weapons. Within an instant, hawk-like, the fuses opened their eyes.

Each second, they flew twenty-six hundred feet and spun over two hundred and fifty times. Inside each fuse, the wet batteries illuminated the vacuum tubes like light bulbs. Glowing tungsten filaments heated, bent, and expanded. Through metal antennas embedded in the nose cones, radio signals broadcast at the speed of light pierced the air like low-pitched sirens, scanning the emptiness. When they found the underbelly of a rattling Nazi drone—which matched the dimensions of Section T's mockup in New Mexico to roughly a foot—the beams ricocheted off the fuselage. Outgoing radio signals mixed with the incoming ones, causing an electronic "ripple" pattern. The radio emissions grew shrill like a child's toy: a high-pitched gravity tube.

The mixing radio waves passed a triggering threshold within two-tenths of a second. In the language of electrons, the message was boosted through the amplifying tubes to triggering tubes to squibs, which set off the high explosives as little as twenty-five feet from the V-1s, the most lethal distance. The drones' thin steel siding was sliced by shrapnel, and flaming wreckage fell to the beach.

Three buzz bombs blew up midair, Salant said, "with a bright yellow flash and fine bang. The fourth belched smoke and fire and dived off its course."

It was "a nice show for a man from Section T."

With increasing precision, the fuse mastered the V-1s.

With every ghost plane stopped, eighteen hundred pounds of explosives did not hit London. Sixty-two American troops did not die. Nor did one hundred and twenty-one parishioners at church. Nor did twenty-two infants at a day care. With each target gunners felled, Londoners did not dive into doorways for cover. Children did not stop playing soccer. Rescue workers did not dig tunnels to find survivors buried under heaps of rubble. Surgeons did not operate desperately on the wounded. Bewildered family members did not wander hospital hallways searching for loved ones until they were sent home.

By the middle of August, the percentage of buzz bombs stopped by the coastal guns had nearly doubled yet again, from 24 percent to 46 percent. General Pile was ecstatic. He penned an effusive letter to the U.S. Army chief of staff George Marshall, on August 12, thanking him for the fuse "which is so secret that I can only describe it by its nickname in this country, 'BONZO.' You will know all about it . . . when I tell you that Dr. Selant [sic] is advising us how best to use it." With the fuse, his gunners were downing V-1s with *less* than a hundred rounds.

Salant estimated that the gadget was delivering one hundred rounds per bird, a figure five or six times better than the standard fuses could deliver. A British analysis agreed that the fuses were five and a half times deadlier, effectively transforming 776 heavy guns defending London into

4,268. And those were just averages. As General Pile noted, "the best batteries are actually getting one bomb for every forty rounds."

Ten times better than regular fuses.

When wide-eyed observers saw an American ack-ack crew effortlessly shoot down four buzz bombs with fewer than twenty shells, the U.S. commander was asked how he had done it. He laughed and said that he'd picked good soldiers—Tennessee hillbillies who were crack shots from hunting squirrels.

When the radar and aiming devices developed by Bush's Office of Scientific Research and Development could get smart fuses close enough— within 180 feet—the rounds per bird for V-1s was a stunning 8.1.

Fifty times better than the old fuses.

Mary Soames, whose battery was sited on an exposed clifftop overlooking the city of Hastings, marveled at the new American equipment, at the proximity fuses, and at the results that soon grew "clear for all to see." (Soames also happened to be Winston Churchill's youngest child.) When she was off duty, she later recalled, she would "sit and watch with relish our guns destroy the divers."

"I'm really happy here," she wrote in her diary, "really happy in my work."

By the third week of August, the success rate for shooting down Hitler's secret weapon had ballooned again, from 46 percent to 67.

Now it was the gunners, not the Royal Air Force fighters chasing drones over the water, who were taking down the vast majority of the V-1s, a reversal that could not have been imagined at the start of the war. Ack-ack crews were downing some two-thirds of the buzz bombs. The planes claimed 17 percent.

"It was quite uncanny," a British colonel recalled. "As an officer on the site you ordered 'cease loading' before you saw any rounds burst." That kind of confidence was unheard of. "And then suddenly, in front of the target, some four rounds [per] gun—thirty-two rounds—appeared in the sky" in front of the drone and "either hit it, blew it up, or knocked it down . . . It was fascinating to watch."

"It was marvelous."

The fuse *looked* different in combat. Whereas time-fused ammunition exploded in a scattered screen of black puffs, like pieces on a vertical chess board, Section T's gadgets did not detonate at all unless provoked.

Instead of a wide veil, the result was often a small cluster of detonations around a drone, a display that should have caught the eye of careful German observers on the French coast.

Salant believed that the unique burst patterns had likely alerted Nazi high command of a new weapon. Over the course of a single day, he wrote, "the enemy might have seen 88 out of 93 of his flying bombs knocked down before his eyes by these suspiciously behaving bursts." After that, surely, "he must have done some inquiring." In Berlin, a German newspaper reported that the Allies had unleashed a "godsend" of a new countermeasure against the pilotless aircraft.

Sometimes V-1s fell after only two or three shots.

One British gunner reported that accuracy had improved so much that it "reached the stage where the gun crews took it in turn to 'have a go.'"

It soon "became one shell, one [drone]."

At the end of August—as Paris was liberated and the Allies captured Colonel Wachtel's westernmost launching sites—the gunners' 67 percent success rate soared to 79 percent. It climbed up to 82 percent. On one particularly fine day when 104 drones were fired at London, only 4 made it through. On August 28, one British gun site took down every single V-1 engaged: 100 percent.

For the Americans in the Battle of the Flying Bombs, the fuse was not their most vibrant memory. They were not told how it worked. (One GI imagined that it "steered" toward the V-1s.) They did not know the fight as a struggle between scientific research labs, between Merle Tuve and Frederick Gosslau, or Section T and Peenemünde, or American and German universities or American and German industry. The gunners never heard of Amniarix or R. V. Jones.

They would recall the people, like a British woman on a beach thankful for an orange, a fruit she had not seen in ages. The British dockmen complaining that the seagulls covered them in excrement whenever the guns fired. The local children who called out "Any gum, chum?" Their fellow soldiers with whom they hauled ammunition, endured the pyrotechnic nights, huddled near the hot metal goliaths, sniffed the ammonia of gunpowder, and dodged shrapnel and falling V-1s.

Some of their most enduring recollections were of the British women of the Auxiliary Territorial Service: the ATS or "ack-ack girls" nearby.

With the Nazis "having a tough time getting any V-1s through to London," one U.S. gunner recalled, "we earned ourselves a holiday every fifth day." The 125th was stationed next to British mixed batteries in Romney Marsh. It wasn't long before "a good portion" of the GIs were "making nightly excursions to the vicinity of these batteries." According to the 125th's unit history, the ATS women "really took to the ack-ack boys." At night, during off-hours, "with all the social activity going on," the American camp grew strangely empty.

Dances on Fridays, Wednesdays, and Saturdays were well attended. Signs warned the eager Americans: NO JITTERBUGS. With several Royal Artillery soldiers, they formed a band, the "Ameri-Anglos." The GIs peopled the local pubs, and savored Highland scotch and ale. "Fond memories of those gallant ATS girls" would linger in the minds of the Americans, as would nights off "and the names of towns such as Lydd, New Romney, Ashford, Folkestone, and Dymchurch."

There was "always a high community spirit."

The 134th recalled the ATS women as "damn friendly and entertaining."

By the end of the summer, every gun site in the Diver Belt had a wireless radio set for entertainment, and books and newspapers to read. "From somewhere," a British historian wrote, "came 1,000 packs of playing cards, 250 sets of darts, 15,000 paperback books and 1,000 folding garden chairs." Concert parties, plays, and movies were put on by a roaming military entertainment service.

By September, the V-1 attack on England was effectively over.

In total, antiaircraft gunners downed 1,550 drones intended for London. When General Pile sent Van Bush a copy of his final report of the battle, he appended a short note of gratitude to Bush's Office of Scientific Research and Development. "With my compliments to OSRD who made the victory possible." (Churchill later asked Roosevelt to pass his thanks along to "the Tuve establishment.")

"More was learned about the potentialities of anti-aircraft work in 80 days," Pile wrote later, "than had been learned in the previous 30 years." He thanked Ed Salant personally.

"Our reputation in the knowledgeable circles here is very high!" Salant wrote, in a letter back to Section T on September 5, 1944. "You can be sure that the [fuse] has saved the lives of thousands here. I do not believe we are through with the flying bomb, but I do not believe it will be a serious menace to London anymore." Morale had already surged. In early August, "there was the nearest thing to hysteria I ever sensed in England," but now "the sting of terror was drawn."

Londoners were "still aware of danger," he wrote, "but they were no longer oppressed by it." Within days, across the country, blackout restrictions were relaxed for the first time in years. Children too young to have ever seen working street lamps gathered to see them illuminated.

30
A COLD WINTER

ou wait until this war is over," Van Bush promised a colleague, in late 1944. "I am going to take a vacation that will make you envious."

He dreamed of returning to his former life. He imagined that after the war, one of his contraptions, a solar-powered pump, might usher in a new phase of irrigation. He puzzled over how to build a cheap solar "collector," and even a solar-powered refrigerator. He mailed his sister a new volume on knot-tying which he firmly believed was "the knot book to end all knot books." He drafted a slim, poetic manifesto about science and "the process by which the boundaries of knowledge are advanced." He pondered a book deal with Houghton Mifflin.

Bush was now a national celebrity. In 1942, *Collier's* magazine had already dubbed him "the man who may win" the war. By 1943, the *Wall Street Journal* was reporting on the engineer who "looks like a Connecticut farmer" and who "rose from a sick bed" to secure funding for his scientific outfit, which was "admittedly more secretive than the military." That year, Bush won an Edison Medal for his career achievements in electrical engineering. As the *New York Times* put it, noting Bush's wartime role, the public had finally "learned to appraise him at his worth."

Appearing before Congress for appropriations, he was referred to as "Doctor" and treated with deference. By 1944, he was gracing the cover of *Time* as the "General of Physics," commander of a "Secret Army" of anonymous American scientists, boss to a legion of six thousand brains who were paid no bonuses or royalties for their covert inventions and whose labs were duly swept for listening devices. The "Tinkering Yankee" with the "unprecedented job," *Time* reported, was "lean, sharp, salty," "shrewd, imaginative," and "self-effacing." OSRD's budget was now one hundred and thirty-five million dollars a year. It had delivered over two hundred new devices.

Lunch with Franklin Roosevelt was a normal occurrence. He had grown close enough to the president that FDR was now inviting Bush and his wife, Phoebe, to join him in prayer services in the East Room of the White House.

Bush, his biographer wrote, was "at the peak of his power."

His reputation had also risen among military brass, thanks in part to the success of the fuse against the V-1s. General Pile's praiseful letter was making the rounds, and Bush briefed FDR on the triumph of the smart weapon that, he told the president, was "the most extraordinary development of this war in many ways."

The Army, in particular, was keen to employ its stockpile of over 1.75 million Section T howitzer fuses. If their field guns were allowed to fire the smart weapons, gunners could ensure that their shells detonated at a perfect height above German troops, over unseen hills, through fog, and across great distances.

Bush hoped to see the fuse deployed over land. And unlike Merle Tuve, who did not enjoy the thought of his defensive weapon being used as an offensive one, Bush had no reservations unleashing it on ground troops. Squeamishness, to OSRD's director, was "logically untenable," a stance equivalent to moral grandstanding that would leave American GIs vulnerable to Axis scientists. The alternative, in his eyes, was to let the enemy run over the Allies with no resistance at all.

"You may be able to join Gandhi in such a point of view," Bush once told one of his scientists, who had quit OSRD on moral grounds, "but I am not."

But the Joint Chiefs of Staff were unsure about the antipersonnel fuse. If the Nazis recovered a dud, which they would, they could reverse-engineer the invention and use it against the Allies — or share it with the Japanese. The Army Air Forces didn't want their planes to face the smart weapon. But the biggest obstacle in Bush's path was his old nemesis Ernest King, the admiral who had resisted the offensive use of radar in the Atlantic. King didn't want his own aircraft to face the fuse.

The gadget worked *too well*, in other words, to chance using.

To assuage the Joint Chiefs, Bush convened a panel of scientists, including Tuve, to estimate how long it would take to reverse-engineer a fuse from a dud. As Merle reported, shell fuses would take some twenty months to replicate, and the Germans would need another seven to ten months to set up a factory line to mass-produce them.

The gamble seemed worth the risk.

"I have agreed to meet with you," King told Bush curtly, when they met to discuss the antipersonnel fuse in late October, 1944, "but this is a military question, and it must be decided on a military basis." Scientists had no say in the matter.

"It is a combined military and technical question," Bush replied coolly, "and on the latter you are a babe in arms and not entitled to an opinion."

Bush considered this a "good start" to their discussion.

By the end of their meeting, King had seen the logic, and on October 25, the Joint Chiefs approved the release of the fuse against troops. Within weeks, Bush himself was bound for France to shepherd the fuse into the Army's field arsenal.

His itinerary took him from New York, to London, to liberated Paris, to Verdun. On November 26, donning a service uniform with no insignia and a noncombatant emblem on his sleeve, he rode an Army jeep to the small French town of Vittel, near the front and some seventy-five miles west of the Rhine River and the German border. That evening, he gathered with a few Army ordnance officers in a little inn that was damaged by a shell "but not quite knocked over." One end of the structure was off its foundation, and the building itself was aslope. The evening faded and the lights went out and stayed out. There was no furniture inside the inn, and Bush sat with the officers on the floor in a circle and shared a bottle of Hennessy brandy. It was as close as the Massachusetts Yankee would get to seeing combat.

Night fell, and they passed the bottle in darkness.

An atmosphere of ecstasy and jubilation followed the Allied advance through France, Luxembourg, and Belgium. The march into Paris, on August 25, 1944, one GI said, was "fifteen solid miles of cheering, deliriously happy people waiting to shake your hand, to kiss you, to shower you with food and wine." French colors flew once again on the Eiffel

Tower. "La Marseillaise" was sung impromptu in plazas, avenues, and movie houses. Journalist Ernie Pyle called it "a pandemonium of surely the greatest mass joy that had ever happened." Jeeps were bejeweled with flowers. One American soldier received an astounding sixty-seven bottles of champagne. As the last pockets of the German army were cleared from the city, GIs already decorated with lipstick fired shots in between kisses from Parisian women.

In London, the British War Cabinet settled on December 31 as the likely end date of the war in Europe. According to American intelligence, the "August battles have done it, and the enemy in the West has had it." Military brass began to refocus on the Japanese, shifting Allied troops to the Pacific. Production of tanks and ships was allowed to dip, and contracts for artillery shells were canceled.

When the U.S. First Army and the British XXX Corps entered Belgium in September, they were met with rapture. "What with champagne, flowers, crowds, and girls perched on the top of wireless trucks," a British lieutenant said, "it was difficult to get on with the war."

The port of Antwerp was seized so quickly that the Germans had no time to sabotage it. Thirty miles of wharfs, over six hundred operating hoists, oil storage capacity for a hundred million gallons, and over eight million square feet of warehouse space were captured intact, a strategic coup that left Berlin newly vulnerable. As a final flourish, a captured German garrison was locked in empty animal cages at the local zoo. Nazis squatted glumly on straw in the lion cage.

The march toward Berlin seemed well on schedule.

In December, men of the 101st Airborne were enjoying Paris when military trucks roaming the cobblestone streets with P.A. systems recalled them. A champagne party was interrupted by news of a breakthrough in the front. The Nazis had counterattacked in the Ardennes forest, along the German border with Belgium and the Netherlands. Flabbergasted GIs were dispatched to Belgium.

When the reinforcements arrived outside Bastogne, on December 19, they encountered a shocking spectacle. Like a scattering mob, retreating American soldiers with no rifles or coats were fleeing the front. Staggering, exhausted grunts shouted out frenzied warnings—the babblings of broken men.

"Run! Run! They'll murder you! They'll kill you!" they told the reinforcements. "They've got everything—tanks, machine guns, air power, everything!"

The German counterattack had begun on the morning of December 16. It was a complete surprise and overwhelmingly violent. No one had expected a winter advance through the Ardennes. American troops called it a "ghost front." One division had reported only two casualties in two months' time. "There's nothing out there," one GI told a newly arrived soldier. Life had been quiet, cold, and dull.

Under a cloudy sky, across an eighty-mile front, German artillery and mortars unleashed thunderclaps over the frigid, misty forests. The assault marked the onset of the Battle of the Bulge, the largest and bloodiest action fought by the United States during the war. For ninety minutes, nineteen hundred German artillery pieces fired barrages at Allied positions. American troops, in the words of a Nazi general, were "pulverized by this hail of steel." One GI recalled how his battalion chaplain was "decapitated by an exploding shell as if by the stroke of a guillotine."

German troops—Waffen-SS, Hitler's very best soldiers, and his "people's army" of young boys, convicts, and grandfathers—followed the barrage.

German infantry approaching one American position, a survivor said, resembled "a bunch of wild cattle." They looked like "wild men" who didn't care if they died—as if they were doped. American infantry companies suffered losses as high as 90 percent. Allied headquarters were abandoned. Fleeing troops burned confidential documents and destroyed their weapons. Others simply deserted their artillery pieces, trucks, jeeps, and supplies and ran away on foot.

Some seventy-five hundred American troops were captured, the biggest U.S. surrender in the European war. Within days, the Nazis had knocked right through the Allied front, smashing through the lines in the shape of a "bulge" fifty miles deep.

It was an intelligence failure to rival the Nazis' surprise on D-Day. The Third Reich had managed, over the final months of 1944, to produce their highest output of wartime armaments. Thousands of new fighting vehicles were sent to the Ardennes rather than to the Eastern Front. Within weeks, the Germans had secretly increased their troop strength to a quarter of a million men. In some of the assault areas, the Nazis held a stunning ten-to-one advantage in troops.

The Allies, now hoping to seize the German invasion by the shoulders and push the front back to the "start line," had more planes and plenty of artillery. They also had the fuse. Field artillery crews had been authorized to start using the smart weapon on Christmas. But after the attack, it was cleared for immediate release.

Beginning on December 18, the U.S. First, Third, and Ninth Armies unleashed the most devastating artillery fire German troops had ever encountered. Near the Belgian city of Lutrebois, a German reconnaissance platoon was hit with a barrage of smart fuse "air bursts" and lost half its men. In the Hürtgen Forest, the Eighth Infantry unleashed a so-called Time-on-Target attack, coordinating dozens of rounds from multiple guns to detonate above the Nazi troops simultaneously. A summary of the incident noted that "96 dead Germans were found and their bodies were reported to have looked as if they had been put through a meat grinder."

General George Patton's Third Army, east of Bastogne, let a German tank crew dismount for the night and then "saturated" the area with salvos of exploding shells detonating at the statistically ideal height. Seventeen tanks were found among the dead. At night, Patton's men fired Section T fuses at a Nazi battalion trying to cross the Sauer River. The gadgets' radio signals reflected off the blackened water, detonating and killing, according to Patton, 702 German troops.

A column of several hundred Germans goose stepping along a highway was decimated. As the new year dawned, the 167th Volksgrenadiers were "cut to pieces." The First and Third Armies converged on the tip of the German incursion, and as their artillery crews covered supply routes with smart fuse ammunition, the Nazi invasion receded. Across the Ardennes, the fuse became the weapon of choice, particularly at night, in the open, and through fog. Shells were showered on German troops crossing critical road junctions, bridges, and highways.

German snipers, nestled in the upper stories of buildings, were usefully "cleaned out" from their perches by smart fuse shells that detonated perfectly above the rooftops.

On January 6, the *Washington Post* reported that a "new secret American artillery weapon" had shattered German counterattacks in Belgium and left the snow "blanketed with Nazi dead." The mysterious weapon was left undescribed. Forward observers reported "that these shells drive the Jerries crazy. After a short barrage our infantry usually captures those that are left for they are in too dazed a condition to even fire their rifles." Prisoners of war described daylight fuse attacks as "huge puffs of black smoke that spread death in all directions" and the nighttime assaults as explosions of fire that sent shrapnel to "hell and heaven."

The fuses exploded as low as ten feet in the air, and high explosives sent shrapnel through logs a foot thick. Captured German prisoners complained that open foxholes and slit trenches no longer afforded protection. "PWs agree that it is practically impossible to take cover against these 'Shrapnels,'" an intelligence report from the First Army summarized. "They expressed amazement that soldiers in foxholes up to 50 [meters] away" were wounded. Describing American artillery as "devastatingly accurate," the Germans wondered why the Wehrmacht did not retaliate in kind. The strangely bursting shells seemed to leave "every yard . . . covered by shrapnel." A German captain remarked that he had served four years on every major front, and yet "had never experienced such devastating artillery fire anywhere."

Rumors spread among German troops of a novel "electro shell" or some new sighting technology. They dubbed it "Doppelzünder," or the "Double Detonator," and described it as a terror weapon. They imagined the Americans had invented some "igniter" set off by the earth's magnetism. German infantrymen unwilling to go on patrol and face the deadly fuse were executed for insubordination. Neither side's veterans had ever seen anything remotely like it. "Nearly all [prisoners] are much perturbed by the 'inhumane' new artillery shells," an Army report summarized, "which they think must be illegal because they are so terrible."

Louis Azrael, a journalist from Baltimore, was stationed at the headquarters of an infantry battalion when the fuse was fired. He recalled captured prisoners "coming back in droves, looking absolutely shattered and stunned." A group of German officers wearing campaign ribbons from the brutal Eastern Front insisted that the volleys of shrapnel were impossible to survive. "To a man," Azrael wrote, "they thought that the terrible beating they had taken was due to some new, unbelievably efficient method we had discovered to train our artillery-men."

Brigadier General Edward Ott described the fuse as the greatest ever development in field artillery. "After twenty-seven years of trying to teach my officers to get the proper height of burst," he said, "we have the answer." General Patton declared that the "funny fuse" would revolutionize combat. "When all armies get this shell," Patton wrote, on December 29, "we will have to devise some new method of warfare."

General Eisenhower's headquarters called it "unprecedented."

The brutal struggle, endured by American and German teenagers, extended into January through an insufferable freeze. Some nineteen thousand American troops paid with their lives, and the Nazis lost over twelve thousand. Artillery caused most of the deaths, but the weather could be just as deadly. Near-zero temperatures, heavy snow, ice, sleet, pneumonia, trench foot, frostbite, and amputations were facts of life. Nights, their stillness cushioned by snowfall, were filled with paranoia. Stumps seemed to move in the dark. Snow slabs falling off pines sounded like footsteps. GIs might doze off on a log and wake up with their hair frozen to the timber. Trucks were chipped out of the iced mud every morning. Tanks skated sideways on frosted roads.

Section T's fuse was shot amidst pearly evergreens, virgin snow stained in blood, and old foxholes that resembled butcher shops. (The ground was so firmly frozen that GIs could not dig new ones.) Soldiers never quite slept, exactly. They gripped each other in trenches or stood up in teams of three and leaned on each other, human tripods drifting into a haze. The dead froze and were covered in snow until the spring thaw or else were loaded away onto trucks like cords of wood.

The fuse, an ordnance report summarized, "had a large share in halting the German drive." By February, the Allies had beaten back the counterattack and were poised to enter mainland Germany and march to Berlin. Section T's smart device, General Patton later told Ed Salant, "won the Battle of the Bulge for us."

In desperation, Colonel Max Wachtel attacked the vital port of Antwerp with his remaining arsenal of V-1 buzz bombs. The assault, which began in October, peaked in February after the Nazis lost the fight in Belgium. That month, as many as one hundred and sixty drones were launched daily at the nerve center of the Allied supply lines. But gunners had the tools to stop them. Of 2,394 V-1s that could have hit the crucial port area, American and British antiaircraft guns armed with the fuse shot down 2,183 — a tally of 91 percent. The American 407th AAA Gun Battalion dubbed themselves the "Buzz Bomb Kings."

Near the end of the engagement, which was finished by April, gunners achieved a record success rate of 97 percent against the V-1s. The port of Antwerp was kept safe and humming.

A month later, the war in Europe was over.

31
TUG OF WAR

Colonel Max Wachtel no longer had any desire to pose as the bearded "Martin Wolf." Instead, he got hold of an Opel P4 — a cheap, sturdy workhorse from 1935 with a slogging top speed of fifty-three miles an hour. Disguising it as a refugee car, he stacked beds and household items atop the roof, wrapped a scarf around his jaw, and on May 9, 1945, two days after the surrender of Germany, posing as an emigrant with a sore tooth, he slipped past a British roadblock outside Hamburg.

He could not predict his fate if he were caught, but he did know that in the final months of the war, Allied agents sought to capture him dead or alive. Wachtel's love interest, a Belgian woman, was waiting in Antwerp, the city he bombarded despite knowing he might cause her death. He did not know she was still alive, or that Section T's fuse might have saved her life. For the foreseeable future, he would hide in the Hamburg suburb of Hummelsbüttel. He bought a mobile home from the local circus, a horse, a sheep, a chicken, and a dog, and became a "farmer."

Wernher von Braun, Peenemünde's aristocratic wunderkind rocket scientist, was not captured either. He was so sure that the Americans would want to employ his engineering services that he turned himself in. As Wachtel prepared for a life on the lam, von Braun and other Peenemünde engineers were holed up at a remote resort in the Bavarian Alps, living "royally in a ski hotel on a mountain plateau" and calmly plotting their next moves. After the V-1 was bested by the fuse, von Braun's rocket, the V-2, had been launched at London. While less deadly than the V-1 bombardment, V-2s flew at supersonic speeds — ten times faster than an airplane—and were nearly impossible to stop. His role in the attack was ignored. Von Braun negotiated his surrender with the U.S. Forty-Fourth Infantry Division. In "captivity," he behaved like "a celebrity" or even "a visiting congressman."

His arrogance turned out to be well placed.

Nazi scientists had special value.

Across the Third Reich, Germany's top scientists were tracked down and arrested. Allied intelligence had compiled a blacklist of one thousand specialists they hoped to interrogate. A virus expert's apartment had been raided in Strasbourg. A doctor suspected of biological-weapon work was detained at a U.S. checkpoint. British troops found an aeronautical engineer hiding under a bridge.

As attention turned to the unfinished war in the Pacific, Allied intelligence sought to uncover what German weapons might have been shared with the Japanese. The Germans were far ahead of the Japanese in military technology. But the Japanese were keen to harness what they could of Nazi science, and had assigned "Scientific Intelligence Officers" to liaise with their European confederates. The Nazis had already given the Japanese guided bombs, listening receivers for submarines, and their early Würzburg and Lichtenstein radar units.

What else did the Japanese have? Equipment had to be confiscated and studied, documents recovered before they were burned, and secrets pried from witnesses. Hundreds of officers in dozens of intelligence programs invaded Germany to pillage Nazi science. One such program was the Alsos Mission, sprung out of the Manhattan Project, which aimed to discover the truth of the German nuclear, chemical, and biological warfare programs. Alsos scientists also sought to find out about a potential Nazi proximity fuse. Army Air Forces commander "Hap" Arnold suspected that the Germans might have already used one.

Section T's Ed Salant was assigned the job.

In February 1945, Salant began his new line of work. As the shy James Van Allen became a sheriff and Navy lieutenant, the snappy Salant became a scientific intelligence agent tasked with finding and interrogating Nazi fuse experts. The reason for his unprecedented mission was simple: only another expert could decipher what the Germans were up to. His military commander was Colonel Boris Pash, a Russian-American and staunch anticommunist. Salant joined Pash in Paris, and promptly gained a reputation for indulging in what Pash jokingly called "applied mathematics," or poker, and for knowing his way around the city.

"To see Paris right," Pash said, "just pick Ed as your guide."

Salant and his fuse team, Pash complained, turned out to be an "even wilder bunch" than he was used to. "They're running our operational men off their feet. We call them Salant's circus. . . . He doesn't care how he travels, what he eats or where he sleeps as long as he can hunt clues to the new Nazi proximity fuse."

Salant interrogated hundreds of scientists over several months.

At the start of the war, he and his team learned, the Germans had perhaps fifty projects devoted to building a smart fuse, or "Abstandzünder." The Nazis had actually produced some rocket fuses that never saw service. They experimented with a barometric fuse that detonated from changes in altitude, and tested an acoustic fuse at Peenemünde. There was even work begun on a radio fuse for bombs as early as 1937.

But not a single project on smart fuses for shells was even in production. The microphonics problem had hampered progress badly. They were unable to develop rugged amplifier tubes and, in fact, declared the task impossible. "It must be apparent," Salant noted dryly in an Alsos report, "that we do not have a high opinion of the German attempts to develop proximity fuzes."

"Every one was a fizzle," he said.

The Third Reich made many mistakes that Tuve, Bush, and the Office of Scientific Research and Development had not. Salant found that there was "too little liaison between laboratory and factory and between technician and military, practically no employment of pure scientists, dispersal of effort in too many directions, dissension, distrust . . . and little sense that the country's war needs were primary." Apart from their aeronautical research, another report concluded, the Germans "failed miserably in availing themselves of their scientific manpower."

Salant found Hitler's fuse scientists utterly depraved. None of them believed that the horrors of the concentration camps, freshly exposed for the world to see, were morally wrong. "Their only objection," Salant said, "was that these practices turned people against Germany." And yet, despite being for the most part "incompetent and dishonest," the Nazi scientists "were ready to tell everything they knew. They were convinced that they were ahead of us and were eager to work for us. . . . They would sell to the highest bidder."

"They had the very same attitude toward the Russians."

The Soviets, in fact, *were* the other bidders. After Germany fell, the "gentlemanly collaboration among Allies," a historian noted, "quickly transformed into one of the greatest competitions for information about weapons-related research in the history of war."

The Allied investigation of German scientific laboratories, one Section T physicist suggested, was most aptly described as "looting." Various intelligence missions would arrive at a target, simply "take whatever they could lay hands on," and then move on to the next mark. There seemed to be no liaison with the Russians at all. On the contrary, "there seemed to be a race." The end of the European war kicked off a chaotic sprint for the secrets of Nazi science.

The Alsos Mission revealed that Hitler's nuclear program was not close to producing an atomic weapon. But Boris Pash still raced to find Germany's nuclear physicists and their documents before they fell into Soviet hands.

According to a Section T report, American and British attitudes toward the German experts were split. Some would have been content to simply destroy the Nazi laboratories and shoot the scientists. Others were satisfied to interrogate the specialists and let them go. A third camp "insisted on shipping the German scientists to America and putting them to work in American laboratories."

As early as May 25, 1945, Van Bush and his OSRD confidant, MIT president Karl Compton, were contemplating the delicate problem vis-à-vis Russia. "At least some of the prominent German scientists under Russian control," Compton reported, "are being taken to Russia, treated with even elaborate hospitality and given opportunities to do scientific work." By July, the Joint Chiefs of Staff had chosen their poison. They established a secret program, later called Operation Paperclip, to recruit the "chosen, rare minds whose continuing intellectual productivity" was useful to American national security. Wernher von Braun would go on to a new life across the Atlantic, as would Robert Lusser, one of the designers of the V-1, and Walter Dornberger, a key commander at Peenemünde. The Soviets would do their own "recruiting," forcibly relocating thousands of German experts. The frenetic race was becoming a tug of war, and German scientists were the rope.

While Nazi intelligence failed to unearth the mysteries of the American smart fuse, the Russians were trying to uncover the very same secrets. Section T's device was deliberately kept from the Soviets. When a batch of Russian officers toured the Diver Belt on the British coast, to inspect the Allied defenses against the V-1s, General Frederick Pile ensured that the gadgets were tucked away beforehand and that no gunnery officers made any mention of a special new fuse.

On Christmas Eve, 1944, as the Battle of the Bulge erupted, a Soviet spy with a small mustache, round glasses, and a plump, unassuming face waited at the Horn and Hardart cafeteria on Thirty-Eighth Street in Manhattan, just west of Broadway. As customers perused evening snacks in the vending machines, a Russian spymaster by the name of Aleksandr Feklisov arrived to find Julius Rosenberg sitting by the window. The men placed two packages on the windowsill, like a pair of holiday gifts.

Feklisov had brought an Omega stainless-steel wristwatch and a crocodile handbag for Rosenberg and his wife. Feklisov did not open Rosenberg's "gift," which weighed some fifteen pounds, until he had returned safely to the Russian consulate. Inside he discovered a proximity fuse, "in working order, brand new, smelling of metal and oil." Rosenberg had somehow assembled the secret gadget from spare parts and then smuggled it out of the Emerson Radio and Phonograph manufacturing plant, where he was working as an inspector for the Army Signal Corps.

It wasn't a Section T fuse for shells. Emerson manufactured smart fuses for rockets and bombs for the splinter group working out of the National Bureau of Standards. But the Russians would learn enough from the device to produce their own fuse, which in 1960 would shoot down the American pilot Francis Gary Powers, ripping the wings off his U-2 spy plane some thirteen miles in the sky.

The Cold War had already begun.

For R. V. Jones, the end of the European war brought on a "year of madness" at MI6. He was worn out, and no longer seemed to have the giddy sense of humor that sustained him at the start of the war. But scientific intelligence was a new field of its own, poised to accelerate as the

Cold War began. Six days before the fall of the Reich, he submitted a proposal, "An Improved Scientific Intelligence Service." He stressed that the service should be designed to resemble "a single perfect human mind, observing, remembering, criticizing, and correlating." But his ideas were ignored. A new, convoluted oversight committee would control scientific intelligence. Only one other of its members had any experience.

Jones was outnumbered.

To colleagues, it was a travesty. To Yves Rocard, a friend, professor of physics at the Sorbonne, and former secret agent, it was a catastrophe. The "little British service which has, one can say, won the war," Rocard wrote later, had lost its power. Its leader, the "man who was alone," the Frenchman continued, "matured by six years of harsh experiences, is now 'drowning' among thirteen others, all new and naïve. Far from being the chief and able to lead them, he is only one among thirteen, and can scarcely speak." The newborn field was overrun by opportunists.

Marginalized, Jones left MI6 for good in 1946.

32
DOWNTIME

On May 8, 1945, for a few short minutes, the fingers of the stenographers inside Section T's Applied Physics Laboratory stopped typing. The tools in the machine shop ceased grinding. The loudspeakers of the public address system crackled to life to celebrate a historic moment. The European war had been won.

Tuve's speech, broadcast to staff, was short on bravado.

"This is not a time when words or phrases are helpful," he said. "For most of us, the succession of frightful possibilities and events, crisis after crisis, day after desperate day, which has finally brought us to this passing moment of military victory, has left us almost without the ability to rejoice or to be glad over anything."

"The losses we have sustained," the speakers blared, "not only in killed and wounded, and in material destruction beyond all comprehension, but in human values, in faith, in confidence that civilization has developed social values which transcend the animal in us . . . have left us somber and somewhat confused."

"We all have close blood relatives in the battle zones, and countless friends there. Some of you have suffered losses which are beyond all words; others of us have been spared the sharp pain of a casualty notice, but all of us have sacrificed the finest and best things in our lives to meet the brutal challenge of this war."

The lesson from Europe, he continued, was that they could not relax. "Too many manufacturers and farmers and professors and engineers and businessmen are not in a position to know, as we do, that the German war just missed having a very different length and ending, through mistakes of theirs and not through virtues of our own. Just one team like Section T in Germany, starting as we did in 1940 . . . with just five hundred workers on pilot lines, would have altered the entire air war over Europe." Had Germany fielded guided missiles a few months earlier, the outcome might have been different. Japan's ferocious use of kamikaze tactics at Okinawa in April, he said, proved that Section T's work remained unfinished.

"I am sure you all join me in the conviction that it is best to celebrate this day by being hard at work as a member of this team." The "insane battles of the Germans" were over, he said, but the peace could not be won until Japan fell.

The sober message, which was preceded by a short prayer, was meant to honor the day and keep staff focused. Publicly, Merle couldn't afford to blink.

Privately, he was jubilant.

In one memo, he outlined staffing plans in case of a possible "EMER-GENCY PEACE." It seemed unreal that the war in the Pacific was nearing an end.

What if "Japan quits in summer (or early autumn) of 1945"?

"Wow!" he wrote. "Nobody is ready for such a crash!"

Merle's hairline was receding. He had gained some fifteen pounds since his prewar days. He had worked himself to exhaustion. His "personal life in all areas" had been "second priority for so long," he noted in late 1944, that he found it "difficult to be even embarrassed" about failing to keep up with old friendships.

His daughter, Lucy, was six and finishing first grade. She had drawn a crayon portrait of him in the doorway, returning home perhaps, his smiling head askew on a square torso. After the Nazis surrendered, she delivered a note stating that she made a booth for selling war bonds, and hoped for her parents' business.

Merle filed away these vital documents. Both Tryg, now eleven, and Lucy, he bragged, had "grown into quite competent individuals, each in their own right."

He received a letter from an Italian physicist whose eight-year-old daughter had died of malaria during the war, who hoped that Tuve had never considered him an enemy, and who now sought help installing a cyclotron particle accelerator near his university in Rome. Merle heard from Odd Dahl, his old friend from the Department of Terrestrial Magnetism, who had spent the war years in Norway. Tuve cabled him promptly: "SPLENDID NEWS THAT YOU ARE ALIVE AND WELL." After five years, he wrote Dahl, he found himself "entirely out of touch with our former . . . scientific interests" and unable to speak of postwar plans he had not yet made.

The fate of the Applied Physics Laboratory was another open question. Staff on loan from the Bureau of Ordnance and institutes like the Department of Terrestrial Magnetism would probably return to their rightful jobs. But the lab—which began as five men pulping tiny glass tubes with a handmade cannon and the wrong gunpowder on a Virginia farm—still had a slew of Navy contracts.

APL was a community now, with a vibrant social life.

By June of 1945, the laboratory had its own first-aid center and nurse. There was not only a softball team but a bowling league. An "Applied Physics Employees Association" formed to decide how to spend the money earned from the Coke machine and the sandwich bar. A "safety committee" ensured that the machine work performed in the shop was not overly hazardous. An APL choral club was planning a production of Gilbert and Sullivan, and needed bass and tenor voices. The lab had thrown a formal gala enlivened by newlyweds. Regular dances were held on Fridays, not Saturdays, because of D.C.'s blue laws restricting drinking.

The "Apple Juice" newsletter would even publish a drink recipe for an "APL Special" cocktail, a refreshment of "one jigger of clear, pure unfermented apple juice" and one jigger of Southern Comfort. Readers were advised not to use an "antenna protector cap" for portioning because "it will dissolve" and that, for mixing, a fuse sleeve of cracked ice "is handy, IF you can waterproof the lower end."

The newsletter was also stuffed with diverting gossip: How a new statistician "who has a lot upstairs" was, unfortunately, married, how "one of our gals borrowed an engagement ring from Alice Brown to scare away an old flame!" and how the "stork department continues to do a roaring business around the Lab."

One staffer had "so many women," the newsletter reported, "he just can't choose among them—horrible situation in these days of man shortage."

Romance was a fine distraction from warfare.

By the spring, James Van Allen was back from the South Pacific. He promptly met his future wife. It was a strange courtship, as his biographer put it, with "war fever still in the air and young couples heading into marriage after only a few month's [*sic*] acquaintance." En route to APL, Van Allen got into an accident. A car had backed into his when the light changed—the impact was no more than a bump, with no need to check for damage. But he still glared at the driver as he passed her car. It must have come as a surprise when he arrived at the lab and saw the stranger at the security checkpoint.

"Who do you think you are throwing dirty looks at me?" she said.

Van Allen was smitten. Here was "a gal with real spunk."

Young couples were not the only ones planning their new postwar lives. In early August, an internal survey of one hundred and fifty of the top laboratory personnel discovered that seventy-five of them had already left, or were thinking of leaving. Tuve himself, his vice chairman Larry Hafstad, and the brash stalwart Dick Roberts were all considering exiting within the next one to four months. There was no longer a doubt in anyone's mind that Japan would lose the war. The question was merely how and when.

The pivotal moment arrived on July 16, 1945. That day, Merle was reading passages from Van Bush's report to President Truman on the future of science and government—a booklet entitled "Science: The Endless Frontier." Bush himself was in New Mexico with his chief deputy James Conant. The men lay on a tarp over the damp sand, gripping dark pieces of glass meant to protect their eyes.

A loudspeaker started a countdown to the atomic age.

Ernest Lawrence, the other Boy Scout from Canton, had succeeded in helping build the nuclear weapon that Merle deeply opposed. Bush and Lawrence gawked at the blinding explosion of light, the gigantic rolling fireball flashing scarlet and green, and the eerie purple afterglow. Shock waves rippled across miles, and a flashing heat filled the cold desert morning, as from suddenly "opening a hot oven."

The test, one witness said, was "a foul and awesome display."

Several weeks later, on August 6, Section T's old ally, the Navy's Deak Parsons, pulled on a flak suit aboard the *Enola Gay* high above the Japanese archipelago. As "weaponeer," Parsons armed "Little Boy" before it was dropped on Hiroshima.

To ensure that the nuclear bomb detonated at an ideal height above the ground, it was equipped with a smart fuse built by Section T's Bob Brode.

On September 2, 1945, Japan surrendered.

After the war, over thirteen hundred Section T workers received Certificates of Merit or Navy ordnance awards. Bush, Tuve, and Ray Mindlin, the designer of the rugged tubes, received Presidential Medals for Merit, then the highest civilian decoration in the United States. Merle and Ed Salant were granted the Order of the British Empire, or OBE. (Salant's nickname, after the war, was "Old Bastard Eddie.")

All told, Tuve's group reported 162 inventions and filed 31 patent applications, including one for the radio proximity fuse. Following OSRD policy, staff were awarded crisp one-dollar bills for the rights to each patented invention.

Over the fall of 1945, the smart fuse enjoyed a short time in the national spotlight. Newsreels from Universal, Paramount, and Metro-Goldwyn-Mayer celebrated its feats in theaters. The rush of publicity was newsworthy in itself. Silver Spring, Maryland, home of the Applied Physics Laboratory, was suddenly "on the front pages of newspapers all over the country." Dr. Merle Tuve, "one of the foremost physicists in the world," was now "being billed as one of the nation's No. 1 heroes." Even Lucy and Tryg were thanked in print for playing "their part by understanding that their father could not give them all the attention he wished."

Hundreds of magazines and dailies devoted ink to Section T's ultramodern smart weapon for shells. The fuse was "one of America's greatest secret weapons of the war," a thinking machine that ranked "next to the atomic bomb as the most revolutionary development" in military science. It was an artilleryman's dream that "eliminates human error" in defending against planes, an "ack-ack secret" that "blew down" the enemy, a "mighty midget" that the Nazis sought and failed to uncover and that "greatly shortened the war."

The story became a quaint piece of Americana. Its legend included a secret building "disguised as an old auto paint shop," and a covert range in the desert outside Albuquerque. There were knotty scientific puzzles, and the "marshalling of 100,000 workers—most of them women" who built little gadgets with an untold purpose. There was the winning record against the kamikazes in the Pacific, Nazis in the Ardennes, and V-1s in England. There was the odd device itself, with its "magic eye," radio set smaller than a man's watch pocket, and tubes tinier than thimbles. Gadgetry packed in a case "no larger than a pink milk bottle."

A sturdy thing a baby could lift that licked the buzz bombs.

EPILOGUE

As the decades passed, Section T's contributions gradually faded from collective memory. Overshadowed by the atomic bomb, and then muddled by a messy patent suit, the story of the fuse receded. Based on the early British work on rocket fuses, which were not effective, a fiction took hold that the British had "invented" the smart fuse and merely passed it along to Merle Tuve to manufacture.

Historians implied the hard work was done in England.

Johns Hopkins University hired a reporter to assemble a complete account of that work. But, Larry Hafstad recalled, it was too "embarrassing to publish . . . all complimentary with no historical detail or technical facts." Several later attempts were begun and "somehow or other they all died." Hafstad said he had "long been disappointed" that the full story never saw daylight. Van Bush, too, regretted that Section T's triumphs had never been properly told. (An insider's history published in 1980 did not draw on available archives and was not widely read.)

The characters of the V-1 drama withdrew to their normal lives.

Ed Hatch and the other survivors from the 130th Chemical Processing Company received Purple Hearts. He returned home to Massachusetts, where he lived out the rest of his years. He found a career in broadcasting, hosting a children's TV show called *Uncle Ed's Fun Club*. On the twenty-first anniversary of the July 3, 1944, tragedy at Sloane Court, he became father to a newborn son, David.

Colonel Max Wachtel was found in hiding in 1946, and was not charged with war crimes. He went on to become manager of the Hamburg airport.

Amniarix survived the concentration camp at Ravensbrück. Rousseau was close to death when she was rescued by the Swedish Red Cross, and met her future husband while being treated for tuberculosis. He was a survivor of Buchenwald and Auschwitz. She died in 2017, after a long, full life, at the age of ninety-eight.

R. V. Jones became a professor at Aberdeen University.

Dick Roberts, as a biophysicist, named the ribosome. Larry Hafstad ran research and development for General Motors. Ray Mindlin explored how crystals generated electric charges under stress. Ralph Baldwin studied lunar craters.

Ed Salant went on to a fine research career at the Brookhaven National Laboratory, in Long Island. He divorced his wife in 1946. She remained, until the 1970s, a psychiatric patient at a hospital in Vermont—apathetic, indifferent, and practically mute. He remarried and moved to East Eighty-Seventh Street, in Manhattan.

James Van Allen and Abigail Halsey were married over fifty years, and had five children and six grandchildren in their "squadron." He would recall his Navy service as "far and away, the most broadening experience of my lifetime."

Van Allen remained at the Applied Physics Laboratory exploring "Altitude Research." He was involved with testing the American versions of the V-1 drone and the V-2 rocket developed at Peenemünde. He devised measuring instruments for the first U.S. satellites, became known as the "father of space science," and discovered the belts of radiation encircling the Earth: "Van Allen belts." Working with Wernher von Braun, Peenemünde's rocket expert, he helped birth the space age.

Van Bush stayed on as president of the Carnegie Institution. Not long after the end of the war, he imagined briefly that he had a brain tumor—a "psychosomatic difficulty." He spruced up his basement machine shop on Hillbrook Lane and fiddled with engines, solar power, an original sailboat design, and even a new technique for applying paint. Retiring in 1955, he settled down in his farmhouse in New Hampshire. There, the "super-gadgeteer," as *Newsweek* dubbed him, invented a novel car engine, a military boat powered by hydrofoil, and tiny medical valves for repairing heart ailments. He also raised turkeys and fashioned pipes.

Bush's Office of Scientific Research and Development was dissolved, as planned, in 1947. The military-industrial complex rose directly from its ashes.

Defense work would not be relaxed. The war proved that nightmares were real. In 1946, during the Nuremberg trials for Nazi war criminals, Hitler's minister of armaments Albert Speer warned that within a decade, "ten men, invisible, without previous warning" would be able to "destroy one million people in the center of New York City in a matter of seconds." There was little reason to doubt him.

The Applied Physics Laboratory today employs over six thousand people. With clients ranging from the Department of Defense to NASA, it is America's largest university-affiliated research center. APL scientists, engineers, and technicians contribute to missile defense, naval warfare, computer security, and space exploration.

Scientific intelligence became a major focus of espionage. Soon after its founding, the CIA opened its own department in the field. In 1993, the agency honored Amniarix and Jones personally at a ceremony in Langley, and endowed an award in his name for "scientific acumen applied with art in the cause of freedom."

A number of Section T "graduates" went on to engage in scientific intelligence work. Ed Salant scouted for Russia's nuclear secrets, and Van Allen advised the CIA during the Cold War. Bush was consulted, throughout his career, on secret government projects. Both he and Merle Tuve would have their names publicly linked to Roswell, New Mexico, Area 51, and the discovery of UFOs.

In 1946, Tuve became director of the Department of Terrestrial Magnetism. With Bush's backing, he returned to the pure science research that was his life's true calling. He scrounged surplus explosives off the Navy and set off mammoth terrestrial detonations to study seismic waves. His researchers fitted out Air Force B-29s to fly through thunderstorms and measure atmospheric electricity. They devised a new telescopic instrument to peer vastly deeper into space.

Merle finally bought a house in the countryside, in rural Maryland. With his wife, Winnie, he planted a large vegetable garden. He loaded bales on the tractor and gave the children hay rides. Weekends, he delved "into the dirt" and trimmed the trees in the pasture. The Tuves kept guinea hens, cattle, and even a pig named Phoebe whom Merle "took a shine to" and who was raised as a pet on bread and jam. When the pig was finally butchered, Merle could not bear to eat any of the pork.

He gave up cursing entirely after the war.

He did not reminisce much about the fuse.

When he was interviewed in 1967 about his wartime work, he found it difficult to recall the precise details of those "five or six years of real stress."

"I've lived another life since then," he said.

In the intervening years, the clear-eyed patriotism of the national emergency had retreated somewhat in Tuve's mind. The urgency of the era was dimmed, sheared of context, and a slight sadness had taken root in its place. Merle blamed himself for pushing Section T staff so hard. He spoke to the interviewer, Al Christman, of remorse. Bush, he said, had nightmares over the 1945 firebombing of Tokyo with napalm. Deak Parsons grieved over his role in dropping the bomb on Hiroshima. For his part, Tuve regretted the brutality of the fuse.

"Our job was antiaircraft," he told Christman. Using the fuse against German troops in the Ardennes was never the intention. "I've never visited Germany," Merle said. "There are too many — too many orphans over there on my account."

Christman held a different opinion about the Battle of the Bulge. He was a veteran of it. The interviewer fought amidst the insane Nazi shelling, foxholes, frostbite, amputations, crimson snow, and frozen GIs whose dead bodies lifted like wooden slabs. He knew exactly how the fuse saved thousands of American lives.

"We killed —" Tuve said.

"Well," Christman replied, "let me tell you this —"

He switched off the tape recorder, and told Merle what he'd seen.

ACKNOWLEDGMENTS

First and foremost, this book is a tribute to the men and women of Section T, the scientists and experts of the Office of Scientific Research and Development, and the scientific intelligence officers and agents involved. They were pioneers who struggled and sacrificed during an impossible period, and it has been humbling and gratifying to tell their stories. I hope this book serves as some small memorial to their accomplishments and also as a reminder of their contributions.

It would have been impossible to draw together such a wide variety of archival sources, many of which have never been written of previously, without the guidance, dedication, and good humor of a great number of archivists and librarians. They include Shaun Hardy at the Department of Terrestrial Magnetism, who hosted me graciously, for many weeks, as I dug through DTM's significant holdings. I'd also like to thank Martha Stum, at the Johns Hopkins University Applied Physics Laboratory, for compiling (and releasing) a set of key proximity fuse reports, and Nicole Horstman, for providing me access to APL photographs. Gene Morris at the National Archives and Records Administration and Patrick Kerwin at the Library of Congress were of particular help. Thanks also go to Director Melanie Mueller, Audrey Lengel, and Sarah Weirich at the Niels Bohr Library and Archives, American Institute of Physics; Nora Murphy at MIT; Dena Adams at the International Spy Museum; Sandi Fox at Navy History and Heritage Command; Marlea Leljedal at the U.S. Army Heritage and Education Center; Moira Zelechoski at the U.S. Army Center of Military History; David McCartney and Denise Anderson at the University of Iowa, Special Collections Department; Andrew Webb at the Imperial War Museum; and Claire Corkill at the Council for British Archeology.

Family members of the book's subjects also helped immeasurably. I'm especially grateful to Alex Schneider, grandson of Samuel "Ed" Hatch, who sparked my interest in the 130th Chemical Processing Company. Schneider's tribute to the 130th (londonmemorial.org) includes details of the V-1 strike that were unavailable elsewhere. Contemporary reports of the incident were censored, and Schneider's accounting represented the first carefully done research on the 130th. He also shared several of his grandfather's wartime letters. In addition, I'd like to thank for their help Eric Stern's daughter Carole Switzer; Al Christman's son Neil Christman; Ed Salant's relatives Patsy and Gregor Asch; the son of Jeannie Rousseau (later Jeannie de Clarens), Pascal de Clarens; James Van Allen's children Margot, Cynthia, Sally, and Peter; R. V. Jones's daughter Rosemary Forsyth; and Merle Tuve's daughter, Lucy Comly.

My agent, Eric Lupfer, helped shape the narrative arc of the book, on a structural level, from the outset. I suspect that very few agents go to such lengths. Thanks, Eric. I've enjoyed our creative partnership and friendship immensely. I'm also indebted to Susan Canavan at Houghton Mifflin Harcourt for believing in the project and acquiring the book. My editors at HMH, Bruce Nichols and Olivia Bartz, have both been fantastic and a pleasure to work with. Discreetly and skillfully, they improved the manuscript in ways too numerous to count.

To my friends and family: Thanks for your love and support.

NOTES

ABBREVIATIONS

AIP	Niels Bohr Library and Archives, American Institute of Physics
APLA	Applied Physics Laboratory Archives
DTM	Department of Terrestrial Magnetism Archives
E	entry
JAVA	James Van Allen Papers
LOC	Library of Congress
MAT	Merle Antony Tuve Papers
NARA	National Archives and Records Administration, College Park, Maryland
NDRC	National Defense Research Committee
n.d.	no date
OH	Oral History
OSRD	Office of Scientific Research and Development
RG	Record Group
TNA	The National Archives, United Kingdom, Kew
UI	University of Iowa Archives
VB	Vannevar Bush Papers

PROLOGUE

page

ix *On May 7, 1944:* Chemical Warfare Service, "A Unit History of the 130th Chemical Processing Company," n.d., NARA, RG 407, E NM3-427. Hereafter referred to as 130th Unit History.

had no speedometer: Information on the specific train line that the 130th used, the LMS, can be found in Samuel Edward Hatch, "Of Sorrow and Joy: July 3, 1994," London Memorial, June 16, 2004, http://www.londonmemorial. org/bombing/testimonials/testimonial-hatch.html; David Wragg, *The LMS Handbook: The London, Midland, and Scottish Railway 1923–47* (Stroud, UK: History Press, 2016), chapters 12 and 19.

This London felt otherworldly: Felicity Goodall, "Life During the Blackout," *Guardian,* November 1, 2009; Geoff Manaugh, "How London Was Redesigned to Survive Wartime Blackouts," *Gizmodo,* January 6, 2014; Leonard James, *The Blackout 1939–1945* (Surrey, UK: Bretwalda Books, 2013).

x   *"He comes when he wants":* Mackay, *Half the Battle,* 76.

alter its overwhelming reliance: Jacobsen, *The Deadly Fuze;* Dobinson, *AA Command,* 231.

twenty thousand to one: Maurice W. Kirby, *Operational Research in War and Peace: The British Experience from the 1930s to 1970* (London: Imperial College Press, 2003), 94. Kirby also discusses British efforts to address the rounds-per-bird problem, as does Stephen Budiansky, *Blackett's War: The Men Who Defeated the Nazi U-Boats and Brought Science to the Art of Warfare* (New York: Vintage Books, 2013).

"It would be just a sheer": Douglas Birch, "'The Secret Weapon of World War II': Hopkins Developed Proximity Fuse," *Baltimore Sun,* January 11, 1993.

xi   *highlighted on the targeting maps:* Sam Tonkin, "Hitler's Plans to Destroy London: Rare Map Revealing Germany's WWII Bombing Targets Is Discovered After 75 Years in an Attic," *Daily Mail,* February 24, 2017.

Samuel Edward Hatch: Hatch, "Of Sorrow and Joy."

"D+7": Jones, *Most Secret War,* 417.

with chemical weapons: See chapter 23; Zachary, *Endless Frontier,* 177; Samuel Edward Hatch, "Service Above All," London Memorial, December 28, 2018, http://www.londonmemorial.org/service-above-all-samuel-edward-hatch-in-his-own-words/; "Here Are the Four Main Gases Used in World War I," *Business Insider,* May 21, 2014; Johannes Preuss, "The Reconstruction of Production and Storage for Chemical Warfare Agents and Weapons from Both World Wars in the Context of Assessing Former Munitions Sites," in *One Hundred Years of Chemical Warfare: Research, Development, Consequences,* ed. Bretislav Friedrich et al. (New York: Springer, 2017), 289–333.

xii   *In the first war in history:* Stewart, *Organizing Scientific Research for War,* ix.

American science and Nazi science: The Allies' greatest achievements in military science were the atomic bomb, advances in radar, code-breaking, and the proximity fuse; the Nazis' greatest advancements came in the field of aeronautics. The battle between the fuse and the V-1 marked the clearest tête-à-tête between the rival technologies. Radar, as will be noted, also played a significant role in defeating the V-1.

4,900-pound "robot bomb": V-1s weighed roughly 4,900 pounds, with some variation between different models. Hölsken, *V-Missiles of the Third Reich,* 335.

twelve seconds of silence: Ramsey, *The Blitz,* vol. 3, 386; Longmate, *The Doodlebugs,* 191. There were actually two quite different ways in which V-1s fell on their targets. As noted in D. G. Collyer, *Buzz Bomb Diary,* 131: "Some of these V-1s glided on, with a distinctive whistling sound, others came straight down." The flying bomb was originally designed to go into a power dive that could last well over a minute; a design flaw caused many of them to cut out suddenly and fall in mere seconds.

more than seven thousand: The number was 7,500, according to data from the U.S. Strategic Bombing Survey. Roy Stanley, *V-Weapons Hunt: Defeating German Secret Weapons* (Barnsley, UK: Pen and Sword Aviation, 2010), 207.

faster than its airplanes: See chapter 29.

morale to its lowest point: Mackay, *Half the Battle,* 135.

xiii   *rounds-per-bird ratio of a hundred to one:* E. O. Salant to L. R. Hafstad, memo, "Status of VTF in the U.K.," September 5, 1944, NARA, RG 165, E NM84-421. As Salant noted, this ratio held for "both visual and blind fire."

scientists made shooting planes: As emphasized in chapter 29, the SCR-584 radar and the M9 gun director also helped shoot down the V-1s.

By September, the V-1s: There were three phases of the V-1 attack on England. See Pile, *Ack-Ack,* 286. The main phase, as Pile explained, ran from roughly June 13 to September 1. Pile also testified that the fuse was ultimately responsible for the 100 percent success rates against the V-1s and was the "final answer to the flying bomb." Frederick Pile, "U.S. and Flying Bombs: Use of 'Proximity Fuses,'" *London Times,* April 5, 1946. Capturing the launching sites helped, but as Max Wachtel himself later wrote, London was not out of reach of the V-1s until November 19, 1944. Wachtel, "Unternehmen Rumpelkammer," 110. Also see Dobinson, *AA Command,* 438: "In the period from 13 June to 5 September the Germans had sent 8,617 ground-launched [V-1] missiles toward Britain, together with about 400 delivered by aircraft, the vast majority in both categories toward London. This was by far the greater proportion of all flying bombs directed at Britain during the war . . . comprising 97 per cent of the ground-launched type." Also see Salant et al., "Part Twelve," 16: "In September, after the flying bomb had been overwhelming defeated." Salant is listed as the main "contributor" to this report by Archambault, but Salant drafted it himself.

Known as the world's first: Baldwin, *They Never Knew What Hit Them,* xiii, describes it as the "world's first 'smart' weapon." Foerstner, *James Van Allen,* 51, describes the shell fuse as "the first 'smart' precision weapon in the history of warfare." Christman, *Target Hiroshima,* 98, describes the fuse as the "first mass-produced smart weapon." Christman's and Foerstner's qualifications speak to one of the principle sources of confusion regarding the proximity fuse: There was never a single fuse. The British developed smart fuses for bombs and rockets prior to the American work on its most important application, in shells, and the earliest ideas for a photoelectric smart fuse appear to originate in Sweden in 1935, as evidenced by a patent from L. M. Ericsson. (Earlier iterations, obviously, did not weigh five pounds.) The list of those claiming to have invented the fuse, one might note, is comically long. For more, see R. W. Burns, "Early History of the Proximity Fuze (1937–1940)," *IEE Proceedings* 140, no. 3 (1993): 224–36.

fuse (or fuze): The purer term is *fuze,* meaning "a mechanical or electronic device to detonate an explosive charge, especially as contained in an artillery shell, a missile, projectile, or the like." The term *fuse* is broader, meaning "a tube, cord, or the like, filled or saturated with combustible matter, for igniting an explosive" but also including *fuze* as a secondary definition. While most of the documentation regarding the fuse employs the spelling *fuze, fuse* also appears frequently in Section T wartime records and was used in nearly all the news coverage in 1945. In addition, *fuse* appears in the *New York Times, Washington Post,* and many other contemporary news sources.

five-pound marvel: Different models had different weights. The Mark 32 fuse, designed for Navy 5"/38 guns, weighed 6.81 pounds. The Mark 45, which was used in numerous variants in England and the Ardennes, weighed as little as 2.4 pounds. The Mark 53 weighed 4.28, 4.79, or 4.91 pounds. United States Navy Bomb Disposal School, *U.S. Navy Projectiles and Fuzes* (Washington, D.C.: U.S. Government Printing Office, 1945), 305–17; Bureau of Ordnance, *VT Fuzes for Projectiles and Spin-Stabilized Rockets,* 9–16.

twenty-two million fuses: Baldwin, *They Never Knew What Hit Them,* 116.

among the most decisive: The fuse's contribution to Allied victory in World War II may be judged, in part, by the totality of this book. As to its technological rank, see Baxter, *Scientists Against Time,* 222, which states of the fuse: "Except for the development of the atomic bomb this constitutes perhaps the most remarkable scientific achievement of the war." Baxter's book, a history of OSRD, won the Pulitzer Prize for History in 1947.

The fuse has its accolades: Baldwin, *They Never Knew What Hit Them,* 116; Salant, "Status of VTF in the U.K."; Jacobsen, *The Deadly Fuze.*

xiv *spoke five languages:* Ramsey, *The Blitz,* vol. 3, 296. Other sources report that Rousseau spoke four languages, not five.

"the most politically powerful inventor": Zachary, *Endless Frontier,* 2.

"battle of scientific techniques": Sean Lawson, *Nonlinear Science and Warfare: Chaos, Complexity and the U.S. Military in the Information Age* (London: Routledge, 2014), 41.

I. ZING-A-ZING!

3 *They lived on a pleasant block:* Tuve to county clerk of Lincoln County, April 11, 1941, LOC, MAT; Tuve to commissioner of motor vehicles, February 17, 1941, LOC, MAT. The Tuve family also owned a Peerless automobile at this time: Comly, interview. Tuve lived on 135 Hesketh Street.

his home was about to fill: See chapter 15.

"Viking" blood: "In Memoriam: Merle Antony Tuve," 207.

4 *Merle bought firecrackers:* Robert D. Lusk, "Their Curiosity Produced Atomic Power and Radar," *Sioux Falls Argus Leader,* March 22, 1958. By the time Merle was eleven, he was buying BB pellets; Merle Tuve, "Accounting Book" (manuscript entries, 1913–1917, LOC, MAT), 24.

phase of starting fires: Whitman, interview. The skunk story is also from this source. The tale of Merle putting his brother in the oven is mentioned here and also relayed in Comly, interview. According to Comly, her father was fourteen years old at the time of the oven incident.

born at home: Tuve to county clerk of Lincoln County.

loving but strict: Comly, interview.

"He's awfully inquisitive": Cornell, "Merle A. Tuve and His Program of Nuclear Studies," 33.

5 *"Can we have these batteries":* Tuve, interview by Cornell, session 2.

Merle and Ernie threw trash: Childs, *An American Genius,* 29–30.

Merle found real trouble: Whitman, interview.

6 *dull because it was newborn:* Holand, *Norwegians in America,* 207–11; Olaf Morgan Norlie, *History of the Norwegian People in America* (Minneapolis: Augsburg, 1925), 148.

These included algebra: Cornell, "Merle A. Tuve and His Program of Nuclear Studies," 4.

Tony read Shakespeare: Evans, *Rosemond Tuve,* 7.

7 *Ernie's parents were also worried:* "Scout Training Keeps Boys Out of Mischief," *McKinney Daily Courier-Gazette,* February 15, 1954.

disassembled a lumber wagon: Tuve, interview by Cornell, session 2.

"sparkling in the morning sun": Merle Tuve, "Two Points of View," 1919, LOC, MAT.

"Five thousand light years": Tuve, interview by Cornell, session 2.

Scouts learned knot tying: Tuve, "1913 Boy Scouts' Diary"; Tuve, "1914 Boy Scouts' Diary."

8   *Hugo Gernsback's Electro Importing Company:* Tuve, interview by Cornell, session 2.

"Let's see what they've got anyhow": Ibid.

"Have the folks got there yet": W. C. Gemmill, "A Great Accomplishment: Local Boys Construct and Have in Operation Wireless Outfit Between Canton and Beloit," *Dakota Farmers' Leader,* October 15, 1915.

buried with them: Whitman, interview.

"Merle, now you stop this": Tuve, interview by Cornell, session 1.

9   *Merle's voice turned to bass:* "Memorial Services," *Sioux Falls Argus Leader,* May 31, 1918.

He began a love affair: Whitman, interview. The voice teacher was Etta Ingbertson.

"I haven't observed that": Lew Tuve to Merle Tuve, February 14, 1916, LOC, MAT.

on the gorgeous lawn: Evans, *Rosemond Tuve,* 10–11.

2. EMERGENCY MODE

10   *"I am neither insane":* Merle Tuve, "Notes," April 6, 1925, LOC, MAT.

11   *Weakened by despair:* "Professor Tuve Dies in Minneapolis," *Sioux Falls Argus Leader,* July 22, 1918.

cold funeral procession: Evans, *Rosemond Tuve,* 13–14. Evans also discusses Ida's reaction, Merle's job at the dry-goods store, Lawrence's return to Canton at Christmas, and Merle's desire to marry his voice teacher and join an opera company. By "over the Sioux River" I mean "overlooking" or "above," not "across."

hiding a textbook: Cornell, "Merle A. Tuve and His Program of Nuclear Studies," 44.

"you can go to college": Tuve, interview by Cornell, session 2.

"I am sorry": Merle Tuve to Carl Lawrence, October 18, 1927, LOC, MAT.

12   *"Life job, I guess":* I'm skipping ahead a bit chronologically to give a general sense of Tuve's finances in the 1920s. Merle Tuve, "Debit Statement," March 21, 1925, LOC, MAT.

How to Study: George Fillmore Swain, *How to Study* (New York: McGraw-Hill, 1917). LOC, MAT.

"Don't be scared": Tuve, interview by Cornell, session 2.

"Can you come down": Tuve, interview by Weiner. The quotes "Hold on a minute" and "There's another way" also come from this source and are spliced together from two meetings. For details of this episode, see also Cornell, "Merle A. Tuve and His Program of Nuclear Studies," 151–62; Merle A. Tuve, "Radio Echoes (the Origin of Radar)," in Haskins, *The Search for Understanding,* 73–76. By "short waves," Tuve was referring to waves under one meter long.

some ninety miles up: Tuve and Breit found that the conductive layer varied between fifty and one hundred miles up, depending on the time, according to Cornell, "Merle A. Tuve and His Program of Nuclear Studies," 160. Tuve and Breit finally reported the variation as extending between fifty and one hundred and thirty miles; see G. Breit and M. A. Tuve, "Existence of the Conducting Layer," *Physical Review* 28, no. 3 (1926): 554–75.

13 *won him a Nobel Prize:* Vannevar Bush stated: "He should have won a Nobel." Bush OH, reel 2-B, 131. Edward Victor Appleton received the Nobel Prize in 1947 for similar work. According to a note scrawled at the bottom of Whitman, interview, the Nobel committee apparently told Tuve that since Appleton was much older, this was his last chance, while Tuve was still young and "would have many chances later." Tuve was twenty-five when the results were published.

"only damn trouble": Tuve, interview by Christman.

also the birth of radar: Edward A. Gargan, "Merle Tuve Dies; Pioneer in Radar," *New York Times,* May 21, 1982. Radar did not actually have a single "birth." As the historian Henry Guerlac wrote: "Radar was developed by men who were familiar with the ionospheric work. It was a relatively straightforward adaptation for military purposes of a widely-known scientific technique, which explains why this adaptation—the development of radar—took place simultaneously in several different countries." David Kite Allison, *New Eye for the Navy: The Origin of Radar at the Naval Research Laboratory* (Washington, D.C.: National Research Laboratory, 1981), 58.

detecting aircraft by radio waves: Tuve, "Radio Echoes," 76. See also Tuve, interview by Christman; Christman, *Target Hiroshima,* 40–56.

"Your account with the Bank": Park Savings Bank to Merle Tuve, November 9, 1931, LOC, MAT.

DTM scientists were spoiled rotten: "Empire & Emperor," *TIME,* January 1, 1940. For the scope of DTM projects, see the annual reports at DTM of the Director of the Department of Terrestrial Magnetism, Carnegie Institution of Washington, for the 1930s.

14 *DTM stood proudly atop:* Brown, *Centennial History of the Carnegie Institution*, 7.

"*one of these contented professors*": Cornell, "Merle A. Tuve and His Program of Nuclear Studies," 117.

"*against bits of iron, metals and crystals*": "Michelson, at 76, Outlines 3 New Tasks," *New York Times*, January 18, 1929.

"*Jack-of-All-Trades*": Budiansky, *Blackett's War*, 43.

15 *Merle squatted atop:* These images are drawn from photos at DTM, some of which appear in magazine articles; see "Light on Heavens' High Places: Washington Scientists Propose Flashing 'Finger-Printed' Beam into Skies; Stratosphere—Unknown World Beckoning Explorers," *Literary Digest*, December 21, 1935; "Shattering the Atom: Scientists Bombard Atomic Nucleus with High Voltage Projectiles," Carnegie Institution of Washington press release, November 11, 1928, DTM.

It blew up: Brown, *Centennial History of the Carnegie Institution*, 49.

"*gambling all the time*": Tuve, interview by Cornell, session 1.

fifty-five-foot-high: "Why Is an Atom?," *Upward*, May 15, 1938.

16 *Their boy, Tryg:* Lucy Tuve Comly, interview with the author, June 24, 2019. She was born June 26, 1938; Tryg was born September 29, 1933.

"*The 'feel' of the people*": Cornell, "Merle A. Tuve and His Program of Nuclear Studies," 518.

"*All ok but lonesome*": Winifred Whitman to Merle Tuve, telegram, June 29, 1934, LOC, MAT. The handwritten letters from Winnie and Tryg mentioned are undated but, based on Tryg's very young-looking handwriting, appear to be from the late 1930s.

17 *Tuve on atomic physics:* See, for example, Associated Press, "Blast Powerful Enough to Raze Skyscraper Pictured in New Atomic Experiment," *Ithaca Journal*, December 16, 1939.

"*Free speech and the radio*": Cornell, "Merle A. Tuve and His Program of Nuclear Studies," 517.

3. SCIENTIFIC SPIES

18 *On October 17, 1939:* Duffy, *Double Agent*, 126; Sherryl Connelly, "New Yorker Risks Life as Double Agent in Nazi Underground, Brings Down Duquesne Spy Ring," *New York Daily News*, July 12, 2014.

19 "*He is in fear*": Duffy, *Double Agent*, 127.

to visit his mother: Alan Hynd, *Passport to Treason: The Inside Story of Spies in America* (New York: Robert M. McBride, 1943), 15.

Consolidated Aircraft Corporation: Michael Sulick, *Spying in America: Es-*

pionage from the Revolutionary War to the Dawn of the Cold War (Washington, D.C.: Georgetown University Press, 2012), chapter 18.

"He did not particularly": Duffy, *Double Agent,* 124.

20 *a criminal record:* Rhodri Jeffreys-Jones, *The FBI: A History* (New Haven, CT: Yale University Press, 2007), 104.

Even his Nazi code name: F. H. Hinsley and C.A.G. Simkins, *British Intelligence in the Second World War,* vol. 4 (Cambridge: Cambridge University Press, 1990), 157.

chief of air intelligence: Ladislas Farago, *The Game of the Foxes: The Untold Story of German Espionage in the United States and Great Britain During World War II* (New York: David McKay, 1972), 325.

"I am your uncle Hugo": Duffy, *Double Agent,* 133.

21 *Nazi recruit pondered:* Ibid., 138.

FBI office in Foley Square: Art Ronnie, *Counterfeit Hero: Fritz Duquesne, Adventurer and Spy* (Annapolis, MD: Naval Institute Press), 228.

Hoover would personally brief: Duffy, *Double Agent,* 140.

back of Sebold's watch: Hynd, *Passport to Freedom,* 42–43. The original text was later released as part of the FBI FOIA file on Frederick Duquesne, available at scribd.com. See section 6, 146. Hereafter cited as Duquesne File.

military secrets the Luftwaffe craved: The Abwehr's list contained questions that would be of value to other branches of the Wehrmacht as well.

22 *brass plaque identified:* Roy Berkeley, *A Spy's London* (London: Leo Cooper, 1994), 7–8.

"Here is a present for you": Jones, *Reflections on Intelligence,* 255.

twenty-eight-year-old harmonica enthusiast: Tim Weiner, "R. V. Jones, Science Trickster Who Foiled Nazis, Dies at 86," *New York Times,* December 19, 1997; M.R.D. Foot, "Obituary: Professor R. V. Jones," *Independent,* December 19, 1997.

also a prankster: Jones, *Most Secret War,* 23–25.

23 *Daisy Mowatt:* In *Most Secret War,* Jones spells her name Daisy Mowat. But in Goodchild, *A Most Enigmatic War,* 139 and elsewhere, her name is spelled Mowatt. Despite the fact that she was Jones's secretary, I've used the latter. Jones, *Most Secret War,* 427, also misspells Salant.

playing a trick: Brian J. Ford, *Secret Weapons: Death Rays, Doodlebugs, and Churchill's Golden Goose* (Oxford: Osprey, 2013), 115.

"A proximity fuse is being": Jones, *Reflections on Intelligence,* 267.

24 *"there was now a scientist"*: Jones, *Most Secret War,* 67.

25 *"first watchdog of national defense"*: Reginald V. Jones, "Scientific Intelligence," *Studies in Intelligence* 6, no. 3 (1962): 58.

scientific intelligence did not exist: Goodchild, *A Most Enigmatic War,* 81.

With Jones in place as MI6's "scientific liaison officer," as Goodchild writes, "The British Government could now act upon gathered and considered scientific knowledge of their enemies for the first time in history."

submitted a bold proposal: Jones, *Most Secret War,* 72.

cultivate sources within each stage: Jones suggested an initial focus on accidental leaks and more passive intelligence-gathering techniques.

John Buckingham, deputy director: Ibid., 75.

not receive the staff: Jones, "Scientific Intelligence," 58.

"Peenemünde, at the mouth": Jones, *Reflections on Intelligence,* 333, reproduces the Oslo Report.

"A German scientist who wishes": Williams, *Operation Crossbow,* 21.

26   *copies of the Oslo Report:* Jones, *Most Secret War,* 69–70.

4. THE WIZARD

27   *not supposed to testify:* U.S. Congress, House of Representatives, Second Deficiency Appropriation Bill for 1939, 29; U.S. Congress, Senate, Second Deficiency Appropriate Bill for 1939, 75.

"science has produced a weapon": Allan A. Needell, *Science, Cold War and the American State: Lloyd V. Berkner and the Balance of Professional Ideals* (New York: Routledge, 2012), 68.

"hit a hard punch": Bush, *Pieces of the Action,* 239.

a bit of a hypochondriac: Zachary, *Endless Frontier,* 55.

28   *Carnegie Institution had an endowment:* Ibid., 83, lists the endowment as thirty-three million dollars. To calculate its value today, I converted 1939 dollars to 2018 dollars (the latest available) using the inflation calculator available at measuringworth.com. The measure I used is "relative output," defined as "the amount of income or wealth relative to the total output of the economy. When compared to other incomes or wealth, it shows the relative 'influence' of the owner of this income or wealth has in controlling the composition or total-amount of production in the economy. This measure uses share of GDP."

Bush oversaw investigations: Carnegie Institution of Washington, *Exhibition Presenting Results of Research Activities of the Carnegie Institution* (Washington, D.C.: 1939).

romp through the Yucatán: Vannevar Bush to Humberto Canto Echeverría, LOC, VB.

fast-moving nebulae: Gale E. Christianson, *Edwin Hubble: Mariner of the Nebulae* (Chicago: University of Chicago Press, 1995), 248.

a particular lifestyle: Zachary, *Endless Frontier,* 91.

talked gadgets with Orville Wright: Bush, *Pieces of the Action,* 150.

presidents would call him Van: Bush OH, reel 6-A, 370.

"Wizard": Fred Fassett Jr. to Vannevar Bush, February 6, 1945, LOC, VB.

29 *"Germany is building better planes":* U.S. Congress, House of Representatives, Second Deficiency Appropriation Bill for 1939, 34. In the following dialogue, not all quotes are sequential. For the quote "Germany has now at *one* station," the emphasis is mine.

"You are figuring on war": Ibid., 40.

30 *scientists at NACA felt:* Ibid.

"Why could you not do": Ibid., 66. This is a quote from Congressman George Johnson on February 25, a few days after the first hearing. Bush was not present.

self-described "Cape Cod Yankee": Bush OH, reel 1-A, 1.

Bush met the senators: The congressional record indicates they met in "the committee room." I've assumed the meeting took place in the main committee room, S-127. *Committee on Appropriations, United States Senate, 138th Anniversary: 1867–2005* (Washington, D.C.: U.S. Government Printing Office, 2005), 222.

"Why is it impossible to extend": U.S. Congress, Senate, Second Deficiency Appropriate Bill for 1939, 78.

"Why cannot they be located": Ibid., 79.

"Do you think": Ibid., 94.

31 *Committee members claimed:* Bush OH, reel 10-A, 631.

Bush returned home to his wife: Ibid., reel 11-A, 697A.

"I made a mess of it": Ibid., reel 1-B, 72. For details of the funding debacle, see Alex Roland, *Model Research: The National Advisory Committee for Aeronautics, 1915–1958* (Washington, D.C.: National Aeronautics and Space Administration, 1985), 158–60.

NACA did get the money: Zachary, *Endless Frontier,* 99.

Bush never forgot: Bush OH, reel 1-B, 72.

"asleep on the technical": Zachary, *Endless Frontier,* 108.

defensive weapons were far stronger: Hermann Balck, *Order in Chaos: The Memoirs of General of Panzer Troops Hermann Balck,* ed. David Zabecki and Dieter Biedekarken (Lexington: University of Kentucky Press, 2015), 284, 331; Victor Davis Hanson, *The Second World Wars* (New York: Basic Books, 2017), 468.

civilian deaths were 14 percent: Yahya Sadowski, *The Myth of Global Chaos* (Washington, D.C.: Brookings Institution Press, 1998), 134.

"lightning war": G. Jones, *The Secret War.* I don't intend to imply that Polish cavalry often charged Germany tanks.

32 *largest air raid:* Overy, *The Bombing War,* 62.

a stunning 67 percent: In addition to bombing, genocide, disease, and starvation were obviously also major contributors to this figure.

Committee on Scientific Aids: Bush, *Pieces of the Action,* 33; National Research Council, *Broadcast Receivers and Phonographs for Classroom Use* (New York: Committee on Scientific Aids to Learning, 1939); National Research Council, *Central Sound Systems for Schools* (New York: Committee on Scientific Aids to Learning, 1940); Zachary, *Endless Frontier,* 107.

Army and Navy research laboratories: G. Pascal Zachary, "Vannevar Bush Backs the Bomb," *Bulletin of the Atomic Scientists,* December 1992, 25–26.

"When military men": Vannevar Bush to former president Herbert Hoover, April 10, 1939, LOC, VB.

33 *"the mysterious ways":* Bush, *Pieces of the Action,* 34.

passed his memo: Zachary, *Endless Frontier,* 109.

were also outdated: Bevin Alexander, *Inside the Nazi War Machine: How Three Generals Unleashed Hitler's Blitzkrieg Upon the World* (New York: New American Library, 2010), chapter 2.

"strange defeat": Paul Kennedy, "Piercing the Fog of War," *Los Angeles Times,* September 24, 2000.

34 *four thirty p.m. on June 12:* Zachary, *Endless Frontier,* 112.

an old legal authority: Stewart, *Organizing Scientific Research for War,* 7.

June 14, as Paris fell: Franklin Roosevelt, "Press Conference 652: Press Briefing by President Franklin Roosevelt," White House, June 14, 1940, http://www.fdrlibrary.marist.edu/_resources/images/pc/pc0102.pdf.

35 *"an end run, a grab":* Van Bush, *Pieces of the Action,* 32–33.

a visiting scientist: Alexander King, *Let the Cat Turn Round: One Man's Traverse of the Twentieth Century* (London: CPTM, 2006), chapter 12. King recalls the phrase as *Illigitimis non carborundum.* Another indirect rendering of the plaque has it as *Illegitimos;* see Ray Ewell to Vannevar Bush, October 28, 1948, LOC, VB. Probably it was *Illigitimis;* see Laura Bradley, "Handmaid's Tale: The Strange History of 'Nolite te Bastardes Carborundorum,'" *Vanity Fair,* May 3, 2007.

"distinguished in a learned profession": Cosmos Club, "Extracts from Bi-Laws Relating to New Members," n.d., LOC, VB.

"Van, you talk": Zachary, *Endless Frontier,* 114.

5. SECTION TUVE

36 *W. T. Ensign marched:* The address is 1530 P Street; the grand entrance is on Sixteenth Street. Zachary, *Endless Frontier,* 119; memo, May 16, 1940, LOC, VB; Memo, June 21, 1940, LOC, VB.

Bush had already abandoned: Zachary, *Endless Frontier,* 120.

37 *Bush discovered:* Vannevar Bush to Gano Dunn, memo, June 11, 1940, LOC, VB.

claimed his new propeller: In Bush OH, Bush discusses how he was approached with such ideas throughout the war. These examples come from 1939 and 1941: Dana W. Drury to Vannevar Bush, September 23, 1939, LOC, VB; Arthur Fuchs to Vannevar Bush, January 21, 1941, LOC, VB.

Military leaders had urgent questions: Vannevar Bush, memo, "Pressure Compensation for Aircraft Personnel," July 16, 1940, LOC, VB; Letter to Rear Admiral Harold G. Bowen, February 19, 1940, LOC, VB.

38 *"must distinguish the really practical":* Zachary, *Endless Frontier,* 122.

NDRC met for the first time: NDRC, "Minutes of the First Meeting," July 2, 1940, NARA, RG 227, E NC-138 3.

Bush recruited his friends: Stewart, *Organizing Scientific Research for War,* 10.

Tolman had been so eager: Richard Tolman to Vannevar Bush, May 11, 1940, LOC, VB; Vannevar Bush to Richard Tolman, May 20, 1940, LOC, VB; Vannevar Bush to Frank Jewett, June 5, 1940, LOC, VB.

Conant, president of Harvard: I. Bernard Cohen, "James Bryant Conant," *Proceedings of the Massachusetts Historical Society* 90 (1978): 112.

Division C fell to Jewett: Oliver E. Buckley, "Frank Baldwin Jewett, 1879–1949," *Biographical Memoirs of the National Academy of Sciences* 27 (1952): 240.

Compton, president of MIT: Daniel J. Kevles, *The Physicists: The History of a Scientific Community in Modern America* (New York: Alfred A. Knopf, 1978), 253.

39 *"Committee on Uranium":* Stewart, *Organizing Scientific Research for War,* 12; Richard G. Hewlett and Oscar E. Anderson Jr., *The New World, 1939/1946: Volume I of a History of the United States Atomic Energy Commission* (University Park: Pennsylvania State University Press, 1962), 24.

"Let's not do": Tuve, interview by Christman.

headlines now feeding: Otto D. Tolischus, "Finns Smash a Red Division; Halt Move to Cut Country; Nazi Drive on Britain Seen," *New York Times,* January 1, 1940; K. J. Eskelund, "50 Finns Killed, 100 Hurt, in Wide Russian Air Raids; British Down 3 Nazi Planes," *New York Times,* February 2, 1940; wireless service, "Nazis Stress Plan of 'Total Warfare,'" *New York Times,* February 8, 1940; Hanson W. Baldwin, "Sea Power and Air Power Content for the Control of a Continent," *New York Times,* April 14, 1940.

a scientific attaché: Baxter, *Scientists Against Time,* 116.

"Don't kid yourself": Tuve, interview by Christman.

at Van Bush's request: Richard Tolman to Vannevar Bush, memo, "List of Personnel Prepared by Dr. Merle Tuve," July 5, 1940, LOC, MAT.

a quick ranking: Ibid.; Merle Tuve, memo, "An Immediate Preliminary List of Personnel Suggested for Consideration for Section Leaders and Technical Aides, NDRC," July 2, 1940, LOC, MAT.

40 *his thirteen-page list:* Tuve delivered it through Tolman.

he did some reconnaissance: Tuve, interview by Christman.

41 *William Blandy:* James P. Rife, *The Sound of Freedom: Naval Weapons Technology at Dahlgren, Virginia 1918–2006* (Washington, D.C.: Department of the Navy, 2007), 65, 72.

"who knew something": Tuve, interview by Christman.

"chew nails and": Ibid.

airpower now posed: Norman Friedman, *Naval Anti-Aircraft Guns and Gunnery* (Barnsley, UK: Seaforth, 2013), chapter 1.

Germany had tested: Ibid., 31.

only airplanes could possibly protect: Douglas Smith, ed., *One Hundred Years of U.S. Navy Air Power* (Annapolis, MD: Naval Institute Press, 2013), 173; Norman Friedman, *Fighters over the Fleet: Naval Air Defence from Biplanes to the Cold War* (Barnsley, UK: Seaforth, 2016), 98.

42 *pressurized catapult systems:* Paul E. Fontenoy, *Aircraft Carriers: An Illustrated History of Their Impact* (Oxford: ABC-CLIO, 2006), 42; William F. Trimble, *Wings for the Navy: A History of the Naval Aircraft Factory, 1917–1956* (Annapolis, MD: Naval Institute Press, 1990), 174; Thomas C. Hone et al., *Innovation in Carrier Aviation* (Newport, RI: Naval War College Press, 2012), 29.

carriers operated behind: Friedman, *Naval Anti-Aircraft Guns and Gunnery,* 16.

the Mitsubishi Zero: James D'Angina, *Mitsubishi A6M Zero* (Oxford: Osprey, 2016), 4.

The 1938 specifications: United States Navy, *Ordnance Specifications: Manufacture and Inspection of 45-Second Mechanical Time Fuze Mark XVIII-2* (Washington, D.C.: Bureau of Ordnance, 1938), LOC, MAT.

thirty grains of explosive: The silk bag was inside the fuse itself. It initiated the high explosives in the shell, which then exploded the round into shrapnel.

within a window of one-fortieth: This figured is derived from Dick Robert's calculation for a shell moving 2,000 feet a second past a 50-foot target; see Roberts, "Development of the Proximity Fuze," 1.

"bombarded and barraged": Baldwin, *The Deadly Fuze,* 60.

43 *ideas were practical:* I'm talking about fuses for shells, not bombs or rockets.

Sunday in August 1940: Philip H. Abelson, "Merle Antony Tuve," *Biographical Memoirs of the National Academy of Sciences* 70 (1966): 411.

space shuttle during launch: National Research Council, *Preparing for the High Frontier: The Role and Training of NASA Astronauts in the Post-Space Shuttle Era* (Washington, D.C.: National Academies Press, 2011), 84.

Rockets had to tolerate: Applied Physics Laboratory, *The "VT" or Radio Proximity Fuze,* 23.

44   *up to twenty thousand g's:* Boyce, *New Weapons for Air Warfare,* 122.

"had little or no prospect": Baldwin, *The Deadly Fuze,* 57–58; Gilbert C. Hoover, "Subject Memorandum for Chief of Bureau: Conference with Dr. Tolman and Dr. Lauritsen of National Defense Research Committee, 12 August, 1940," August 17, 1940, NARA, RG 227, E NC-138 1.

formally established on: Applied Physics Laboratory, *The "VT" or Radio Proximity Fuze,* 19.

Bush had been sick: Zachary, *Endless Frontier,* 117; W. M. Gilbert to Theodore Francis Green, August 14, 1940, LOC, VB; W. M. Gilbert to Earl O. Benson, August 19, 1940, LOC, VB.

Nobel Prize–winning chemist: Harold Urey to Richard Tolman, July 31, 1940, LOC, VB.

Merle had a different: Roberts, "Autobiography," 1.

On September 3, Bush: Vannevar Bush to Merle Tuve, September 3, 1940, DTM.

6. CHOICE OVERLOAD

45   *"let's shoot them straight up":* Tuve, interview by Christman.

An old howitzer: Brown, *Centennial History of the Carnegie Institution,* 111; R. C. Tolman to M. A. Tuve, memo, "Authority to transfer one 37 mm gun to Section S, Division A," April 15, 1941, LOC, MAT. The gun was a 37 mm M1916.

fired a one-pound round: War Department, *Basic Field Manual Volume III, Basic Weapons Part Four: Howitzer Company* (Washington, D.C.: United States Government Printing Office, 1932), 2, LOC, MAT. According to the manual, rounds weighed slightly more than one pound, but of course Tuve and Section T were firing modified rounds.

faintest of nods: Baldwin, *The Deadly Fuze,* 71.

Gardiner Means: Rick Allen, "The Garden View from Meadowland Park," *Washington Post,* April 9, 1992; Henry Porter to John Fleming, memo, "T-4 Proving Ground on Dr. Means' farm," January 29, 1941, LOC, MAT.

"Quite frankly": Bush OH, reel 7-A and 7-B, 438.

46 *Tuve's team consisted:* Applied Physics Laboratory, *The "VT" or Radio Proximity Fuze,* 26.

Roberts, in mid-August: Roberts, "Autobiography," 1; Roberts, "Development of the Proximity Fuze," 3.

godfather to his child: Dick Roberts to Merle Tuve, undated, LOC, MAT.

Roberts ventured down: Roberts, "Autobiography," 1.

47 *"the duration of the acceleration":* Ibid., 2.

"put a vacuum tube": Tuve, interview by Christman.

looked up Dynamite: Baldwin, *The Deadly Fuze,* 69.

"Blasting and Sporting Powder": Baldwin, *The Deadly Fuze,* 70–73. Baldwin discusses the florist on pages 72 and 73 as if they were different suppliers. Shenk was the florist.

48 *physicists built their own:* Roberts, "Autobiography," 2.

"powdered glass": Tuve, interview by Christman.

common Navy gun barrel: I'm referring to the 5″/38 Navy guns.

clock fuses were three: I'm referring to the Mk 18 mechanical time fuse, fit for 5″/38 guns; United States Navy Bomb Disposal School, *U.S. Navy Projectiles and Fuzes,* 279.

smaller than a paper clip: Applied Physics Laboratory, *World War II Proximity Fuze,* 181; "Sylvania Subminiature Tube Program," *Signal* 6, no. 4 (1952): 25.

49 *In early September, Tuve:* Baldwin, *The Deadly Fuze,* 62; Roberts, "Autobiography," 3.

assist from London: R. C. Tolman, "Memorandum by R. C. Tolman on Discussion of Proximity Fuses at a conference with members of the British Scientific Mission and representatives of the Army and Navy, held at 10:00 A.M. September 17 in the Munitions Building," September 18, 1940, NARA, RG 227, E NC-138 1; W. R. Furlong, memo, "Notes on Conference with British Mission September 17, 1940," December 19, 1940, NARA, RG 227, E NC-138 7. The British also announced that they had developed a thyratron trigger tube that could withstand eighteen thousand g's. Apparently on this basis, claims have been made that the British originated rugged tubes. However, thyratrons did not present the same microphonic problems and were not nearly as fragile as amplifier and oscillator tubes. According to Baldwin, *The Deadly Fuze,* 113, "the oscillator and amplifier tubes had about the same failure rate and the thyratron about one-fourth of the rate." In short, the trigger tubes were four times sturdier. This was why the German attempts to build a smart fuse, uncovered later by the Alsos Mission, focused on radio-sensitive thyratrons and avoided amplifiers entirely: it would allow them to skirt the microphonics problem and much of the rugged-tube problem. The British Pye Radio design for a pulse fuse, which turned out to be unrealistic, likewise skirted much of the size and microphonics problems by excluding the oscillator from the design. Another point about the British thyratron: It may have been based on the tube delivered to R. V. Jones with the famous Oslo Report. The sample Jones received was a German-developed thyratron tube for shells, and as he notes in *Most Secret War*, 70, "by now [Admiralty Research Laboratory] had found that the electronic tube was much better than anything that we had made in the same line." Since the German tube arrived in November of 1939, it seems reasonable to speculate that by September of 1940, the British thyratron was at least partly based on the superior German tube. As noted in Applied Physics Laboratory, *The First Forty Years*, 2, "As early as May 1940, the British had fashioned a radiosensitive thyratron trigger." For further details of how early British efforts were inspired by German designs, see Jones, *Reflections on Intelligence*, 308. For a description of Pye Radio's early work on rugged tubes, see Mark Frankland, *Radio Man: The Remarkable Rise and Fall of C.O.* (London: Institute of Engineering and Technology, 2002), 106–11. As Frankland notes on page 109, Pye Radio and GEC appear to have been unable to get their rugged pentodes to work reliably in the British pulse fuse until August 6, 1941. Section T's successful oscillator tests occurred earlier, on April 20 and May 8, 1941.

Two days later: Applied Physics Laboratory, *The First Forty Years*, 2.

an Australian physicist: William Butement, who was born in New Zealand.

50   *saved Section T time:* I'm skeptical of claims, such as those made Frederick Pile, that Section T would not have been able to develop the fuse without British assistance. In Tuve, interview by Christman, Tuve says: "We heard some rumors of circuits they were using in the rockets over in England, then they gave us the circuits, but I had already articulated the thing into the rockets, the bombs and shell." Also see Applied Physics Laboratory, *The "VT" or Radio Proximity Fuze*, 23, where it notes that before British information arrived, Section T "re-invented several British technical developments before they learned anything except the fact that the British were working on a radio fuze of some kind which was reported to depend on the Doppler effect." As far as the circuitry itself, as Tuve wrote in 1944: "The one outstanding characteristic in this situation is the fact that success of this type of fuze is not dependent on a basic technical idea — all of the ideas are simple and well known everywhere." Of cooperation with the British, he wrote: "We have given them complete information on our entire experience since early 1941, with British

scientific and industrial men stationed here in our laboratory continuously but they cannot yet make successful AA fuzes." M. A. Tuve to Charles F. Fell, September 28, 1944, NARA, RG 165, E NM84 421. One Section T report regarding the "Radio Proximity Fuze for AA Shell" stated that "the status of development is similar to that of the corresponding British work." Tuve, in a handwritten note in red pencil dated February 11, 1942, added: "Like hell it is, or ever has been!" Richard Tolman to Section Chairmen, memo, "Project Summaries," February 9, 1942, DTM. Vannevar Bush likewise dismissed British claims of having "invented" the fuse and merely delivered it to the U.S. to produce: "Churchill in his book says that the British invented the proximity fuse and turned it over to the Americans to manufacture. Churchill knew better than that. That kind of statement from Churchill was the kind of thing that makes the chaps in this country who really worked hard on the thing a bit furious." Bush OH, reel 7-A, 436-A. Also, in the OSRD London report of fuse efforts, it reads: "Since the Section T program developed so much more rapidly than that of the British, the technical contributions were almost entirely eastward and liaison was centered primarily in British agencies in the United States." Salant et al., "Part Twelve," 5. Many similar examples could be cited. The lawsuit referenced in the epilogue has only clouded matters. Clearly it was an excellent result for the U.S. Navy that patent rights were assigned to Butement et al., as it saved them millions in royalty payouts. The details were explained by a Navy patent lawyer in Brennan, "The Proximity Fuze." The gist is this: A U.S. inventor, Russel H. Varian, claimed to have conceived of a radio fuse before Section T. Since the U.S. government owned the rights to the British patent of Butement et al. (conceived still earlier), then if Tuve claimed that Section T's design was based on Butement's, Varian would lose and the U.S. Navy would save millions of dollars. Tuve clearly believed this to be a righteous outcome and worked to make it reality. Remember, Butement's wiring (for rocket fuses) itself was rudimentary, based on widely known concepts. By claiming that Section T's fuse circuitry "descended" from Butement's, Tuve ensured Varian would lose and the U.S. Navy would be protected from financial liability. This is the basic story explained by Brennan, minus a clear distinction between rocket and bomb fuses, which were relatively unimportant, and shell fuses, which were successfully developed only by Section T—and also minus an understanding of the simple nature of the circuitry as compared to the engineering challenges of the shell fuse. Tuve also left clues as to his view of the patent dispute and the true story. In a note on the patent interference, beside a line describing how the Tizard mission delivered information on the radio fuse to Section

T, Tuve wrote: "but not the one we made to work!" In the section regarding how Cockcroft delivered a version of Butement's wiring to Section T, Tuve wrote: "but not the one we used! This one <u>from</u> England was <u>not</u> successful, but the result (against Varian) is correct." United States Patent Office, *Butement et al. v. Varian,* Patent Interference Hearing No. 86,648, April 19, 1964, DTM.

Section T ran preliminary tests: Applied Physics Laboratory, *World War II Proximity Fuze,* 29.

Their budget was twenty-five thousand dollars: R. C. Tolman to V. Bush, memo, August 27, 1940, 9, NARA, RG 227, E NC-138 3.

"reasonable balance": John Fleming to Merle Tuve, September 5, 1940, DTM.

51   *All Merle had:* Merle Tuve to John Fleming, September 9, 1940, LOC, MAT; John Fleming, memo, November 4, 1940, DTM.

"full time, if required": John Fleming to Merle Tuve, September 25, 1940, LOC, MAT.

wish list of 31 scientists: Merle Tuve, memo, "Suggested for Clearance," September 13, 1940, LOC, MAT.

"Our own men (all)": Tuve, "Notebook II," September 23, 1940.

"busy days and nights": Merle Tuve to Ed Condon, October 23, 1940, LOC, MAT.

six men, full-time: John Fleming, memo, November 4, 1940, DTM.

wrong kind of powder: Roberts, "Autobiography," 2.

52   *"record for a fifty-yard dash":* Baldwin, *The Deadly Fuze,* 72.

a briar patch: Boyce, *New Weapons for Air Warfare,* 129.

Merle had thirteen projects: Merle Tuve to Director Fleming, November 4, 1940, DTM.

by way of a rumor: Foerstner, *James Van Allen,* 53; "Model Airplane Tube Led to Proximity Fuse," *Boston Globe,* January 5, 1964.

experts at Raytheon: Applied Physics Laboratory, *The "VT" or Radio Proximity Fuze,* 28.

53   *"Important use your services":* Tuve, "Notebook IV," November 6, 1940.

Van Bush gave Section T: Applied Physics Laboratory, *The "VT" or Radio Proximity Fuze,* 25.

"structurally special tubes": Tuve, "Notebook IV," November 12, 1940.

instructions to Raytheon: Ibid., November 6, 1940.

four hundred revolutions a second: Boyce, *New Weapons for Air Warfare,* 114.

experts thought the microphonics: Ibid., 122.

his fifteen or so scientists: Applied Physics Laboratory, *The "VT" or Radio Proximity Fuze*, 29; Merle Tuve, "Notebook V," December 12, 1940.

54 *leaving behind a couplet:* Boyce, *New Weapons for Air Warfare*, 168–69.

7 · UNCANNY DAYS

55 *R. V. Jones expected:* Jones, *Most Secret War*, 154.

exceedingly clever physicist: Goodchild, *A Most Enigmatic War*, 398.

"irretrievably ugly": Young, *Enigma Variations*, 74.

"ugliest country house": Russell-Jones and Russell-Jones, *My Secret Life in Hut Six*, 116. This is a secondhand summary of an overheard remark.

56 *modest structure:* Gordon Welchman, *The Hut Six Story* (New York: Mc-Graw-Hill, 1982), 10.

"operators [were] constantly": Young, *Enigma Variations*, 74. This refers to circa 1942.

codebreakers passed on: Jones, *Most Secret War*, 92.

series of vague clues: G. Jones, *The Secret War*.

57 *"One cannot possibly get":* Jones, *Most Secret War*, 96.

"For twenty minutes": Churchill, *Their Finest Hour*, 340.

another system of beams: Jones, *Most Secret War*, 135–45.

58 *good enough airborne radar:* David T. Zabecki, ed., *World War II in Europe: An Encyclopedia* (New York: Taylor and Francis, 1999), 1436.

guns stationed throughout London: Mortimer, *The Longest Night*, 17, 109; Paul Rabbitts, *London's Royal Parks* (London: Shire Publications, 2014), chapter 6.

"largely wild and uncontrolled": Mortimer, *The Longest Night*, 142.

flat-out terrifying: See Mackay, *Half the Battle*, as well as Pile, *Ack-Ack*, 107: "We were open to air attack in a way we had never dreamed of."

some five billion dollars: Gerhard L. Weinberg, *A World at Arms: A Global History of World War II*, 2nd ed. (Cambridge: Cambridge University Press, 2005), 143.

thousands of large blimps: Ramsey, *The Blitz*, vol. 2, 153.

Germany lost just 1.5 percent: Robert Tombs, *The English and Their History* (New York: Alfred A. Knopf, 2015), 712.

within seventy miles: Overy, *The Bombing War*, 143.

fourteen hundred shelters were declared: Ibid., 144–45.

59 *had no latrines:* Amy Helen Bell, *London Was Ours* (London: I. B. Tauris, 2008), 59.

hundred thousand people sheltered: Overy, *The Bombing War*, 148.

Twenty thousand children: Ibid., 136, 142.

It was four thirty p.m.: Jones, *Most Secret War,* 154.

60   *"bandaged and with pronounced":* Ibid., 155.
Jean and Marie Rousseau: Sebba, *Les Parisiennes,* 86; Robert Gildea, "Jeannie Rousseau Obituary," Guardian, September 6, 2017.

61   *near-photographic memory:* William Grimes, "Jeannie Rousseau de Clarens, Valiant World War II Spy, Dies at 98," *New York Times,* August 29, 2017.
The hospitals in Dinard: D'Albert-Lake, *An American Heroine in the French Resistance,* 42–43.
twenty-one thousand British troops: J. E. Kaufmann and H. W. Kaufmann, *Hitler's Blitzkrieg Campaigns: The Invasion and Defense of Western Europe* (Conshohocken, PA: Combined Books, 1993), 281.
Clocks in Dinard: D'Albert-Lake, *An American Heroine in the French Resistance,* 48.

62   *"We are becoming 'Nazified'":* Ibid., 52.
under Nazi control: Some of the censors were sympathetic to the Resistance. See Hugh Schofield, "'All Honor to You'—the Forgotten Letters Sent from Occupied France," *BBC News,* May 10, 2014.
mayor of Dinard: Rossiter, *Women in the Resistance,* 129; Sebba, *Les Parisiennes,* 86.
There had been rumors: D'Albert-Lake, *An American Heroine in the French Resistance,* 64. At least, this was the rumor in the nearby town of Pleurtuit.
"She doesn't want anything": Oliver Holmey, "Jeannie Rousseau, Spy for the French Resistance," *Independent,* August 29, 2017.
a crisp white shirt: David Ignatius, "After Five Decades, a Spy Tells Her Tale," *Washington Post,* December 28, 1998.
"wanted to be liked": Ibid.
Her German was fluent: de Clarens, interview by Ignatius.
Dinard was the headquarters: Brown, *"C": The Secret Life of Sir Stewart Menzies,* 521.

63   *embark from Cherbourg:* McKinstry, *Operation Sea Lion,* chapter 13.
In September of 1940: Ignatius, "After Five Decades, a Spy Tells Her Tale."
Jacques Cartier Prison: Corinne Jaladieu, *La Prison Politique Sous Vichy: L'Exemple des Centrales d'Eysses et de Rennes* (Paris: L'Harmattan, 2007), 47; Geoffrey P. Megargee and Joseph R. White, *The United States Holocaust Memorial Museum Encyclopedia of Camps and Ghettos, 1933–1945,* vol. 3: *Camps and Ghettos Under European Regimes Aligned with Nazi Germany* (Bloomington: Indiana University Press, 2018), 210–11.

"Nothing, Papa": Ignatius, "After Five Decades, a Spy Tells Her Tale."

64 *"whether the dangerously obtained"*: Jones, *Most Secret War*, xiv.

8. PEENEMÜNDE

65 *most advanced weapons research:* Georg, *Hitler's Miracle Weapons*, 12.

Pomeranian deer: Middlebrook, *The Peenemünde Raid*, 15.

three bridges: Petersen, *Missiles for the Fatherland*, 66.

Since 1936, colossal efforts: Campbell, *Target London*, 40.

German military engineers: Petersen, *Missiles for the Fatherland*, 60.

"a forgotten paradise": De Maeseneer, *Peenemünde*, 49.

66 *four hundred and forty-seven inhabitants:* Petersen, *Missiles for the Fatherland*, 60.

In 1937, the first three hundred and fifty: Walpole, *Hitler's Revenge Weapons*, 15.

To reach it by train: De Maeseneer, *Peenemünde*, 72–73.

At its height: Petersen, *Missiles for the Fatherland*, 90.

four movie theaters: There were four cinemas on Usedom.

"to spiritually unite German": Petersen, *Missiles for the Fatherland*, 90.

67 *conscripted Polish prisoners:* Campbell, *Target London*, 44.

an estimated five hundred and fifty million: De Maeseneer, *Peenemünde*, 79. The exchange rate for reichsmarks and dollars between 1936 and 1940 ranged from 2.48 to 2.5. I used a rate of 2.49, then used a "real cost" calculator to adjust that two hundred and twenty million dollars in 1940 currency to 2018 dollars, the latest year data was available. More information can be found at www.measuringworth.com and in R. L. Bidwell, *Currency Conversion Tables: A Hundred Years of Change* (London: Rex Collings, 1970), 23.

died by hanging: Petersen, *Missiles for the Fatherland*, 85.

twenty-eight-year-old rocket expert: Middlebrook, *The Peenemünde Raid*, 21.

Hitler's SS: Andrea Sommariva, *The Political Economy of the Space Age: How Science and Technology Shape the Evolution of Human Society* (Wilmington, DE: Vernon Press, 2018), 15.

68 *"exerted a strange":* King and Kutta, *Impact*, 58.

Hitler's "superweapons": The V-weapons are distinct from but are usually included in lists of Hitler's "Wunderwaffe." See, for example, Arnold Blumberg, "Hitler's Super Weapons, Explained: Jets, Bombers, Rockets and Deadly Tanks," *National Interest*, August 18, 2018. The V-3 was not developed at Peenemünde.

The Paris Gun: De Maeseneer, *Peenemünde*, 52–53.

advanced experimental wind tunnel: King and Kutta, *Impact*, 50.

one and a half million: De Maeseneer, *Peenemünde,* 55. The wind tunnel was estimated to cost three hundred thousand reichsmarks. See the earlier notes on currency conversion.

69   *reach speeds of Mach 4.4:* King and Kutta, *Impact,* 50.

German chemists had devised: De Maeseneer, *Peenemünde,* 69.

In February of 1940: King and Kutta, *Impact,* 80.

70   *peculiar and completely novel:* Paul Schmidt had begun experimenting with a pulsejet engine as early as 1928; see Hölsken, *V-Missiles of the Third Reich,* 50.

a spark plug ignited: The spark plug was used only for starting the engine, after which point the engine was "self-igniting."

fifty times a second: Many sources claim the figure is around fifty; some say several dozen times a second. Others report it as between forty and forty-five times a second.

Argus began flight tests: Hölsken, *V-Missiles of the Third Reich,* 49–50.

9. DON'T SLOW DOWN

71   *At 2:45 a.m., March 15:* Merle Tuve to John Fleming, March 15, 1941, LOC, MAT.

"Confidential and secret": Merle Tuve, memo, "Secrecy Notice," March 28, 1941, LOC, MAT.

72   *enter the Cyclotron Building:* Merle Tuve, memo, "Notice," February 10, 1941, LOC, MAT; John Fleming, "Memorandum for Miss Evans, Miss Griffin and Messrs. Price and W. W. Wright," February 8, 1941, LOC, MAT.

Acoustics in the Cyclotron: Merle Tuve to Raymond Burrows, February 17, 1941, LOC, MAT; Henry Porter, memo, "Immediate Needs in Connection with Realignment," May 10, 1941, LOC, MAT; Merle Tuve to John Fleming, memo, "Some Items Needed in Cyclotron Building," February 10, 1941, LOC, MAT.

conceal their workbenches: Merle Tuve to John Fleming, memo, "Some Items Needed in Cyclotron Building," February 10, 1941, LOC, MAT.

to hang their coats: Merle Tuve to John Fleming, December 26, 1940, LOC, MAT.

Ceiling sprinklers: Henry Porter to John Fleming, July 28, 1941, LOC, MAT.

"adequate supply of pencils": Merle Tuve to John Fleming, memo, "Stationary [*sic*] Supplies," February 14, 1941, LOC, MAT.

73   *Sears, Roebuck workbenches:* Merle Tuve to Section T File, memo, "Sears Roebuck Benches, April 17, 1941," April 17, 1941, LOC, MAT.

poles for opening: Merle Tuve to John Fleming, memo, "Window Poles," February 15, 1941, LOC, MAT.

meetings around a lathe: Merle Tuve to John Fleming, memo, "Additional Furniture Need for Defense Group," April 5, 1941, LOC, MAT.

By February of 1941: Tuve, "Notebook VII," January 22, 1941; Merle Tuve, memo, February 9, 1941, LOC, MAT; Merle Tuve to John Fleming, memo, "Personnel Situation as of February 17, 1941," February 18, 1941, LOC, MAT. Tuve's notebook entry lists twenty-seven men. The February 9 memo lists more but many of these are clearly aspirational, as clarified in Tuve's February 18 memo.

Equipment to build and test fuse parts: John Fleming to Richard Tolman, August 2, 1941, LOC, MAT. The letter encloses three lists of Section T property and equipment.

the "Powder Room": Merle Tuve, memo, "Notice Regarding Explosives," April 3, 1941, LOC, MAT.

"liable to throw particles": Merle Tuve, memo, "Following Are Two Rules That Must Be Enforced," June 17, 1941, LOC, MAT.

"of considerable size": Merle Tuve to John Fleming, memo, "First Aid Kit for Cyclotron Building," February 10, 1941, LOC, MAT.

"a <u>not too violent</u> form": Merle Tuve, memo, "Baseball," April 14, 1941, LOC, MAT; Merle Tuve, memo, "Baseball Notice," April 16, 1941, LOC, MAT.

74 *aptly named Jack Workman:* His legal name was Everly John Workman; he also went by E.J. His bio is available at timeline.unm.edu/item/ej-workman.html.

Through a military liaison: Richard Tolman to General R. H. Somers, January 15, 1941, LOC, MAT.

experiments at the Albuquerque airport: E. J. Workman and R. E. Holzer to Merle Tuve, memo, "Preliminary Report, Electrostatics of Airplanes," rec. January 28, 1941, LOC, MAT.

too weak to trigger: Section T, *Final Report of the Development of the Radio and Other Proximity-Fuzes,* 31; C. C. Lauritsen to R. C. Tolman, March 14, 1941, LOC, MAT; Navy Bureau of Ordnance, memo, "Division Monthly Report," June 28, 1941, DTM.

required Merle to order: John Fleming to A. L. Williams, January 28, 1941, LOC, MAT; A. L. Williams to John Fleming, January 17, 1941, LOC, MAT.

"two photographs of poor quality": J. A. Bearden, *National Defense Research Committee Report No. A-12—Summary Report: Studies of Acoustic Proximity*

Fuzes (Washington, D.C.: NDRC, 1944), 49, NARA, RG 227, E NC-138 29.

"flying toilet bowl": Dennis, "A Change of State," 273.

the bombs' "self-noise": Section T, *Final Report of the Development of the Radio and Other Proximity-Fuzes,* 29.

75   *James Van Allen:* Van Allen, "What Is a Space Scientist?," 3.

he failed the physical: George H. Ludwig, "James Alfred Van Allen: From High School to the Beginning of the Space Era, A Biographical Sketch," October 9, 2004, UI, www-pw.physics.uiowa.edu/van90/VanAllenBio _LudwigOct2004.pdf.

a promising young Fellow: Van Allen, interview, 82.

"fiddling while Rome burned": Ibid., 88.

two thousand times brighter: By "in darkness," I mean dawn, when the illumination level is about one lumen per square foot; Larry Hafstad, *National Defense Research Committee Progress Report No. A-1: The Photoelectric Proximity Fuze as of November 4, 1940* (Washington, D.C.: NDRC, 1940), 1, NARA, RG 227, E NC-138 29.

76   *"If we had the whole place":* Van Allen, interview, 90.

The "rudimentary" radio circuitry: Van Allen, "My Life at APL," 174.

"if you brought a fly swatter": Van Allen, interview, 90.

"away from the sun": J. E. Henderson to Merle Tuve, memo, "Dahlgren Test for June 5, 1941," May 27, 1941, DTM.

thirty-four-year-old assistant professor: H. Deresiewicz, ed., *The Collected Papers of Raymond D. Mindlin* (New York: Springer, 1989), xiv.

had a taste for Chopin: Bruno A. Boley, "Reminiscences of Raymond D. Mindlin on the Occasion of the Hundredth Anniversary of His Birth" (speech, Boulder, CO, June 26, 2006), University of Mississippi, olemiss .edu/sciencenet/mindlin/Boley_Mindlin_Speech.pdf.

77   *"Fair to Mindlin":* Ibid., 4.

"torsion of structural beams": Deresiewicz, *Collected Papers,* xiv.

He was recruited in January: Dana Mitchell to Raymond Mindlin, telegram, January 10, 1941, DTM.

Mindlin had just finished: Henry Porter, "Recollections on the Development of Radio-Controlled Proximity Fuzes," *Johns Hopkins APL Technical Digest* 4, no. 4 (1983): 297.

"like popcorn": "Big Tacoma Bridge Crashed 190 Feet into Puget Sound," *New York Times,* November 8, 1940.

"excessive oscillations": Othmar H. Ammann et al., *The Failure of the Tacoma Narrows Bridge: A Report to the Honorable John M. Carmody* (Washington, D.C.: Board of Engineers, 1941), 1.

Section T lied to him: Baldwin, *The Deadly Fuze,* 109. This was the recommended story for all consultants without clearance; C. C. Lauritsen to Merle Tuve, September 9, 1940, DTM.

a bevy of centrifuges: Henry Porter et al., "T-4 Report No. 1," January 18, 1941, DTM.

including a Canadian tube: Applied Physics Laboratory, *The "VT" or Radio Proximity Fuze,* 28–29.

In early 1941, Mindlin produced: R. D. Mindlin to H. H. Porter, memo, "Summary of My Work Since My Association with Section T (T-4), February 1, 1941," February 27, 1941, LOC, MAT.

miniature cantilevers: The tubes had cantilevered springs.

78 *"Mindlin appeared one morning":* Henry H. Porter, "Recollections on the Development of Radio-Controlled Proximity Fuzes," 297.

inch long and 0.00075 of an inch: Baldwin, *The Deadly Fuze,* 113.

teams of women: Vinton K. Ulrich to Dana Mitchell, June 4, 1941, DTM.

a very high tension: Applied Physics Laboratory, *World War II Proximity Fuze,* 182–83.

reached out to manufacturing: G. Scott to Robert Brode, February 20, 1941, DTM; E. V. Sundt to Robert Brode, February 21, 1941, DTM.

of all metals: I'm referring to natural metals.

"Tuve said don't slow down": Tuve, "Notebook X," March 19, 1941.

floors in a high-rise: Henry H. Porter, et al., Rugged Vacuum Tube, U.S. Patent 113,235, filed January 24, 1944, and issued December 3, 1963.

79 *Gardiner Means was worried:* Gardiner Means to John Fleming, May 2, 1941, DTM.

falling from fifteen thousand feet: Roberts, "Autobiography," 2.

known as Stump Neck: Klingaman, *APL—Fifty Years of Service to the Nation,* 11.

four-inch plate of steel: Baldwin, *The Deadly Fuze,* 74.

nearly a minute: T. M. Perry to L. M. Mott-Smith, June 23, 1942, DTM.

a Chesapeake retriever: Boyce, *New Weapons for Air Warfare,* 129.

80 *A "Star Shell":* W.H.P. Blandy and Gilbert C. Hoover, memo, "5/25 Mark XXVII-1 Illuminating Projectiles Modified for Recovery—Visit of Special Representatives to Army Proving Ground," April 18, 1941, LOC, MAT.

a March test shoot: Memo, "Section T-4 Proof Report," March 20, 1941, LOC, MAT.

On April 20, Jack Workman: Applied Physics Laboratory, *World War II Proximity Fuze,* 182; Boyce, *New Weapons for Air Warfare,* 109; Applied Physics Laboratory, *The "VT" or Radio Proximity Fuze,* 30.

On May 8, Merle traveled: Applied Physics Laboratory, *The "VT" or Radio Proximity Fuze,* 30.

"a new breed of officer": Christman, *Target Hiroshima,* vii.

"the finest technical officer": Ibid., 1.

81 *stretching twenty miles:* Rusty Dennen, "Dahlgren Firing Range Has Been a Fixture on the Potomac River Since Turn of the Century," Fredericksburg.com, July 20, 2003.

five miles downstream: Section T, *Progress Report as of April 19, 1942,* 9.

signals in three of seven: Tuve notebook lists the third round as "probably" heard; Tuve, "Notebook XII," May 8, 1941.

"Success is now in sight": Vannevar Bush to President Franklin Roosevelt, July 16, 1941; OSRD, *Report of the National Defense Research Committee for the First Year of Operation,* 25–27.

"It is now clear": Merle Tuve to Richard Tolman, June 11, 1940, DTM.

drafted a plan: Merle Tuve, memo, "Verbal Comments by Dr. Tuve, June 11, 1941, regarding Section T Work at the Department of Terrestrial Magnetism, Carnegie Institution of Washington," June 11, 1941, DTM.

had to order more desks: Memo, "Immediate Needs in Connection with Realignment," May 10, 1941, LOC, MAT.

10. PRYING EYES

82 *"Where are the mics":* Duffy, *Double Agent,* 215; Duquesne File, section 8, 236.

"x-ray mirror": Duquesne File, section 6, 13.

the FBI had rented: Ibid., section 2, 303.

largest successful espionage case: Baime, *The Arsenal of Democracy,* 213.

83 *Once declared dead:* Sam Roberts, "Nazi Spies and New York Perspectives," *New York Times,* August 8, 2014.

Three days later, on June 28: Duffy, *Double Agent,* 219.

Across America, German operatives: Ibid., 220.

84 *"Newspapers report arrest here":* Duquesne File, section 8, 240.

Tuve filed away a clipping: Associated Press, "Nazi Spies Told to Get Secrets of New Weapons, Jury Hears," September 15, 1941, LOC, MAT.

"unable to find anyone": Duquesne File, section 6, 288.

hoped to sabotage: Ibid., section 1, 71.

85 *"Suggestions for Protection of Industrial":* J. Edgar Hoover to Henry Porter, August 12, 1941, LOC, MAT.

already had a copy: Duquesne File, section 7, 242.

On August 7, 1941: "Memorandum of Conference," August 7, 1941, NARA, RG Group 227, E NC-138 169.

NDRC had been restructured: Zachary, *Endless Frontier,* 129–30. FDR approved the change in May; it was not formally restructured until June 28, 1941.

87   *building a nuclear weapon:* Zachary, *Endless Frontier,* 189–217.

"Section on Uranium": Stewart, *Organizing Scientific Research for War,* 121.

one side of the debate: Tuve, interview by Cornell, session 1; Zachary, *Endless Frontier,* 195.

"The Germans can't afford": Tuve, interview by Cornell, session 1.

"Oh fudge": Zachary, *Endless Frontier,* 193.

Lawrence startled Compton: Hiltzik, *Big Science,* 226.

88   *"was running the show":* Zachary, *Endless Frontier,* 193.

the Cyclotron Building register: Section T, "Cyclotron Building Register," March 23, 1941 (register book, February 1941 — May 1942, DTM).

"gone under so many different": Cornell, "Merle A. Tuve and His Program of Nuclear Studies," 517.

Merle abandoned the "uranium affair": According to Tuve's recollection, he quit. But it's clear that whatever Tuve's stance (and provocations), Vannevar Bush also eased him out the door; Van Bush to Merle Tuve, August 14, 1941, LOC, MAT.

a group of French industrialists: Sebba, *Les Parisiennes,* 86–87.

the old Hôtel Majestic: Christophe Chommeloux and Caroline Laleta Ballini, *The Peninsula Paris: The Making of a Parisian Masterpiece* (Hong Kong: International Publishing Concepts, 2015).

II. ESCALATION

90   *extended its combat zone:* Olson, *Those Angry Days,* 295.

sunk by a German submarine: Duffy, *Double Agent,* 213.

"to steer a course between": Steven Casey, *Cautious Crusade: Franklin D. Roosevelt, American Public Opinion, and the War Against Nazi Germany* (Oxford: Oxford University Press, 2001), 30.

vote was 203 to 202: J. Garry Clifford and Robert H. Ferrell, eds., *Presidents, Diplomats, and Other Mortals* (Columbia: University of Missouri Press, 2007), 99.

legislature even reduced: Zachary, *Endless Frontier,* 195.

91   *stopgap loan:* Karl Compton to John D. Rockefeller, May 8, 1941, LOC, VB; John D. Rockefeller to Vannevar Bush, July 24, 1941, LOC, VB.

more than three million troops: Duffy, *Double Agent,* 214.

swaths of American industry: Baime, *The Arsenal of Democracy,* 106.

Radio engineers: L. M. Mott-Smith to F. E. Terman, July 29, 1941, LOC, MAT.

"*Who are you to telegraph Harris*": Merle Tuve to P. M. Morse, May 22, 1941, LOC, MAT.

"*very, very few diplomatic*": Zachary, *Endless Frontier*, 123–24.

"*very brilliant, not inclined*": Vannevar Bush, "Memorandum for Dr. Conant," September 19, 1941, NARA, RG 227, E NC-138 1.

92 *he relegated "Group B*": Merle Tuve, memo, "Space and Room Assignments," May 9, 1941, DTM. The shell fuse group was designated "Group C." There doesn't appear to have been a Group A.

nine times more valuable: Baldwin, *The Deadly Fuze*, 99.

Group B's main office: J. E. Henderson to H. H. Porter, memo, "Office Chairs for Room 205," June 19, 1941, LOC, MAT.

was ten degrees hotter: J. E. Henderson to H. H. Porter, memo, "Roof-Sprinkling System for Cyclotron Building," June 25, 1941, LOC, MAT.

A hot spell in D.C.: Ibid. I've verified the archival references with the historical weather data for DCA airport for June 21 and 22, 1941, available at wunderground.com.

sunbeams poured unmercifully: Henderson, "Roof-Sprinkling System for Cyclotron Building."

Tuve's planned expansion: M. A. Tuve to John A. Fleming, memo, "Approval of Tentative Plans for Expansion of Section T," July 25, 1941, LOC, MAT.

Group B was removed: Baxter, *Scientists Against Time*, 227. The Bureau of Standards had also been working on bomb and rocket fuses since 1940. Ibid., 291.

steps away from the White House: L. Grant Hector to Dr. M. Benjamin, August 22, 1941, DTM.

"*balance wheel*": "Transcript of Meeting on VT (Proximity) Fuze," 70.

visited England in June: Larry Hafstad, memo, "Diary of Larry Hafstad," June 1941, NARA, RG 227, E NC-138 176.

"*If this works*": Ibid., June 27, 1941.

93 *the pulse design:* Ibid., June 1–4, 1941. The British believed, even in June of 1941, Hafstad wrote, that "any complete proximity fuse requiring both oscillator and detector would be too bulky for use in a shell."

abandon the design: The pulse design was abandoned by the end of October 1941; Wallace T. Baker, memo, "Flow Sheet of Section T Fuze Milestones from 19 Aug 1940 to Jul 1942," July 29, 1942, transcribed from the original, APLA.

lost their juice: Memo, "Extract from J. P. Teas letter to O. W. Torreson dated August 28," n.d. (1941), DTM.

English-made tubes: M. A. Tuve to L. G. Hector, memo, "Possible American Reproduction of the GEC Tube," July 13, 1941, DTM; M. A. Tuve to R. C. Tolman, memo, "British Liaison," October 22, 1941, LOC, MAT.

demonstration of rocket fuses: Hafstad, "Diary of Larry Hafstad," June 6, 1941.

Benjamin arrived in Buffalo: L. G. Hector to M. A. Tuve, memo, "Visit of Dr. Benjamin on Tour for August 17, 18, 19, and 20th," August 21, 1941, DTM.

at the General Electric Company: Irene Glausiusz and Leslie Rübner, "Pamela and Raymond Foreman," *Kingsbury Courier,* 2015, 17.

summer-evening scenic tour: Hector, "Visit of Dr. Benjamin."

94 *room 102 of Philosophy Hall:* H. H. Porter to M. A. Tuve, memo, "Program for the Study of Tube Noise and Microphonics," April 27, 1941, LOC, MAT.

discovered the main culprit: R. D. Mindlin to L. Grant Hector, memo, "Microphonics of British Tubes," October 5, 1941, LOC, MAT.

again that Friday: Section T, "Cyclotron Building Register," August 22, 1941; Tuve, "Notebook XVI," August 21 and 22, 1941.

Ten days later, heading home: "Missing Atlantic Ferry Bomber Crashes in Britain: Strikes Hill on Reaching Old Country," *Lethbridge Herald,* September 2, 1941.

special prototype tubes: L. Grant Hector to W. L. Webster, memo, "American Tubes for Experimental Use by the British," September 30, 1941, DTM.

It took five days: Glausiusz and Rübner, "Pamela and Raymond Foreman," 21.

should buy about two thousand: Raytheon Production Corporation, "Section T Contract Questionnaire," April 16, 1942, LOC, MAT.

95 *"You know, Tuve":* Tuve, interview by Christman.

"just the opposite of what": "Transcript of Meeting on VT (Proximity) Fuze," 2.

Deak Parsons first saw: The description of the event comes from Baldwin, who presumably heard it from Roberts, but the date comes from the fuse-milestones timeline I received from the Applied Physics Laboratory; Baldwin, *The Deadly Fuze,* 88; Baker, "Flow Sheet of Section T Fuze Milestones from 19 Aug 1940 to Jul 1942."

96 *at least 50 percent reliability:* Baxter, *Scientists Against Time,* 229.

Brothers Joe and John Erwood: Baldwin, *The Deadly Fuze,* 89–90.

on West Erie Street: W.H.P. Blandy, "Memorandum of Telephone Conversation Between Lieut. (jg) E. Earle, U.S.N.R., and Lieut. V. Hicks, U.S.N.R., of December 31, 1941. File No. L24," January 16, 1942, LOC, MAT.

of at least fifty: Section T, *Progress Report as of April 19, 1942,* 73.

97 *had some five hundred components:* Foerstner, *James Van Allen,* 53. I've used 5 percent instead of 4, which Foerstner uses, since $0.99^{300} = 0.049$.

90 percent of the time: L. M. Mott-Smith to L. Grant Hector, memo, August 4, 1941, DTM.

On September 8, 1941: Section T to Vannevar Bush, memo, September 8, 1941, LOC, MAT.

borrow a hiring technique: "Transcript of Meeting on VT (Proximity) Fuze," 82.

98 *"If we reach out":* Tuve, interview by Christman.

"Roberts triumvirate": "Transcript of Meeting on VT (Proximity) Fuze," 81.

Tuve drafted: Charles Lauritsen was vice chairman of Division A, but early on he worked with Tuve and Section T. Foerstner, *James Van Allen,* 52–53; Associated Press, "Thomas Lauritsen, Nuclear Physicist," *New York Times,* October 17, 1973; John Fleming to B. P. Seaman and C. V. Broadley, July 1, 1941, enclosure A, LOC, MAT.

Bright, youthful faces arrived: Section T, "D.T.M. Photo Album XXVIII" (photo album, 1941, DTM).

Evelyn Wood: "Evelyn Wood Résumé," n.d., LOC, MAT; John Fleming to Marvin L. Faris, October 22, 1941, LOC, MAT.

99 *"During the last six weeks":* M. A. Tuve to J. A. Bearden, memo, "Formulation of Plans for Emergency Ordnance Research Based on Experience with Section T," November 26, 1941, DTM.

"Emergency Group X": Memo, "Emergency Staff Assignments," October 27, 1941, LOC, MAT; Memo, "Weekly Section T Officers' Meeting," October 27, 1941, DTM.

a new type of spring: Foerstner, *James Van Allen,* 54–55; Van Allen, "My Life at APL," 174.

Devised in mid-November: Memo, "Conversation with Van Allen," November 14, 1941, LOC, MAT.

Seventy-two tubes: Lynn G. Howell to L. M. Mott-Smith, memo, "Tube Shoot of December 3, 1941," DMT.

Later patented: Van Allen, interview, 95–96.

test of ten units: Memo, "Abstracted Report of Group Supervisors: November 1, 1941," November 5, 1941, DTM.

Section T's Secret Assembly Shop: F. M. Walter, memo, "Report of Secret Assembly Shop," October 4, 1941, DTM.

"dangerous accident": M. A. Tuve to Section T Files, memo, "Personal Report for Week Ending October 4, 1941," October 4, 1941, LOC, MAT.

100 *He penned a nasty letter:* This happened after Pearl Harbor; Warren Weaver to M. A. Tuve, December 19, 1941, NARA, RG 227, E NC-138 8.

"I am frankly shocked": M. A. Tuve to Warren Weaver, December 22, 1941, NARA, RG 227, E NC-138 8.

Weaver griped that Tuve: Warren Weaver to James Conant, December 30, 1941, NARA, RG 227, E NC-138 1.

British were increasingly frustrated: James B. Conant to R. C. Tolman, October 16, 1941, NARA, RG 227, E NC-138 1.

"the assumed way in which": M. A. Tuve to R. C. Tolman, memo, "British Liaison," October 22, 1941, LOC, MAT.

"An absolute master": Zachary, *Endless Frontier,* 124.

"Of course he is a wonderful": Ibid.

"For the last 16 months": M. A. Tuve to J. A. Bearden, memo, "Formulation of Plans for Emergency Ordnance Research Based on Experience with Section T," November 26, 1941, DTM.

"rambling memorandum": M. A. Tuve to Section T Files, memo, "Personal Report for Week Ending November 29, 1941," December 6, 1941, DTM.

101 *"TIME IS SHORTER THAN WE THINK":* M. A. Tuve to J. A. Bearden, memo, "Proposal of a Definite Plan for the Future Operation of Section T," December 5, 1941, DTM.

two privates at a radar station: U.S. Congress, *Investigation of the Pearl Harbor Attack: Report of the Joint Committee on the Investigation of the Pearl Harbor Attack* (Washington, DC: United States Government Printing Office, 1946), 140.

12. READY OR NOT

105 *headed to Griffith Stadium:* Details from the game were drawn from S. L. Price, "The Second World War Kicks Off: December 7, 1941, Redskins Versus Eagles on Pearl Harbor Day," *Sports Illustrated,* November 29, 1999; Dan Steinberg, "How the *Washington Post* covered the Pearl Harbor Redskins Game," *Washington Post,* December 7, 2011; Shirley Povich, "At Redskins-Eagles Game, Crowd Was Kept Unaware That War Had Begun," *Washington Post,* December 7, 1991.

Hawaiian sugarcane and pineapple: Shirley, *December 1941,* 138.

107 *broadcasts were interrupted:* Frank Fitzpatrick, "A Quiet Sunday Afternoon Shattered by a News Bulletin: Pearl Harbor Reorders Philadelphia," *Philadelphia Inquirer,* December 7, 2016. Some of the broadcasts are available at otr.com/r-a-i-new_pearl.shtml.

Parsons was late: Christman, *Target Hiroshima,* 83.

"deadly calm": Shirley, *December 1941,* 178.

108 *At Fort Lewis:* Thomas Saylor, *Remembering the Good War* (Minneapolis: Minnesota Historical Society Press, 2005), chapter 1.

some four thousand antiaircraft troops: Shirley, *December 1941,* 179.

MADE IN JAPAN: Ibid., 143.

FBI was tasked: Toshio Whelchel, *From Pearl Harbor to Saigon: Japanese American Soldiers and the Vietnam War* (New York: Verso, 1999), 4.

North Dakota and Montana: Shirley, *December 1941,* 172.

109 *Hatch perched on a staircase:* Hatch, "Service Above All." The speech was at twelve thirty. Hatch misremembers it as being broadcast at nine o'clock in the morning.

ashamed to be crying: Shirley, *December 1941,* 164.

"There is no blinking": Melissa Chan, "'A Date Which Will Live in Infamy': Read Present Roosevelt's Pearl Harbor Address," *Time,* December 6, 2018.

"word came down": Hatch, "Service Above All."

"The Americans are a bunch": Sacks, "Local Man Who Helped Develop 'Best Kept Secret.'"

Clenching the bow: Zachary, *Endless Frontier,* 147–48.

110 *"congenial group":* Bush OH, reel 3-A, 178.

"live a reasonably sane life": Ibid., 177.

updated President Roosevelt: Bush to Roosevelt, July 16, 1941.

authorized 207 contracts: OSRD, *Report of the National Defense Research Committee for the First Year of Operation,* 56–58.

Dartmouth would soon: Baxter, *Scientists Against Time,* 218.

six million dollars: OSRD, *Report of the National Defense Research Committee for the First Year of Operation,* 60.

forty million dollars: Stewart, *Organizing Scientific Research for War,* 200.

new military medicines: Andrus, *Advances in Military Medicine,* vol. 1, xli–liv.

Under OSRD's Division A: Richard C. Tolman, memo, "Brief Report on the Status of the Work of Division A of the National Defense Research Committee as of June 30, 1941," July 2, 1941, NARA, RG 227, E NC-138 2.

head of Division D: Memo, "General Report of Division D, National Defense Research Committee," n.d., NARA, RG 227, E NC-138 2.

111 *the rancher's son:* Jewett's father was also an engineer.

"permanent impairment to": R. W. King, reports for Divisions C-2, C-4, and C-5, July 16, 1941, NARA, RG 227, E NC-138 2.

supervised by Roger Adams: Stewart, *Organizing Scientific Research for War,* 38.

Division B scientists: Division B, memo, "Summary of Progress of A Sections to June 30, 1941," June 23, 1941, NARA, RG 227, E NC-138 2.

OSRD's final meeting: NDRC, "Minutes of Fifth Meeting," November 28, 1941, NARA, RG 227, E NC-138 3.

112 *twenty-three hundred U.S. servicemen:* Matthew Diebel, "Pearl Harbor Remembrance Day 2017: What Happened on That Fateful Day 76 Years Ago?," *USA Today,* December 7, 2017.

Eight battleships: United States Navy, "Navy Statement on the Pearl Harbor Attack," *New York Times,* December 6, 1942.

Bush's family: Zachary, *Endless Frontier,* 141.

"The nature of the world": Ibid., 143.

three Japanese airplanes: Another damaged plane later crash-landed; Martin Middlebrook and Patrick Mahoney, *The Sinking of the* Prince of Wales *and* Repulse: The End of the Battleship Era (Barnsley, UK: Pen and Sword, 2014), chapters 9 and 11.

"most modern British": Bush, *Pieces of the Action,* 77.

Historians later described: Middlebrook and Mahoney, *The Sinking of the* Prince of Wales *and* Repulse; Robert Farley, "This Was the Exact Day the Battleship Became Completely Obsolete," *National Interest,* June 13, 2017.

technology wasn't fully respected: Bush, *Pieces of the Action,* 91; Zachary, *Endless Frontier,* 2.

on the naval docks: Bush, *Pieces of the Action,* 91.

113 *353 Japanese planes:* Martin Gilbert, *The Routledge Atlas of the Second World War* (New York: Routledge, 2008), 56.

Hawaiian streets: Jack G. Henkels, "Civilians Died on Dec. 7, Too," *Honolulu Star-Bulletin,* December 7, 1996; Willis David Hoover, "Remember Pearl Harbor: The Day of Infamy That Changed Honolulu Forever," *Honolulu,* December 7, 2016.

13. NO ALIBIS

114 *Tuve received a call:* Merle Tuve to Richard Tolman, memo, "Report for Week Ending December 20, 1941," December 28, 1941, DTM.

Within days, Van Bush: Ibid.

was freezing: Ibid.

"analytical shoots": For example, see Merle Tuve to Richard Tolman, memo, "Official Report for Week Ending January 17, 1942," January 19, 1942, DTM.

115 *Weather Bureau reported:* Memo, "Abstracted Report of Group Supervisors for Week Ending December 13, 1941," December 18, 1941, DTM.

"almost impossible": M. A. Tuve to R. C. Tolman, memo, "Official Report for Week Ending December 13th, 1941," n.d., DTM.

finding another test site: T. C. Roberts to H. H. Porter, memo, "Report on Proving Ground," January 25, 1942, DTM.

"overwhelming shooting schedule": M. A. Tuve to R. C. Tolman, memo, "Official Report for Week Ending January 3, 1942," January 7, 1942, DTM.

two hundred shots a week: Tuve to Tolman, "Official Report for Week Ending December 13th, 1941."

"Urgent requests for acceleration": Tuve to Tolman, "Official Report for Week Ending January 3, 1942."

December's field tests: Memo, "Abstracted Report of Group Supervisors for Week Ending December 13, 1941"; Memo, "Abstracted Report of Group Supervisors for Week Ending December 20, 1941," December 26, 1941, DTM; Memo, "Abstracted Report of Group Supervisors for Week Ending December 27, 1941," December 31, 1941, DTM; Memo, "Abstracted Report of Group Supervisors for Week Ending January 3, 1942," January 8, 1942, DTM.

double his field crew: Tuve to Tolman, "Official Report for Week Ending December 13th, 1941."

"less vulnerable to new hands": M. A. Tuve to R. C. Tolman, memo, "Official Report for Week Ending December 6, 1941," December 16, 1941, DTM.

116 *"couldn't get 25 units":* "Transcript of Meeting on VT (Proximity) Fuze," 84.

most innocent changes: Ibid., 72, 84.

Even the Crosley Corporation: Memo, "Abstracted Report of Group Supervisors for Week Ending January 24, 1942," January 29, 1942, DTM.

Crosley built refrigerators: McClure et al., *Crosley;* brochure, "Crosley and the Defense Program," Crosley Corporation, 1941, DTM.

over fourteen million dollars: Brochure, "Crosley and the Defense Program."

ran the Cincinnati Reds: "Crosley Again Heads Reds," *New York Times,* November 16, 1937.

Crosley's vice president: Clement's official title was vice president in charge, research and engineering; Lewis M. Clement to P. J. Larsen, December 5, 1941, DTM.

arrived at the Crosley headquarters: "Transcript of Meeting on VT (Proximity) Fuze," 3–4.

"we would be contacted": Ibid., 3.

Clement's sixth-floor office: McClure et al., *Crosley,* 362.

six hundred square feet: Ibid., 366.

"making what are considered": Memo, "Abstracted Report of Group Supervisors for Week Ending January 24, 1942."

117 *"They have achieved":* Merle Tuve, memo, "Notes for Discussion with Dr. Conant, Friday, February 20, 1942," February 19, 1942, LOC, MAT.

"obvious improvements": "Transcript of Meeting on VT (Proximity) Fuze," 83.

"Absolutely only one change": Baldwin, *The Deadly Fuze,* 93; Tuve, "Notebook XX," December 30, 1941.

"to the assembled group": "Transcript of Meeting on VT (Proximity) Fuze," 83.

"like an endless series": Ibid., 81.

core design of the fuse: Applied Physics Laboratory, *World War II Proximity Fuze,* 4; J. J. Hopkins to R. H. Thayer, memo, "Adaptation of the Grounded Grid Oscillator to the Proximity Fuze," November 22, 1944, APLA; Paul J. Larsen, "Early Development of the Radio Proximity Fuze: The Design and Manufacture of the Mk 32 Radio Proximity (VT) Fuze," excerpted from "A Compilation of Status Reports on OSRD/JHU/APL Projects as of November 30, 1944," n.d., APLA.

A month after Pearl Harbor: M. A. Tuve, memo, "Personal Report for Week Ending January 10, 1941," January 12, 1942, DTM.

"Prescription A": M. A. Tuve to R. C. Tolman, memo, "Official Report for Week Ending January 24, 1942," January 25, 1942, DTM.

On January 29, at Dahlgren: Memo, "Abstracted Report of Group Supervisors for Week Ending January 31, 1942," February 5, 1942, DTM.

Of the fifty rounds shot: I've used the figure usually cited for this test, 52 percent, which appears in official histories of OSRD and the above cited memo. But it's interesting that in Tuve's notebook, he records that three of the fuses did not fit in the shells: "Threads too large to go into shells in #27 33 40." For those shells, he wrote, "Not shot." Tuve also calculates in his notebook "26 out of 47 = 55%." The 52 percent figure would be 26/50. See the entry after the January 28 entry in Tuve, "Notebook XXI."

Parsons phoned in: Baldwin, *The Deadly Fuze,* 97.

118 *"This proves you know":* "Transcript of Meeting on VT (Proximity) Fuze," 57–58.

worth eighty million dollars: Baxter, *Scientists Against Time,* 229.

On February 24: "Bond Raised to $1,000 in Ordnance Theft Case," *Evening Star,* February 27, 1942; Clipping, "Worker, Bomb Parts Seized: Navy Employee Also Carried Secret Data," n.d., LOC, MAT.

Navy insisted: E. E. Merkel to L. R. Hafstad, memo, "Navy Department Request for Arming of Section T, Division A, NDRC Representatives," December 26, 1941, LOC, MAT.

appointed as deputy sheriffs: Section T to Victor Hicks, June 29, 1942, NARA, RG 227, E NC-138 8.

"a crusty old drill": Van Allen, interview, 97.

"damage that might arise": G. Pettengill to the Department of Terrestrial Magnetism, February 12, 1942, LOC, MAT.

locking up all confidential: Memo, "Memo for Buckingham, Capron, Field, Lacy, Mott-Smith, Schulte," April 7, 1942, LOC, MAT.

blackout drills: John Fleming, memo, "Notice to All Regular and Temporary Employees," March 13, 1942, LOC, MAT.

119 *"revolvers in their hands":* Victor Hicks to W. S. Parsons, memo, "Security of T-3G Shipments in Cincinnati," June 5, 1942, NARA, RG 227, E NC-138 8.

eighty thousand troops captured: These included Indian and Australian troops.

Battle of the Java Sea: This was a combined American, Australian, British, and Dutch operation.

delays in delivering the fuse: Baldwin, *The Deadly Fuze,* 87–88.

ten times less than: Tuve to Tolman, "Report for Week Ending December 20, 1941."

highest-quality rugged tubes: This was already the case in December of 1941; Tuve to Tolman, "Official Report for Week Ending December 6, 1941."

Hygrade Sylvania built lamps: The lamps were marketed under the Hygrade name; Larry R. Paul, *Made in the Twentieth Century: A Guide to Contemporary Collectibles* (Lanham, MD: Scarecrow Press, 2005), 225.

"A Chat with Roger Wise": "A Chat with Roger Wise," *Sylvania News* 8, no. 1 (1939): 2.

120 *"expert hand-operators":* Section T, *Progress Report: February 19, 1942,* 4.

three thousand tubes a day: Ibid., 5.

adapted for their antiaircraft: R. C. Williams to M. A. Tuve, memo, "Characteristics of Guns of Interest to Section T," April 10, 1942, NARA, RG 227, E NC-138 8.

used rounds as small as: I'm referring to 90 mm shells. The Navy and Army both used shells smaller than 90 mm, but the rounds for the Army's 90 mm and British 3.7-inch guns were considered the smallest potential targets for adaptation.

one and a half inches: Section T, *Progress Report: February 19, 1942,* 8b.

Merle assigned the task: The project was designated "T-81." Memo, "Revised Staff Assignment: Effective Feb. 10, 1942," LOC, MAT.

associate professor of physics: Benjamin Bederson and H. Henry Stroke, "History of the New York University Physics Department," *Physics in Perspective* 13 (2011): 281.

Salant's weekly summaries: E. O. Salant to E. D. McAlister, memo, "Report of Week's Activities Ending February 14, 1942," February 14, 1942, LOC, MAT; E. O. Salant to E. D. McAlister, memo, "Week's Activities," February 23, 1942, LOC, MAT; E. O. Salant to E. D. McAlister, memo, "Report of Week's Activities Ending February 28, 1942," March 2, 1942, DTM.

also appointed Salant: Memo, "Revised Staff Assignment: Effective Feb. 10, 1942."

121 *spare only eight men:* Ibid.

room was meant for two: Memo, "Space Allotments at D.T.M.," n.d., LOC, MAT.

four times the space: M. A. Tuve to E. D. McAlister, memo, "Army-British Schedule," March 27, 1942, DTM.

pleaded with the Navy: M. A. Tuve, memo, "Personal Report for Week Ending January 10, 1942," January 12, 1942, DTM.

pleaded with Director Fleming: M. A. Tuve to J. B. Conant, memo, "Necessity for Prompt Action on Additional Facilities for Section T, Division A," February 11, 1942, LOC, MAT.

"unconscious dislike for unexpected": Merle Tuve, "Notes for Discussion with Dr. Conant, Friday, February 20, 1942."

On March 31, Bush removed: V. Bush to Richard C. Tolman, March 31, 1942, NARA, RG 227, E NC-138 1.

"alibis," as he called: M. A. Tuve to H. H. Porter, memo, "Research Battery Situation at National Carbon Company," September 11, 1942, LOC, MAT.

thirty thousand square feet: M. A. Tuve, memo, "Resumé of Officers' Conference on Hopkins Contract," March 24, 1942, DTM.

Tuve decided to call it: Ibid.; Klingaman, *APL — Fifty Years of Service to the Nation,* 8.

14. DR. JONES'S RAID

122 *On a cold, still morning:* Ford, *The Bruneval Raid,* 35.

ribbed aluminum floors: Millar, *The Bruneval Raid,* 167.

123 *"a violent technical reconnaissance":* Jones, "Scientific Intelligence," 66.

whether German radar existed: R. Jones, *The Secret War of Dr. Jones.*

"a basic intelligence assault": Jones, "Scientific Intelligence," 65.

124 *Würzburg radars, Jones guessed:* Jones, *Most Secret War,* 194.

"Look, Charles": Ibid., 236.

spy named Gilbert Renault: Wolfgang Saxon, "Gilbert Renault, 79; Hero of Resistance in Wartime France," *New York Times,* July 30, 1984.

125 *Roger Dumont:* Millar, *The Bruneval Raid,* 78–79.

"Good morning, Fritz": Ibid., 80.

126 *someone with technical expertise:* Jones, *Most Secret War,* 237.

mechanic named Charles Cox: Millar, *The Bruneval Raid,* 17.

127 *"Don't be worried too much":* Jones, *Most Secret War,* 238.

at 11:55 pm: Ford, *The Bruneval Raid,* 43.

Caught in the slipstream: Millar, *The Bruneval Raid,* 169.

128 *"Like a searchlight on":* Ibid., 172.

129 *ten minutes, Cox:* Brian Austin, *Schonland: Scientist and Soldier* (Bristol, UK: Institute of Physics Publishing, 2001), 239.

On March 2, Jones: Jones, *Most Secret War,* 242–44.

When Hitler heard: Clay Blair, *Hitler's U-Boat War: The Hunters, 1939–1942* (New York: Modern Library, 2000), chapter 9.

a complete success: P. Jones, *The Secret War of Dr. Jones.*

130 *those in Jones's circle:* Jones, *Most Secret War,* 181; Goodchild, *A Most Enigmatic War,* 398.

his most extraordinary source: Jones did not receive the report directly.

traveling by train: Ignatius, "After Five Decades, a Spy Tells Her Tale."

15. THE GARAGE

131 *Rumors spread around:* Klingaman, *APL—Fifty Years of Service to the Nation,* 10.

"They push-a them up": Ibid., 10.

FDR had to personally: Ibid., 8.

upped the insurance cost: D. Luke Hopkins to W. S. Parsons, memo, "Expansion Program," May 1, 1942, RG 227, E NC-138 8.

The area was tranquil: Jerry McCoy, *Downtown Silver Spring, Then & Now* (Mount Pleasant, SC: Arcadia, 2010).

132 *over a hundred thousand dollars:* Invoice, "Condensed Statement of Expenditures," June 1942, NARA, RG 227, E NC-138 8.

"There was rarely a day": "In Memoriam: Merle Antony Tuve," 214.

Supplies poured into: Invoice, "Condensed Statement of Expenditures."

Domestic concerns crept: Tuve, "Notebook XXI," April 1, 1942.

invading the Tuves' home: Comly, interview; Whitman, interview.

"absolute chimney": Comly, interview.

133 *"The military is full"*: Ibid.

probably included: For amusing background, see Rob Chirico, *Damn!: A Cultural History of Swearing in Modern America* (Durham, NC: Pitchstone, 2014).

"It wasn't his nature": Comly, interview.

did not like it: Ibid.

134 *"one of the finest but"*: Klingaman, *APL—Fifty Years of Service to the Nation,* 10.

"Results took precedence": Jacobsen, *The Deadly Fuze.*

pair PhDs and hams: Ibid.

Paul Ertsgaard: Section T Personnel File, Edwin Paul Ertsgaard, LOC, MAT.

John Doak: Section T Personnel File, John Bell Doak, LOC, MAT.

Joseph Teresi: Section T Personnel File, Joseph Anthony Teresi, LOC, MAT.

spate of oil men: Memo, "Personnel at Section T, DTM, on Leave of Absence from Other Companies," February 4, 1943, LOC, MAT.

"The important thing was": Klingaman, *APL—Fifty Years of Service to the Nation,* 10–11.

"Merle was gifted": Jacobsen, *The Deadly Fuze.*

"Time meant nothing": Klingaman, *APL—Fifty Years of Service to the Nation,* 11.

"You were working": Jacobsen, *The Deadly Fuze.*

his rotary telephone: Baldwin, *The Deadly Fuze,* 203.

135 *a set of "Running Orders"*: According to Baldwin, *The Deadly Fuze,* 80, these were widely understood from the beginning and formally listed only gradually. Van Allen had a written copy; memo, "Section T 'Running Orders,'" February 2, 1944, UI, JAVA.

"This is it": Baldwin, *The Deadly Fuze,* 105.

"You never knew what": Klingaman, *APL—Fifty Years of Service to the Nation,* 11.

the end of May: L. W. Baxter to D. Luke Hopkins, memo, "Identification," May 29, 1942, NARA, RG 227, E NC-138 8.

jumped to 196: Applied Physics Laboratory, "Telephone Numbers and Addresses of Employees as of July 2, 1942," NARA, RG 227, E NC-138 8.

"challenge strangers": M. A. Tuve to All Group Supervisors, memo, "New Personnel," April 29, 1942, LOC, MAT.

sophisticated ADT alarm system: L. W. Baxter to Board of County Commissioners, Montgomery County, May 12, 1942, NARA, RG 227, E NC-138 8; D. Luke Hopkins to W. S. Parsons, memo, "Security—American District

Telegraph," June 13, 1942, NARA, RG 227, E NC-138 8; D. Luke Hopkins to W. S. Parsons, memo, "Security," June 22, 1942, NARA, RG 227, E NC-138 8; D. Luke Hopkins to W. S. Parsons, memo, "Security—American District Telegraph," June 13, 1942, NARA, RG 227, E NC-138 8.

136 *Silver Theater:* Klingaman, *APL—Fifty Years of Service to the Nation,* 9.

"Assume always an enemy agent": L. R. Hafstad to All Applied Physics Laboratory Personnel, memo, "Security," July 3, 1942, MAT LOC.

top secret classification: The top secret classification was not introduced in 1944; Security Committee to Unit, Group, and Project Supervisors, memo, "Classification of Material Connection with Section T Work," April 18, 1944, NARA, RG 227, E NC-138 8.

couldn't find his copies: E. O. Salant to L. R. Hafstad, memo, "Lost Document," July 22, 1942, NARA, RG 227, E NC-138 8.

"some very unfortunate": D. Luke Hopkins to Irvin Stewart, memo, "Incident Reported to the F.B.I.," May 23, 1942, NARA, RG 227, E NC-138 8.

a man taking pictures: Hudson Moore Jr. to W. S. Parsons, memo, "Taking a Picture of Building at 8621 Georgia Avenue by Unknown Photographer," June 29, 1942, NARA, RG 227, E NC-138 8.

Even Deak Parsons's wife: Christman, *Target Hiroshima,* 91–92.

137 *near Leonardtown, Virginia:* Memo, "Abstracted Report of Group Supervisors for the Week Ending April 25, 1942," May 5, 1942, DTM.

seven hundred test rounds every day: Baldwin, *The Deadly Fuze,* 75.

after inspecting it himself: Memo, "Outline of Telephone Conversation Between Comdr. Parsons and Mr. Wayne Coy—April 24, 1942—12:35 p.m.," n.d., NARA, RG 227, E NC-138 8.

some submachine guns: Memo, "Mr. Hopkins Telephones the Following Questions for Comdr. Parsons," NARA, RG 227, E NC-138 8.

Richard Crane and David Dennison: For example, see H. R. Crane and D. M. Dennison to D. I. Lawson, "Memo on Problem of Hit Probabilities," January 6, 1942, DTM.

"Vulnerability of Modern Planes": D. R. Inglis to M. A. Tuve, memo, "Preliminary Suggestion of Experiments on Vulnerability of Modern Planes to Shell Fragments," March 18, 1942, DTM.

Roberts and seven other: R. B. Brode to L. R. Hafstad, memo, "Parris Island, Expedition I," April 4, 1942, DTM.

engineless Taylor Cub: Ibid.; Section T, *Progress Report as of April 19, 1942,* 40–41; Brode to Hafstad, "Parris Island Report." About 32 percent of the shells got within thirty feet. On April 13 the gun was 11,150 feet from the airplane; it was 10,200 feet on April 14.

138 *needed to be more sensitive:* Roberts, "Autobiography," 7.

within seventy feet: Buford Rowland and William Boyd, *U.S. Navy Bureau of Ordnance in World War II* (Washington, D.C.: U. S. Government Printing Office, 1953), 295. These distances varied according to which fuse model was being used and the angle of the radio beam off the plane.

proving hard to adapt: Merle Tuve to F. A. Vick, June 29, 1942, NARA, RG 227, E NC-138 8.

damaging vapor residue: Ross Wood to L. G. Hector, February 9, 1942, DTM.

careful study revealed: Baldwin, *The Deadly Fuze,* 92. Baldwin doesn't date this episode but suggests it happened the first summer after Crosley was hired.

"It now appears impractical": M. A. Tuve to W. S. Parsons, memo, "Implied Commitments of OSRD Funds for Section T," after July 1, 1942, NARA, RG 227, E NC-138 8.

"catch flies and put": Baldwin, *The Deadly Fuze,* 98.

the final obstacle: Section T, *Progress Report as of April 19,* 24.

learned a week later: "In Memoriam: Merle Antony Tuve," 212.

no less than five: Section T, *Progress Report as of April 19,* 24.

"If you need thirty clocks": "In Memoriam: Merle Antony Tuve," 212.

139 *during the Parris Island tests:* Section T, *Progress Report as of April 19,* 41.

16. THREE RUNS, THREE HITS

140 *888-foot Navy aircraft carrier:* Keith, *Stay the Rising Sun,* prologue.

eighteen fighters and thirty-six: Bureau of Ships, "Preliminary Report: Loss of U.S.S. *Lexington,* May 19, 1942, Coral Sea," *War Damage Report* 16, June 15, 1944, 2.

Jesse Rutherford Jr.: Keith, *Stay the Rising Sun,* prologue, chapter 9.

twelve-foot, two-ton monster: The *Lexington* had 5"/25 Mk 10 guns; Robert C. Stern, *The Lexington Class Carriers* (Annapolis, MD: Naval Institute Press, 1993), 98. The total gun length is 11.85 feet; I've rounded up.

141 *did not deter:* Richard W. Bates, *The Battle of the Coral Sea, May 1 to May 11 Inclusive, 1942: Strategical and Tactical Analysis* (Washington, D.C.: National Technical Information Service, 1947), 100.

four Mk 19 "gun directors": Stern, *The Lexington Class Carriers,* 98.

were all preset: Gunnery Officer, U.S.S. *Lexington,* to Commanding Officer, U.S.S. *Lexington,* "Action Fought in the Coral Sea with the Japanese Navy, 7–8 May 1942," May 30, 1942, 3.

producing explosions: Keith, *Stay the Rising Sun,* chapter 9.

the formation had broken: Bates, *The Battle of the Coral Sea,* 97.

142 *Like victims at Pompeii:* Keith, *Stay the Rising Sun,* chapter 9.

splintered a storage locker: Bureau of Ships, "Preliminary Report: Loss of U.S.S. *Lexington*," 3.

only six of the fifty-four: Gunnery Officer, "Action Fought in the Coral Sea," 7. The gunnery officer claimed six planes downed, but he might have been overly optimistic in his estimate. According to other Navy reports, the figure might have been four; Commanding Officer, U.S.S. *Lexington,* to the Commander-in-Chief, U.S. Pacific Fleet, "Report of Action—the Battle of the Coral Sea, 7 and 8 May 1942," May 15, 1942, 13.

At 12:47 p.m.: Bureau of Ships, "Preliminary Report: Loss of U.S.S. *Lexington*," 7.

were safely rescued: Keith, *Stay the Rising Sun,* chapter 10.

"Air offense is definitely superior": Commanding Officer, "Report of Action—the Battle of the Coral Sea," 9.

143 *Dick Roberts was impressed:* Roberts, "Autobiography," 7.

August 10, 1942: Baxter, *Scientists Against Time,* 232. Several sources, including Baldwin, put the date as August 12, but Baxter is correct; "Antiaircraft Phototriangulation Worksheets" at DTM list the firing date as August 11.

carried a thousand men: Christman, *Target Hiroshima,* 94–95.

144 *vessel was not designed:* Mark Stille, *US Navy Light Cruisers, 1941–45* (Oxford, UK: Osprey, 2016), chapter 4.

a maze of control rooms: Diagram, "Cutaway of a Typical Cruiser," *All Hands* 498 (1958): 32–33. The caption describes the diagram as a typical cruiser but, as is clear from the outline of the ship, the diagram portrays a Cleveland-class light cruiser.

on a "shakedown" cruise: Roberts, "Autobiography," 7.

145 *Parsons had requested six:* Christman, *Target Hiroshima,* 94.

had never once seen: Baldwin, *The Deadly Fuze,* 144.

would help to aim: Section T, "Quarterly Report as of September 30, 1942," 16.

A standard gun crew: Manual, "5-Inch Twin Gun Mounts Mark 28, 32, and 38 All Mods: Description and Instructions," OP 805, First Revision, Department of the Navy, Bureau of Ordnance, March 19, 1957, 30–47, available at https://maritime.org.

promptly crashed: Baldwin, *The Deadly Fuze,* 142.

from three thousand yards away: Section T, "Quarterly Report as of September 30, 1942," 16.

146 *"You've wrecked two":* Jacobsen, *The Deadly Fuze.*

over an hour later: "Antiaircraft Phototriangulation Worksheet," Run 7, U.S.S. *Cleveland,* August 11, 1942, DTM; "Antiaircraft Phototriangulation

Worksheet," Run 8, U.S.S. *Cleveland,* August 11, 1942, DTM. Where there are minor discrepancies between the worksheets and the summary of the test results in Section T's quarterly report the following month, I've used the quarterly report.

Eight shots and: Section T, "Quarterly Report as of September 30, 1942," 16.

captain came down: Baldwin, *The Deadly Fuze,* 144–45.

canceled their shore leave: Boyce, *New Weapons for Air Warfare,* 121.

"Three runs, three hits": Baldwin, *The Deadly Fuze,* 146.

17. INTO THE FLEET

147 *rear admiral:* Office of Government Reports, *United States Government Manual* (Washington, DC: Office of War Information, 1942), 273; G. F. Hussey, Jr. to Merle Tuve, March 23, 1945, LOC, MAT.

"It'll take an afternoon": "Transcript of Meeting on VT (Proximity) Fuze," 14.

148 *When Shumaker took him:* Ibid., 15.

seven-pound fuses: The Mark 32 Navy fuses, which is what Hussey would have seen, weighed 6.81 pounds; United States Navy Bomb Disposal School, *U.S. Navy Projectiles and Fuzes,* 306.

"I agree": Baldwin, *The Deadly Fuze,* xx.

the footage of the Chesapeake: Another reality surely made the footage look even better: shells that weren't triggered by the drones wouldn't have exploded at all.

Devices made at Crosley: Section T, "Quarterly Report as of September 30, 1942," 16.

149 *Merle formed an emergency:* Ibid., 8.

when the hotshot physicist: Roberts, "Autobiography," 7.

"friction in human transactions": Section T, "Quarterly Report as of September 30, 1942," 5.

By September 19: Memo, "Abstracted Report of Section T Activities for the Week Ending September 19, 1942," September 30, 1942, DTM.

September 26: Memo, "Abstracted Report of Section T Activities for the Week Ending September 26, 1942," October 7, 1942, DTM.

one of Tuve's teams modified: Section T, "Quarterly Report as of September 30, 1942," 13.

ship one hundred thousand: Ibid., 5.

150 *cost only six cents:* "Memo of Telephone Conversation Between Mr. Pollak and Lieut. Comdr. M. M. Mathsen," September 18, 1942, NARA, RG 227, E NC-138 8.

replaced with black lacquer: L. R. Hafstad to M. A. Tuve, memo, "Notes on Our Discussion of Navy Needs," October 20, 1942, NARA, RG 227, E NC-138 8.

Beryllium copper: L. R. Hafstad to M. A. Tuve, memo, "Crosley Engineering Program," October 12, 1942, NARA, RG 227, E NC-138 8.

listed by OSRD: OSRD, "Transition Office Circular No. 3. Critical Materials and Substitutions," November 11, 1942, LOC, MAT.

"drop, jolt, and jumble": W. H. Offenhauser Jr. to Mr. Torreson, memo, "Jolt and Jumble Tests," July 24, 1942, DTM. The standard drop test was actually forty feet, but as suggested later in the chapter, that standard would have been impossible to meet in fuses incorporating the reserve glass battery. The five-foot standard is mentioned in "Conference at DTM November 7, 1942: Porter, Tuve, Torreson," November 7, 1942, LOC, MAT, which quotes the Navy Bluejacket's Manual: "Any loaded shell dropped five feet or more is carefully packed for return to the ammunition depot. Flare shells or other special shells are promptly dropped over the side."

Section T needed to know: M. A. Tuve to L. R. Hafstad, September 30, 1942, NARA, RG 227, E NC-138 8; L. R. Hafstad to O. W. Torreson, memo, "Squibs at reduced air pressure," September 9, 1942, LOC, MAT; L. R. Hafstad to M. A. Tuve, memo, "Crosley engineering program," October 12, 1942, NARA, RG 227, E NC-138 8; Memo, "Abstracted Report of Section T Activities for the Week Ending September 19, 1942," September 30, 1942, NARA, RG 227, E NC-138 8.

no battery existed: "Transcript of Meeting on VT (Proximity) Fuze," 26–36.

151 *way back to 1918:* Department of the Interior, *Annual Report of the Commissioner of Patents for the Year 1918* (Washington, D.C.: Government Printing Office, 1919), 384.

generate the same capacity: See H. F. French, "Improvements in B-Battery Portability," *Proceedings of the I.R.E.* 29, no. 6 (1941): 299–303. As French describes, the big jump in 1938 was due to the fact that the Hygrade Sylvania Corporation and the Philco Radio and Television Corporation introduced 1.4-volt tubes that "offered the possibility of cutting receiver power demand to one half, or even one third, of its prior value."

French's innovative batteries: Ibid.

two inches wide: O. W. Torreson, "Energizer Story," November 1944, DTM.

tiniest battery cells in half: "Transcript of Meeting on VT (Proximity) Fuze," 33.

frozen larger batteries: Baldwin, *The Deadly Fuze,* 199.

Jean Paul Teas: Section T Personnel File, Jean Paul Teas Jr., LOC, MAT.

152 *with the draft board:* M. A. Tuve to Dr. Irvin Stewart, September 30, 1942, LOC, MAT.

The units aged quickly: H. F. French to M. A. Tuve, November 3, 1942, LOC, MAT; Memo, "Abstracted Report of Section T Activities for the Week Ending September 19, 1942," September 30, 1942, NARA, RG 227, E NC-138 8.

a replacement-battery program: Torreson, "Energizer Story."

suggested by Harry French: "Transcript of Meeting on VT (Proximity) Fuze," 34.

sulfuric and chromic acids: Section T to Victor Hicks, memo, "Use of Reserve Battery in T-3 Units," September 19, 1942, NARA, RG 227, E NC-138 8.

a forty-foot plunge: Memo, "Monthly Survey of Section T Activities: October, 1942," November 20, 1942, DTM.

when Section T dropped them: Ibid.

"It is now September": M. A. Tuve to H. H. Porter, memo, "Research Battery Situation at National Carbon Company," September 11, 1942, LOC, MAT.

153 *sat in on meetings:* Memo, "Conference at O. W. Torreson's Office, D.T.M., October 5, 1942," October 6, 1942, LOC, MAT; Memo, "Conference at J.H.U., November 10, 1942," November 12, 1942, LOC, MAT.

"inclined to give various alibis": Tuve to Porter, September 11, 1942, LOC, MAT.

Neither Rear Admiral Hussey: E. D. McAlister to M. A. Tuve, memo, "Items We Need from the Navy," October 20, 1942, NARA, RG 227, E NC-138 8; L. R. Hafstad to M. A. Tuve, memo, "Notes on Our Discussion of Navy Needs," October 20, 1942, NARA, RG 227, E NC-138 8; H. H. Porter to M. A. Tuve, memo, "Navy Action," October 20, 1942, NARA, RG 227, E NC-138 8.

shelf life of merely six months: Section T, "Quarterly Report as of September 30, 1942," 18.

Van Allen was commissioned: Foerstner, *James Van Allen,* 57.

Eleven years had passed: Ibid., 26.

one of three unmarried men: M. A. Tuve to Director of Naval Officer Procurement, October 26, 1942, NARA, Group 227, E NC-138 8.

starved to death: Foerstner, *James Van Allen,* 56.

the recommended gear: W. S. Parsons to Messrs. Dilley, Peterson, and J. Van Allen, memo, "Notes on Procedure After Your Commissioning," October 28, 1942, NARA, RG 227, E NC-138 8.

some fifty-nine hundred: Foerstner, *James Van Allen,* 56, suggests all fuses were delivered already fitted inside shells, as does Christman, *Target Hiroshima,* 95, but Van Allen's diary suggests that there were about 5,420 "loose" fuses not fitted into shells that were instead packed in "skid boxes" of 40. Van Allen, "Field Notes," November 30, 1942.

154 *"steeped in the virtues":* Foerstner, *James Van Allen,* 56–57.

"condensed set of working formulas": Van Allen, "Field Notes," November 23, 1942.

The Thanksgiving menu: Foerstner, *James Van Allen,* 57.

He had already spoken: Christman, *Target Hiroshima,* 95.

Neil Dilley: D. Luke Hopkins to director of Naval Officer Procurement, October 26, 1942, October 28, 1942, NARA, RG 227, E NC-138 8.

Bob Petersen: Section T Personnel File, Robert Petersen, LOC, MAT.

okayed a "hunting license": Christman, *Target Hiroshima,* 95.

155 *The bane of a gunner's job:* Van Allen, interview, 99.

"black box characteristics": Jacobsen, *The Deadly Fuze.*

They didn't believe him: Van Allen, interview, 98.

18. NEW TRICKS

156 *On New Year's Day:* Boyce, *New Weapons for Air Warfare,* 121; Marion R. Kelley, "Chronology—APL History," December 28, 1964, APLA, 1–43. Boyce reported that Parsons attended the meeting, which seems in error; Kelley does not.

in the Cabinet Room: White House Stenographer's Diary, "Franklin D. Roosevelt Day by Day," January 1, 1943. The Pare Lorentz Center, Franklin Delano Roosevelt Presidential Library, available at http://www.fdrlibrary .marist.edu/daybyday/.

Tuve's first encounter: Boyce, *New Weapons for Air Warfare,* 121.

a notorious Luddite: Zachary, *Endless Frontier,* 161.

like Deak Parsons: Rhodes, *The Making of the Atomic Bomb,* 477.

the course of three days: Day log, "Franklin Roosevelt Day by Day," December 30 and 31, 1942, and January 1, 1943, Pare Lorentz Center, Franklin Delano Roosevelt Presidential Library, available at fdrlibrary.marist.edu/daybyday/.

the morning of January 5: Christman, *Target Hiroshima,* 83; Commanding Officer, U.S.S. *Helena,* to Commander in Chief, United States Fleet, "War Diary, U.S.S. *Helena,* period January 1, 1943, to January 31, 1943," January 31, 1943, fold3.com, file A16-3, serial 0011, micro serial 46850, reel A239.

Munda Point airfield: Commanding Officer, U.S.S. *Helena,* to Commander in Chief, U.S. Pacific Fleet, "Munda Airfield—Bombardment of—morning of 5 January, 1943," January 8, 1943, fold3.com, file A16-3, serial 001, micro serial 45433, reel A189.

157 *"Let's go get into a fight":* Christman, *Target Hiroshima,* 96.

was a surprise: Executive Officer (Acting), U.S.S. *Helena,* to Commanding Officer, "After Battle Report, Attack by Japanese Dive Bombers at 0915, January 5, 1943," January 7, 1943, fold3.com, file A16-3, serial 002, micro serial 45685, reel A198.

were out of range: Commanding Officer to Commander in Chief, U.S. Pacific Fleet, "Anti-Aircraft action by U.S.S. Helena against surprise dive bomber attack at 2245 Zebra, 4 January, 1943," January 8, 1943, fold3.com, File A16-3, Serial 002, Micro Serial 45685, Reel A198.

"This beats target practice": Christman, *Target Hiroshima,* 97.

a pistol and killed himself: Ibid.

158 *Admiral Willis Lee:* Joseph Thomas, *Leadership Embodied: The Secrets to Success of the Most Effective Navy and Marine Corps Leaders,* 2nd ed. (Annapolis, MD: Naval Institute Press, 2013), 104–7.

"talk shop and drink": U.S.S. *North Carolina,* "War Diary: 1 December 1944–31 December 1944," December 28, 1944, 8, fold3.com, micro serial no. 103158, reel A1356.

met Lee in his cabin: Van Allen, interview, 98–99; Van Allen, "What Is a Space Scientist?," 12; Foerstner, *James Van Allen,* 59.

shot down five Japanese: Klingaman, *APL—Fifty Years of Service to the Nation,* 13.

chose to remain: Van Allen, interview, 102.

159 *a fuse-instruction manual:* Ibid., 106.

"power of the sea": Foerstner, *James Van Allen,* 66.

learn from the Navy: Van Allen, interview, 101.

"essentially a floating refrigerator": Ibid., 106.

"a way of speaking": Foerstner, *James Van Allen,* 58.

a fifty/fifty mix: Ibid.

By March 31, Section T: Section T, "Quarterly Report: January 1, 1943–March 31, 1943," 13.

traded their stuffy lodgings: Zachary, *Endless Frontier,* 141.

a machine shop: Ibid.

160 *straining to coax color:* Bush, "Laboratory Notebook."

He devised a process: Vannevar Bush, "Memo on a System of Picture Transmission," April 7, 1942, LOC, VB.

manufacturing machine guns: L. H. Adams to R. C. Tolman, October 4, 1941, LOC, VB.

mooring aircraft to ships: H. H. Arnold to Vannevar Bush, November 14, 1940, LOC, VB.

"I have a Saint Bernard dog": V. Bush to Dogs for Defense Incorporated, December 22, 1942, LOC, VB.

"which would incorporate": V. Bush to W. S. Parsons, August 31, 1942, NARA, RG 227, E NC-138 8.

"A radio proximity fuse": Section T, "Quarterly Report as of December 31, 1942," 3.

161 *108 Allied craft:* Zachary, *Endless Frontier,* 166.

"the toughest man": Jonathan W. Jordan, *American Warlords: How Roosevelt's High Command Led America to Victory in World War II* (New York: NAL Caliber, 2015), 202.

"He is the most even-tempered": Zachary, *Endless Frontier,* 161.

"There's too much radar": Buderi, *The Invention That Changed the World,* 161.

"terrible blind spot": Zachary, *Endless Frontier,* 161.

"made the decisions": Bush OH, reel 13-B, 816.

"trying to mess into things": Zachary, *Endless Frontier,* 161.

162 *"not just one way":* Ibid., 164.

offered a new defense: OSRD, *Report of the National Defense Research Committee for the First Year of Operation,* 16.

a breach of channels: Zachary, *Endless Frontier,* 168; White House Stenographer's Diary, "Franklin D. Roosevelt Day by Day," March 24, 1943.

"a struggle between": Buderi, *The Invention That Changed the World,* 161.

"research-based statistical analysis": Zachary, *Endless Frontier,* 171.

163 *over forty-four months:* Bush, *Pieces of the Action,* 88; Buderi, *The Invention That Changed the World,* 167; Naval History and Heritage Command, "German U-Boat Casualties in World War II," n.d., available at history.navy.mil.

the dominant factor: Buderi, *The Invention That Changed the World,* 169.

Germany withdrew their submarines: Ibid., 168.

"superiority in the field": Baxter, *Scientists Against Time,* 43.

19. CHERRY STONE

164 *In April 1943:* Hölsken, *V-Missiles of the Third Reich,* 72.

a Luftwaffe artillery school: Edward B. Westermann, *Flak: German Anti-Aircraft Defenses, 1914–1945* (Lawrence: University Press of Kansas, 2001), 76.

"The Colonel is urgently": Wachtel, "Unternehmen Rumpelkammer," 100.

top Luftwaffe commander: Campbell, *Target London,* 68.

165 *"Have you ever heard"*: Wachtel, "Unternehmen Rumpelkammer," 100.
    The führer sought "revenge": De Maeseneer, *Peenemünde*, 108.
    on May 12: Campbell, *Target London*, 71.
    Twenty-seven feet long: Different models of the V-1 had slightly different lengths. For specifications see Hölsken, *V-Missiles of the Third Reich*, 335.
166 *four hundred miles per hour:* Ibid.
    three German companies: Nijboer, *Meteor I*, 20.
    The initial launching system: Hölsken, *V-Missiles of the Third Reich*, p. 83.
    Wachtel's regiment would fire: Zaloga, *German V-Weapon Sites*, 17; Walpole, *Hitler's Revenge Weapons*, 39.
    force of sixteen g's: Hölsken, *V-Missiles of the Third Reich*, 75.
167 *six thousand workers:* De Maeseneer, *Peenemünde*, 127.
    from Germany's top universities: Middlebrook, *The Peenemünde Raid*, 24.
    Advertisements for various: Petersen, *Missiles for the Fatherland*, 68.
    "We had parties": Ibid., 96.
    eleven thousand foreigners: Middlebrook, *The Peenemünde Raid*, 29.
    as young as sixteen: Ibid., 30.
    village of Trassenheide: For a map of the camp, see Hölsken, *V-Missiles of the Third Reich*, 33.
168 POLES AND DOGS: Middlebrook, *The Peenemünde Raid*, 33.
    largest freestanding structures: Petersen, *Missiles for the Fatherland*, 91.
    twelve hundred more laborers: Middlebrook, *The Peenemünde Raid*, 33.
    "not even a mouse": Ibid., 34.
    Bunkers and launchers: The V-1 sites were not yet under construction, but the first V-2 site was; Zaloga, *German V-Weapon Sites*, 20.
    The V-1 was slated: Hölsken, *V-Missiles of the Third Reich*, 84.
    Hidden microphones adorned: Williams, *Operation Crossbow*, 121.
    General Ludwig Crüwell: Derek R. Mallett, *Hitler's Generals in America: Nazi POWs and Allied Military Intelligence* (Lexington: University Press of Kentucky, 2013), 22.
    General Wilhelm von Thoma: Zabecki, *World War II in Europe*, 520.
169 *"no progress whatsoever can":* Jones, *Most Secret War*, 333.
    "It looks": Ibid., 332.
    "a new factory at Peenemünde": Campbell, *Target London*, 53.
    a German tank expert: Ibid., 56–57.
    an extraordinary hunch: Jones, *Most Secret War*, 336.
170 *the winter snow:* Campbell, *Target London*, 55.
    Frederick Lindemann: Ibid., 335.
    irrigation "mud pumps": Ibid., 339.
    spotted a Nazi rocket: Ibid., 340–41.

"up to four thousand casualties": Ibid., 343.
"living and sleeping quarters": Middlebrook, *The Peenemünde Raid*, 45.
remained unknown to the Allies: Ibid., 46.

20. A LONDON FUSE

172 *simply had to know*: D. Luke Hopkins to W. S. Parsons, memo, "Security," September 16, 1942, NARA, RG 227, E NC-138 8. Theodore Wolfe was the son of John Edgar Wolfe and Vera Fields Wolfe; "In Memoriam: Kenneth M. Wolfe," *Spirit of Jefferson Farmer's Advocate*, December 29, 1988. John Edgar Wolfe was the brother of James William Mason Wolfe, who was Garland W. Wolfe's father. Garland and Theodore also appear together in "Personal and Social Activities of Charles Town and Vicinity: Millville," *Spirit of Jefferson, Charlestown, West Virginia*, July 22, 1942.

The Navy let him off with a warning: D. Luke Hopkins to W. S. Parsons, memo, "Security," October 8, 1942, NARA, RG 227, E NC-138 8.

a squad of women: Applied Physics Laboratory, *The First Forty Years*, 10; Klingaman, *APL — 50 Years of Service to the Nation*, 11. Lucy Trundle apparently went by Lulu or Lula; Herb Rudefer, "Sports," *APL-JHU-S* 1, no. 20 (1946): 4, APLA.

Section T numbered 480: Memo, "Summary of Resignations," June 5, 1943, NARA, RG 227, E NC-138 8.

The Navy Yard complained: L. R. Hafstad to S. J. Ratner, memo, "Disposal of Confidential 'Junk,'" March 26, 1943, NARA, RG 227, E NC-138 8.

gas tank and pump: L. W. Baxter to Brode et al., memo, "Gasoline for APL-JHU Vehicles," January 19, 1943, NARA, RG 227, E NC-138 8.

173 *construction had begun*: Memo, "Additional Information in Support of Request to Erect a Two-Story Building," December 1, 1942, NARA, RG 227, E NC-138 8; D. Luke Hopkins to M. A. Tuve, memo, "Special Authorization for Third Floor — New Building," November 27, 1943, NARA, RG 227, E NC-138 8.

Security incidents had increased: D. Luke Hopkins to W. S. Parsons, memo, "Security," September 16, 1942, NARA, RG 227, E NC-138 8; Lt. H. Moore, Jr. to J. A. Bearden, memo, "Report upon investigation of anonymous phone call to Mrs. James R. Day," November 4, 1942, NARA, RG 227, E NC-138 8; L. R. Hafstad to Lieutenant Hudson Moore, memo, "Security," September 11, 1942, NARA, RG 227, E NC-138 8; L. R. Hafstad to Lieutenant Hudson Moore, memo, "Security," September 16, 1942, NARA, RG 227, E NC-138 8; L. R. Hafstad to M. A. Tuve, memo, "Breach of secrecy by Cincinnati and Centralab," September 10, 1942, NARA, RG 227, E NC-138 8;

H. Moore, Jr. to W. S. Parsons, memo, "Violation of security by Crosley et al.," September 10, 1942, NARA, RG 227, E NC-138 8.

shipped seven hundred and twenty thousand fuses: Section T, "Quarterly Report as of June 30, 1943," 12.

eighty-eight thousand rugged tubes: Ibid., 7.

"Working so fast that observation": Lyman J. Briggs, *NBS War Research: National Bureau of Standards in World War II* (Washington, D.C.: Department of Commerce, 1949), 19.

"Every girl would rap": Baldwin, *The Deadly Fuze,* 116.

174 *"More Women War Workers":* Ad, *Ipswich News and Chronicle,* January 15, 1943.

Sylvania had "feeder plants": "Transcript of Meeting on VT (Proximity) Fuze," 58–59.

fielded a request from: Ibid., 17–23.

Capacitors, rugged tubes: Baldwin, *The Deadly Fuze,* 160.

112 companies would help: Abelson, "Merle Antony Tuve," 411.

new manufacturing contracts: Section T, "Quarterly Report: January 1, 1943–March 31, 1943," 30.

175 *order for antiaircraft fuses:* Kelley, "Chronology—APL History," 3–42.

Its liaison in D.C.: Salant et al., "Part Twelve," 9, MIT; E. O. Salant to M. A. Tuve, "British Army AA Situation," August 31, 1943, NARA, RG 227, E NC-138 8.

April 5, 1943, Ralph Baldwin: Baldwin, *They Never Knew What Hit Them,* 87.

"Ralph," Furuholmen began: Ibid., 87. Baldwin misspells Furuholmen; See R. B. Baldwin to E. D. McAlister, memo, "Meeting at the Army War College September 27, 1943," October 4, 1943, NARA, RG 227, E NC-138 8.

176 *twenty-five times deadlier:* Baldwin, *The Deadly Fuze,* 174.

howitzer shells: Ibid., 168.

weighed only twenty-three pounds: Williams to Tuve, "Characteristics of Guns of Interest to Section T."

On April 29, at Aberdeen: Baldwin, *The Deadly Fuze,* 170–71.

Ione Berkeley: Baldwin, *They Never Knew What Hit Them,* 90.

177 *"harder to swallow":* Tuve, interview by Christman.

90 percent success rate: Baldwin, *The Deadly Fuze,* 177–78.

one million antipersonnel fuses: Ibid., 189. The record is somewhat confused on this point. In Marion R. Kelley, "Chronology—APL History," 3–42, Kelley writes that a secret order for one million Army AA fuses for 90 mm guns was placed by General G. M. Barnes in September of 1942. This suggests that the one million fuses described by Baldwin were already ordered and merely disclosed in September of 1943. But antipersonnel fuses could not have been ordered in September of 1942, which was prior to Baldwin's meeting at the War College. Since Baldwin was the Army liaison, I've relied on his narrative but have also assumed that an initial order of Army AA fuses came in late 1942.

information about new weapons: Bennett Archambault, *Report of OSRD Activities in the European Theater.*

178 *Section T's experts arrived:* Salant et al., "Part Twelve," 23.

Section T's main concern: Section T, "Quarterly Report: January 1, 1943–March 31, 1943," 30–35; E. O. Salant to E. D. McAlister, memo, "Results of 4.5 Trials to Date," February 19, 1943, NARA, RG 227, E NC-138 8; A.F.H. Thomson to E. O. Salant, memo, "Proposed Modification of Reserve Battery," February 26, 1943, LOC, MAT.

Salant was well equipped: Edward Oliver Salant, "Selective Service System: Affidavit—Occupational Classification," March 25, 1943, NARA, RG 227, E NC-138 170.

needed the distraction: Seaburg, "Whatever Happened to Thelma Adamson?," 81. Seaburg notes that Adamson was "institutionalized" in 1941 but was not committed to Brattleboro Retreat until 1942.

tests at Shoeburyness: E. O. Salant to Vannevar Bush, memo, "Activities of Mission from Section T to Great Britain," April 15, 1943, NARA, RG 227, E NC-138 170; E. O. Salant to M. A. Tuve, memo, "Proposed Tests in the U.K.," April 23, 1943, NARA, RG 227, E NC-138 170; Salant et al., "Part Twelve," 9–10.

179 *height of twenty feet:* J. B. Doak to E. O. Salant, memo, "British Drop Tests of Complete 4.5 Rounds Fused with Mark 33 at Woolwich Arsenal, London," August 27, 1943, NARA, OSRD, RG 227, E NC-138 9.

met with Professor John Cockcroft: E. O. Salant, memo, "British Army Request for PF Exclusive of AA Use," August 31, 1943, NARA, RG 227, E NC-138 8.; E. O. Salant to M. A. Tuve, memo, "British Army AA Situation," August 31, 1943, NARA, RG 227, E NC-138 8.

Pile's gunners: Reuters, "Gen. Sir Frederick Pile Dies; Led British Antiaircraft in War," *New York Times,* November 15, 1976.

Known as Tim: Wesley Boyd, "An Irishman's Diary," *Irish Times,* January 8, 2007.

"the leavings of the Army": Ibid.

recruit over seventy thousand women: Dobinson, *AA Command,* 313.

fuses for 3.7-inch guns: Salant to Tuve, "British Army AA Situation." The Army's heavy stationary guns, the 4.5 and 5.25 inch guns, did not require the smaller, Mark 45 fuse.

only twenty-eight pounds: Williams to Tuve, "Characteristics of Guns of Interest to Section T." I'm referring to 4.5-inch Royal Navy shells.

Section T set in motion: E. O. Salant to E. D. McAlister, memo, "Supply of Mark 45 for British Army," December 7, 1943, NARA, RG 227, E NC-138 8.

According to MI6: See chapter 21.

21. AMNIARIX

180 *eighty-five B-17 bombers:* Overy, *The Bombing War,* 560; Allen V. Martini, "Fifteen Minutes Over Paris," *Saturday Evening Post,* November 20, 1943.

watched from Parisian rooftops: Charles Glass, *Americans in Paris* (New York; Penguin, 2010), chapter 34.

four B-17s fell: Martin W. Bowman, *US Eighth Air Force in Europe: Eager Eagles 1941–Summer 1943* (Barnsley, UK: Pen and Sword Aviation, 2012), 117.

parachuting onto French soil: Glass, *Americans in Paris,* chapter 34.

In Paris, at great risk: Ibid.

181 *Parisian newspapers under:* Rosbottom, *When Paris Went Dark,* 290.

Tuberculosis cases were up: Ibid., 291.

Five hundred pieces: Lynn Nicholas, *The Rape of Europe: The Fate of Europe's Treasures in the Third Reich and the Second World War* (New York: Vintage, 1994), 170; Tom Bazley, *Crimes of the Art World* (Santa Barbara, CA: Praeger, 2010), 90.

French Jews were hunted: Renée Poznanski, *Jews in France During World War II* (Hanover, NH: University Press of New England, 2001), 309.

a bleak prison: Rosbottom, *When Paris Went Dark,* 294.

The city's grand hotels: Ibid., 119, 334, 458; Susan Roland, *Hitler's Art Thief* (New York: St. Martin's, 2015), 215.

182 *Rousseau had encountered:* Ignatius, "After Five Decades, a Spy Tells Her Tale."

56 Avenue Hoche: Ibid. The Ignatius piece does not divulge the exact address on Hoche, but Wachtel provides it in Wachtel, "Unternehmen Rumpelkammer," 110.

"a piece of furniture": Ignatius, "After Five Decades, a Spy Tells Her Tale."

She teased them: Ibid.

Raketen was not: P. Jones, *The Secret War of Dr. Jones.*

"Pursue it": Ignatius, "After Five Decades, a Spy Tells Her Tale."

"masterpiece in the history": Grimes, "Jeannie Rousseau de Clarens."

183 *active-duty captain:* Jones, *Most Secret War,* 351.

He made frequent trips: P. Jones, *The Secret War of Dr. Jones.*

"You really think so": Ibid.

at 26 rue Fabert: Ignatius, "After Five Decades, a Spy Tells Her Tale."

"Concentrated on the island": Fourcade, *Noah's Ark,* 258–59; Jones, *Most Secret War,* 351–52.

Lieutenant Colonel Heinrich Sommerfeld: Hölsken, *V-Missiles of the Third Reich,* 202.

184 *found the claims preposterous:* P. Jones, *The Secret War of Dr. Jones.*

added a note: Ibid.

"completely out of the ordinary": Fourcade, *Noah's Ark,* 258.

"Who was Petrel's": Ibid., 259.

The moon over Peenemünde: Middlebrook, *The Peenemünde Raid,* 63.

185 *at least 50 percent incendiary:* Ibid., 68–69.

German planes accidentally: Jones, *Most Secret War,* 346–47.

"clear as a bell": Middlebrook, *The Peenemünde Raid,* 112.

186 *Some 75 percent:* Ibid., 144.

"the scientific heart": Ibid., 150.

All but one of: Jones, *Most Secret War,* 346. As Jones notes, the raid did burn some key documents, but the Germans had copies stored safely away.

The Luftwaffe airfield: Middlebrook, *The Peenemünde Raid,* 153.

Hanna Reitsch slept: Philip Joubert de la Ferté, *Rocket* (London: Hutchinson, 1957), 83.

The camp, Wachtel said: Ibid.

needed a vacation: Jones, *Most Secret War,* 349.

Frank briefed Jones on August 31: Several prominent accounts incorrectly suggest that Amniarix's "Wachtel Report" led to the major bombing of Peenemünde in mid-August 1943. It was delivered after that bombing, and its major contribution concerned the V-1, not the V-2.

187 *Nine days prior, on August 22:* De Maeseneer, *Peenemünde,* 222–23. According to Jones, *Most Secret War,* 349, it was a turnip field.

Danish naval intelligence: King and Kutta, *Impact,* 118.

two major surgeries: Jones, *Most Secret War,* 350.

At first, Jones wasn't sure: Ibid. Jones claims it was an experimental version of a V-1, which seems unlikely. It resembled far more a large HS 293 glider bomb. It appears more likely that the Allies believed the Bornholm prototype was in the same family as the HS 293 and that the Peenemünde 20 was a still larger version of Bornholm prototype. Also see Goodchild, *A Most Enigmatic War,* 489.

For the first time: Jones, *Most Secret War,* 352.

known as the "Wachtel Report": Williams, *Operation Crossbow,* 154.

22. SKI SHAPES

191 *the British government decided:* Jones, *Most Secret War,* 353; Mackay, *Half the Battle,* 191.

giant rabbit pens: Stephen Wade, *Air Raid Shelters of the Second World War* (Barnsley, UK: Pen and Sword Military, 2011), chapter 1.

six-and-a-half feet: Norman Longmate, *How We Lived Then: A History of Everyday Life During the Second World War* (London: Pimlico, 2002), 130.

their ductile frames: Donoughue and Jones, *Herbert Morrison,* 290.

require so much metal: Jones, *Most Secret War,* 353.

192 *"installing, under the cover":* Ibid., 356.

193 *seize plots of land:* Zaloga, *German V-Weapon Sites,* 24.

unusual construction: Campbell, *Target London,* 145.

Michel Hollard: Martelli, *Agent Extraordinary,* 34–35.

a Protestant missionary: Williams, *Operation Crossbow,* 148.

194 *Across two hundred miles:* Martelli, *Agent Extraordinary,* 164.

engineer named André Comps: Ibid., 167. Comps was recruited through one of Hollard's agents, Robert Rubenach.

Jones received it: Jones, *Most Secret War,* 360.

saw from the blueprint: Martelli, *Agent Extraordinary,* 175.

195 *other peculiar features:* Jones, *Most Secret War,* 360.

196 *MI6 now had a robust:* Ibid., 369.

197 *tear gases:* Brian J. Ford, *Secret Weapons,* 32–33.

a strange nighttime broadcast: Australian Associated Press, "Hitler's Speech About Italy Was a Record," *Adelaide Mail,* September 11, 1943.

"The technical and organizational": Longmate, *The Doodlebugs,* 43.

"the theme of 'revenge'": Ibid.

hampered by design changes: Hölsken, *V-Missiles of the Third Reich,* 123.

the Fieseler plant: King and Kutta, *Impact,* 100.

198 *American Ninth Air Force bombed:* Zaloga, *Operation Crossbow,* chapter 4.

Wachtel did not even mention: Ibid.

only fourteen feet wide: Zaloga, *German V-Weapon Sites,* 25.

large-scale daylight raids: Zaloga, *Operation Crossbow,* chapter 4.

Only two ski sites: Ibid.

"sledgehammers for tintacks": Zaloga, *German V-Weapon Sites,* 32.

safely outside of Beauvais: Campbell, *Target London,* 198.

199 *"At the slightest suspicion":* Ibid., 205.

already using a new code name: Jones, *Most Secret War,* 373–74.

23. RANCH COUNTRY

200 *"If they want to make a fountain":* Zachary, *Endless Frontier,* 154.

"give a man the symptoms": Ibid., 157.

Alsos: Thiesmeyer, *Combat Scientists,* 164–65.

201 *On Christmas Eve 1943:* Harvey H. Bundy, "Memo for General Marshall," December 24, 1943, NARA, RG 165, E NM84 498.

"growing mass of intelligence": Memo to the Secretary, Joint Chiefs of Staff, "Implications of Recent Intelligence Regarding Alleged German Secret Weapons," January 4, 1944, NARA, RG 165, E NM84 498.

the new British committee: Pelly was chairman of the Crossbow Sub-Committee of the British Joint Intelligence Committee, which reported to Churchill's Defense Committee through the British Joint Chiefs of Staff.

On oral instructions: Memo, "Meeting No. A, 29 December 1942," RG 165, E NM84 498.

202 *Bush was intimately involved:* S. G. Henry to Vannevar Bush, December 30, 1943, NARA, RG 218, E UD 92.

a pool of scientists: Joint Committee on New Weapons to E. V. Hungerford Jr., January 19, 1944, NARA, RG 218, E UD 92; V. Bush to S. G. Henry, January 24, 1944, NARA, RG 218, E UD 92.

anthrax was too: Memo, "Meeting No. A, 29 December 1942," RG 165, E NM84 498; memo, "Meeting No. B, 30 December 1942," NARA, RG 165, E NM84 498.

delivered in the hot: Homer W. Smith, Division 9 NDRC, "H Vapor: Summary of Data on Toxicology and Casualty Production," March 1, 1944, NARA, RG 218, E UD 92.

At Bush's urging: The same NACA engineers (including Jerome Hunsaker) who analyzed the P-20 were among those Bush recommended to Henry as advisers; Joint Committee on New Weapons to E. V. Hungerford Jr., January 19, 1944, NARA, RG 218, E UD 92; S. G. Henry to V. Bush, "Report of 19 February 1944 by the Interpretations Sub-Committee," February 21, 1944, NARA, RG 218, E UD 92.

the "Peenemünde 20": Harvey H. Bundy, "Memo for General Marshall," December 24, 1943, RG 165, E NM84 498.

203 *a bloated, winged missile:* G. S. Henry to V. Bush, memo, "Periodic Reports," February 29, 1944, NARA, RG 218, E UD 92.

The V-1 design reached: J. C. Hunsaker to C. P. Nicholas, February 16, 1944, NARA, RG 165, E NM84 498.

full-scale replica: S. G. Henry to V. Bush, February 8, 1944, NARA, RG 218, E UD 92.

threaten the D-Day invasion: Burton L. Lucas, memo for the secretary, Joint Chiefs of Staff, "Implications of Recent Intelligence Regarding Alleged German Secret Weapon," January 6, 1944, NARA, RG 165, E NM84 498.

four hundred and fifty miles per hour: E. N. Jacobs, "Secret Preliminary Analysis: Discussion Concerning Intended German Attack on England from "Cross-Bow Ski Sights [*sic*]," February 16, 1944, NARA, RG 165, E NM84 498.

exacerbated the range problem: Memo for the Joint New Weapons Committee, "Release of PD-74 Fuzes for Use Against German Secret Weapons," February 5, 1944, NARA, RG 165, E NM84 498.

Henry's committee asked: Ibid.

So did the British: Combined Chiefs of Staff, memo by the Representatives of the British Chiefs of Staff, "Use of Influence-Fuzed Bombs Over Enemy Territory," January 13, 1944, NARA, RG 165, E NM84 498.

for use over land: Combined Chiefs of Staff, memo by the Representatives of the British Chiefs of Staff, "Use of Influence-Fuzed Bombs Over Enemy Territory," August 13, 1944, NARA, RG 165, E NM84 498.

"as a special case": Joint Committee on New Weapons and Equipment, Memo for Information No. 3, "Use of PD-74 Influence Fuzes Against German Secret Weapon," February 24, 1944, NARA, RG 218, E UD 92.

"where duds cannot be": Ibid.

an emergency request: Memo, "Monthly Survey of Section T Activities, Jan. 15, 1944," January 24, 1944, NARA, RG 227, E NC-138 8; E. D. McAlister to M. A. Tuve, memo, "Army Fuses," February 18, 1944, NARA, RG 227, E NC-138 1.

at the Army War College: Baldwin, *The Deadly Fuze,* 260–61.

204 *"for the particular target":* McAlister to Tuve, "Army Fuses."

patrolling cowboys worked: Boyce, *New Weapons for Air Warfare,* 145.

over twenty-nine thousand desolate acres: William Hume II, "New Mexico Proving Ground: Construction Work Under Contract OEMsr-264 As Amended," June 20, 1943, NARA, RG 227, E NC-138 29. The proving ground was also used for an OSRD Division 1 project.

205 *"shots in the dark":* Ibid.

Section T's satellite team: Boyce, *New Weapons for Air Warfare,* 145; Marx Brook et al., "Final Report Contract OEMsr-264," January 1, 1945, Section T, Group 227, E NC-138 29.

"gasping urgency": Summary of meeting, September 11, 1943, LOC, MAT.

206 *three-day train ride:* Tuve, "New Mexico Trip Notebook," MAT, August 4 and 7, 1943.

"Nicest kind of place": Ibid.

"School of Mines": Comly, interview.

request of Ed Salant: C. C. Haworth Jr. to E. D. McAlister, memo, "New Mexico Program," March 2, 1944, NARA, RG 227, E NC-138 9.

Workman's men hung: C. C. Haworth Jr., to E. D. McAlister, memo, "Burst Pattern Obtained with Mock-Up of Special Glider in New Mexico," March 23, 1944, RG 227, E NC-138 9.

207 *twenty-five feet from the "drone":* S. G. Henry to Vannevar Bush, memo, "Periodic Report: Countermeasures Development Report Dated 3 April 1944," April 8, 1944, NARA, RG 218, E UD 92.

an 86 percent chance: Ibid.

Section T's Jean Paul Teas: Teas, "Trip Report."

hidden in Savernack Forest: Mitchell, "Special Mission."

tested at the Shoeburyness: Ibid.

five American antiaircraft battalions: Ibid.

A super-explosive called RDX: Memo, "Notes on Conversations, Captain Burke, Captain Teas with Colonel Denham," April 28, 1944, NARA, RG 227, E NC-138 8; Baxter, *Scientists Against Time,* 257. RDX was brought to the U.S. as part of the Tizard mission.

American Crossbow Committee reported: War Department Crossbow Committee, "Report on Enemy Preparations for Attack Against England by Long-Range Projectiles Launched from Channel Coast of France," May 12, 1944, NARA, RG 165, E NM84 498.

208 *Allies knew that drone trials:* Jones, *Most Secret War,* 414.

"practicable": War Department Crossbow Committee, "Report on Enemy Preparations for Attack Against England by Long-Range Projectiles Launched from Channel Coast of France," May 12, 1944, NARA, RG 165, E NM84 498.

24. CHEMICAL BOYS

209 *"This past week, we studied":* Samuel E. Hatch to David and Esther Hatch, January 2, 1944, courtesy of Alex Schneider and the Hatch family. *He was of Eastern European:* Hatch, "Service Above All."

fundamentals of gas warfare: Leo P. Brophy and George J. B. Fisher, *The Chemical Warfare Service: Organizing for War* (Washington, D.C.: Center of Military History, United States Army, 1989), 273, 319; Bai Tian, *GIS Technology Applications in Environmental and Earth Sciences* (Boca Raton, FL: CRC Press, 2016), 183.

210 *a cutting-edge technique:* In Hatch's letter to his parents of January 2, 1944, he wrote: "Our processing plant is known as the M2 Water Suspension Plant." See Noyes, *Chemistry,* 207; Leo P. Brophy, Wyndham D. Miles, and Rexmond C. Cochrane, *The Chemical Warfare Service: From Laboratory to Field* (Washington, DC: Center of Military History, United States Army, 1988), 331.

like doing laundry: Brophy and Fisher, *The Chemical Warfare Service: Organizing for War,* 322.

a hodgepodge group: See "130th Chemical" n.d., a list of the 130th, including which men were fatalities from the V-1 strike, available at londonmemorial.org. For Zanfagna, also see the death notice in the *Boston Globe,* July 1, 1948. For Strout, see "Auburn Man Dies in England, Believed Robot Bomb Victim," *Lewiston Daily Sun,* July 17, 1944.

all sorts of professions: Some home states, education levels, and previous employments were found through World War II U.S. Army enlistment records, available at fold3.com.

211 *clerk at a Portland:* "Three Portland Soldiers, Friends for Years, May Have Died Together," *Portland Press Herald,* July 17, 1944. The article misspells Cooke's name.

"In the spring": Hatch, "Service Above All." Cooke may have sung a variant of the standard tune or Hatch may have misremembered the cadence slightly. The standard version runs: "Around her hair she wore a yellow ribbon / She wore it in the springtime, in the early month of May /And if you asked her why the heck she wore it / She'd say she wore it for her soldier who was far, far away."

reminded married GIs: "Sgt. Chester R. Peterson Killed on Duty in Britain: Robot Bomb May Have Caused Death," *Portland Press Herald,* July 7, 1944; "Pvt. Donat Patry Killed in England," *Lewiston Daily Sun,* July 20, 1944; "Rumford Corporal Killed in Action," *Lewiston Daily Sun,* July 17, 1944; "Auburn Man Dies in England."

"zeal and industriousness": 130th Unit History, 6.

Sibert's commander had never: Ibid., 7. The commander referenced was the Commanding General of the Unit Training Center, Camp Sibert.

"I never felt better": Samuel E. Hatch to David and Esther Hatch, May 10, 1944, courtesy of the Alex Schneider and the Hatch family.

212 *"like monkeys in a cage"*: George J. Maskin, "When Yank Was a Novelty to U.K.: Oldest Outfit in 8th Bomber Command Marks Birthday," *Stars and Stripes,* May 17, 1944.

one hundred and thirty thousand American troops: Lynne Olson, *Citizens of London,* 272.

"greatest operating military": Ibid.

full-blown American colony: The living quarters information, including the addresses set aside on Sloane Court, can be found in "Central Base Station History: 1942–1944," *Historical Reports and Monographs — Geographical Command Reports,* RG 498, fold3.com, file 587a, roll MP63-9_0119; "Central Base Station: Installations," RG 498, fold3.com, file 60, roll MP63-9_0006.

spiffy Class A uniforms: S. Hatch to D. and E. Hatch, May 10, 1944.

213 *"The Yanks were the most":* Donald L. Miller, *Masters of the Air: America's Bomber Boys Who Fought the Air War Against Nazi Germany* (New York: Simon and Schuster, 2006), 138.

Red Cross's Rainbow Corner: Olson, *Citizens of London,* 282.

"oversexed, overpaid": David Reynolds, *Rich Relations: The American Occupation of Britain 1942–1945* (London: Phoenix, 1995).

"Let me close with": S. Hatch to D. and E. Hatch, May 10, 1944.

at 4 Crinan Square: 130th Unit History, 9.

"M2 Water Suspension" plants: Ibid.

nearly eight thousand pounds: Brophy et al., *The Chemical Warfare Service: From Laboratory to Field,* 330.

the Stars and Stripes*:* "Crimea Now a Nazi Boast: Sebastopol Called Useless; Reds Cite Huge Enemy Losses; Fronts Quiet," *Stars and Stripes,* May 12, 1944; "Heavies Out 15th Straight Day to Batter 11 Air, Rail Targets Deep Behind Channel Defense," *Stars and Stripes,* May 10, 1944; "Rail Yards in Germany Are Blasted; Ploesti Bombed; Lines Feeding West Wall Hit; Wall Itself Plastered," *Stars and Stripes,* June 1, 1944.

214 *"'D-Day' Fever Has":* United Press, "'D-Day' Fever Has All U.S. Steamed Up," *Stars and Stripes,* May 29, 1944.

"Garlands of colored lights": Allan M. Morrison, "Britain Goes About Its Business Calmly, Come Hell or Invasion," *Stars and Stripes,* May 16, 1944.

"The last breathing space": Associated Press, "Berlin Warns of Allied Blows Inland as Well as on the Coast," *Stars and Stripes,* May 16, 1944.

single-engine Percival Proctor: Jones, *Most Secret War,* 406–7.

a "Baby Blitz": Overy, *The Bombing War,* 120–21.

under the dining-room table: Jones, *Most Secret War,* 396.

remained dangerously dispersed: Ibid., 401.

215 *six hundred Royal Air Force sorties:* Ibid., 408.

six of the ninety-two targeted: Ibid., 411; Buderi, *The Invention That Changed the World,* 214–15.

size of the drone: Another clue to the wingspan was the width of the doorway it would have to exit through; Brian Johnson, *The Secret War* (Barnsley, UK: Pen and Sword, 1978), 147.

continued to feed MI6: Jones, *Most Secret War,* 373–74.

216 *to interview her directly:* de Clarens, interview by Ignatius.

the name Madeleine Chauffeur: Sebba, *Les Parisiennes,* 214.

"just a poor little French": de Clarens, interview by Ignatius.

shipped to Ravensbrück: Sebba, *Les Parisiennes,* 215–16.

25. THINGS CARRIED

217 *Combat engineers:* For a nice account, see Rob Morris, "A Combat Engineer on D-Day," November 3, 2018, warfarehistorynetwork.com.

hundreds of peculiar: In addition to the First, Fifth, and Sixth Special Engineer Brigades, the 131st Quartermaster Battalion planned to bring 102 duck boats ashore on D-Day, and the 280th Quartermaster Battalion planned to bring 43 ashore. The British apparently also ferried several hundred. Eyewitnesses confirm hundreds of DUKWs; Kim Briggeman, "Lolo D-Day Hero Passes Away Days After Commemoration," *Missoulian,* June 10, 2019; memo, "Operation Neptune Monograph, Provisional Engineer Special Brigade Group," 203, fold3.com, file 493F.

backed off the LST: The rear of the DUKWs had more buoyancy, so backing off made them easier to maneuver at launch; "Operation Neptune Monograph," 203.

hundreds of pints of blood: John Boyd Coates, ed., *Blood Program in World War II* (Washington, DC: Office of the Surgeon General, 1964), 484.

called serum albumin: Andrus, *Advances in Military Medicine,* vol. 1, 364.

218 *Two billion pills:* Daniel T. Willingham, *When Can You Trust the Experts? How to Tell Good Science from Bad in Education* (San Francisco: Jossey-Bass, 2012), 108.

wasn't enough penicillin: Andrus, *Advances in Military Medicine,* vol. 2, 717.

its yield by one hundred times: Ibid., 718.

"beach barrage rockets": Baxter, *Scientists Against Time,* 206–7.

hundreds of radar jammers: Compton et al., *Applied Physics,* 69.

keep the Nazis ignorant: Michael J. Donovan, *Strategic Deception: Operation Fortitude* (Carlisle, PA: U.S. Army War College, 2002), 10.

219 *electronic-targeting gadgets:* Buderi, *The Invention That Changed the World,* 214.

"the most sophisticated faking": Baxter, *Scientists Against Time,* 72.

"Boat, machine guns, rockets": For rocketboatmen, see William Howard Palmer, *We Called Ourselves Rocketboatmen* (Atlanta: LitFire, 2012), 66–70.

special navigation equipment: Baxter, *Scientists Against Time,* 69–70.

A medical boat sank: Coates, *Blood Program in World War II,* 699.

220 *additional curious contraptions:* Colin Flint, *Geopolitical Constructs: The Mulberry Harbours, World War Two, and the Making of a Militarized Transatlantic* (Lanham, MD: Rowman and Littlefield, 2016), 5.

M1 mechanical smoke generators: Baxter, *Scientists Against Time,* 287; Brooks E. Kleber and Dale Birdsell, *The Chemical Warfare Service: Chemicals in Combat* (Washington, DC: United States Army Center of Military History, 1990), 327; Brophy and Fisher, *The Chemical Warfare Service: Organizing for War,* 308.

OSRD's David Langmuir: Baxter, *Scientists Against Time,* 286.

ferried Section T fuses: Teas, "Trip Report"; Mitchell, "Special Mission."

in crates of ten: War Department, "Technical Bulletin TB 9X-59: Preliminary Instructions, Fuze T74E6," March 23, 1944, NARA, RG 165, E NM84 421.

221 *at National Carbon:* Baldwin, *The Deadly Fuze,* 262.

hayloft of a French barn: Mitchell, "Special Mission."

"electromagnetically": War Department, "Technical Bulletin TB 9X-59."

within sixty feet: Ibid.

Enlisted men were taught: According to Mitchell, "Special Mission," they were "instructed only in the packaging and marking and also the assembly of the ammunition."

fuses were not needed: According to Teas, "Trip Report," the fuses were set to be used in defense of the artificial harbors but were not actually issued because of "technical restraints." As Mitchell reported, "Colonel Patterson then stated that unless air activity increased decidedly, he did not see any reason for using the T74 Fuze. Only approximately 40 sorties occurred during the nights while I was there and no planes got within range of the harbors . . . plans were in place for the use of the T74 in Cherbourg." Teas added: "No air raids developed over Cherbourg, however, and no operational use of the fuzes occurred."

chief science adviser: Dennis Hevesi, "H. G. Stever, Who Advised Leaders on Science, Dies at 93," *New York Times,* April 14, 2010.

Morning of June 6: Guy Stever, *In War and Peace: My Life in Science and Technology* (Washington, D.C.: Joseph Henry Press, 2002), 43.

landed in a C-47: Ibid., 45–46.

26. WACHTEL IS HIDING

223 *country manor Amniarix discovered:* Jones, *Most Secret War*, 373–74; Wachtel, "Unternehmen Rumpelkammer," 118–19; Walpole, *Hitler's Revenge Weapons*, 198.

He relocated himself: Wachtel, "Unternehmen Rumpelkammer," 110.

Château d'Auteuil: Bibliographie de la France, ou Journal Général de l'Imprimerie et de la Librairie et Des Cartes Géographiques, Gravures, Lithographies et Œuvre de Musique (Paris: Chez Pillet Ainé, 1850), 626.

formally off-limits: Wachtel, "Unternehmen Rumpelkammer," 110.

"Men": Ibid., 108.

224 *supervision of the German military:* Zaloga, *German V-Weapon Sites*, 30.

the Luftwaffe authorized: Ibid.

The new "modified sites": Ibid., 32; Zaloga, *V-1 Flying Bomb*.

prefabricated ramps: De Maeseneer, *Peenemünde*, 289–90.

took just eight days: Zaloga, *V-1 Flying Bomb*, 16.

Detecting them from the air: Campbell, *Target London*, 226; Zaloga, *V-1 Flying Bomb*, 16. The Allies did get better at identifying the sites over time.

On March 5, Hitler: Campbell, *Target London*, 223–24.

Production was ongoing: Hölsken, *V-Missiles of the Third Reich*, 84.

established additional factory lines: Ibid., 102.

225 *over eleven thousand prisoners:* Sellier, *A History of the Dora Camp*, 84.

acetylene lamps: Ibid., 58.

V-1s would soon emerge: Hölsken, *V-Missiles of the Third Reich*, 154.

a series of bunkers: Neliba, *Kriegstagebuch des Flakregiments 155 (W)*, 41; Campbell, *Target London*, 237.

In May, Wachtel completed: Campbell, *Target London*, 234.

"batteries with especially well-trained": Hölsken, *V-Missiles of the Third Reich*, 140.

226 *a few hours' break:* Wachtel, "Unternehmen Rumpelkammer," 114.

headquarters received word: Neliba, *Kriegstagebuch des Flakregiments 155 (W)*, 41.

"The fight for the flying bomb": David Irving, *The Mare's Nest* (London: William Kimber, 1964), 233.

shaved off his beard: Wachtel, "Unternehmen Rumpelkammer," 114–15.

numbering some sixty-five hundred: Zaloga, *Operation Crossbow*, chapter 2. Not all the men would be manning the launching sites. As Zaloga notes, an additional four thousand Luftwaffe troops were also assigned to the program.

across sixty-four camps: Hölsken, *V-Missiles of the Third Reich,* 202. For more details on the different types of launching sites see ibid., 180.

supply depots code-named: Campbell, *Target London,* 239.

"Catastrophic supply conditions": Wachtel, "Unternehmen Rumpelkammer," 115.

227 *"Soldiers":* David Irving, *Mare's Nest,* 234.

At five thirty p.m.: Neliba, *Kriegstagebuch des Flakregiments 155 (W),* 43.

seven of the sixty-four sites: De Maeseneer, *Peenemünde,* 306.

Critical items had still: Neliba, *Kriegstagebuch des Flakregiments 155 (W),* 44.

a map of London: Wachtel, "Unternehmen Rumpelkammer," 115.

radio beacons: Zaloga, *V-1 Flying Bomb,* 33.

at eleven p.m.: Campbell, *Target London,* 241.

228 *ten Nazi drones launched:* Hölsken, *V-Missiles of the Third Reich,* 202; Neliba, *Kriegstagebuch des Flakregiments 155 (W),* 43. According to Zaloga, an additional nine failed launches were fired earlier as part of the "first salvo"; Zaloga, *V-1 Flying Bomb,* 18.

"It did not work": Wachtel, "Unternehmen Rumpelkammer," 116.

27. A COMET GONE WRONG

229 *The early hours:* Longmate, *The Doodlebugs,* 89.

British firemen awoke calmly: Alec Savidge, "V1 Number One, June 13, 144," WW2 People's War, BBC, June 12, 2003.

230 *Witnesses who glimpsed:* Longmate, *The Doodlebugs,* 91.

like burning black crosses: Williams, *Operation Crossbow,* 211.

engines' steel shutters: Hölsken, *V-Missiles of the Third Reich,* 51.

unlike anything British ears: Longmate, *The Doodlebugs,* 91; Campbell, *Target London,* 246–47.

231 *detonated on agricultural land:* Longmate, *The Doodlebugs,* 92.

strawberry field: Campbell, *Target London,* 246.

The fourth V-1 landed: There appears to have been some jurisdictional confusion at the time—which persists today—about whether the V-1 actually landed in Bethnal Green or in Bow. Both locations are regularly cited; Longmate, *The Doodlebugs,* 95.

"a large glass mirror": Ibid.

Six died: Campbell, *Target London,* 247.

Ministry of Information: Haining, *The Flying Bomb War,* 32.

"several people had been killed": Ibid., 248.

"mountain hath groaned": Jones, *Most Secret War,* 417.

Seventy-three drones: Ibid., 418.

232 *"available information does not":* Haining, *The Flying Bomb War,* 53.

London newspapers announcing: Ibid., 54–56; "Our Scientists Will Defeat It," *Evening News,* June 16, 1944; "How to Spot Ghost Planes," *Evening News,* June 16, 1944.

On June 18, a Sunday: Jones, *Most Secret War,* 424.

"They are nuisances": Hatch, "Of Sorrow and Joy."

climbed to the top floors: Ibid.; 130th Unit History, 6.

meet up with the Women's: Samuel Hatch to David and Esther Hatch, June 22, 1944, courtesy of Alex Schneider and the Hatch family.

"It has never awakened": Ibid.

233 *British defenses were threefold:* Pile, *Ack-Ack,* 284; Dobinson, *AA Command,* 433. After the guns were moved to the coast, there was also a pocket of airspace between the guns and the barrage balloons where the fighters re-engaged with the flying bombs.

the Allies estimated that: Memo, "Crossbow Committee: Report as of 0600, 28 June 1944," July 1, 1944, NARA, RG 165, E NM84 498.

the V-1's top speed: Hölsken, *V-Missiles of the Third Reich,* 335.

Gloster Meteor: Nijboer, *Meteor I,* 15. Top speeds depend on altitude.

Of the fastest RAF fighters: Ibid.

"'bits fell off'": D.D.R.E., "Cleaning Up of Fighters Engaged on Operations Against the German Flying Bomb," Scientific Sub-Committee (Flying Bomb) of Crossbow Committee, letters and miscellaneous (German flying bomb, V1), January 1, 1944 to December 31, 1944, TNA, AVIA 6/25668. Archive hereafter referred to as TNA, UKSSCV1.

fitting them to rockets: Letter to W. S. Farren, July 5, 1944, TNA, UKSSCV1.

paint on the Tempests: Memo, "Note on a Visit to Tempest Squadron Newchurch on 24th June, 1944," TNA, UKSSCV1.

Parachutes might be affixed: Memo, "Use of Wire Barrages Against Peenemunde 16," n.d., TNA, UKSSCV1.

Paper bits could: W.G.A. Perring to the Secretary, Ministry of Aircraft Production, memo, "Flying Bomb—Blocking of Grille," July 13, 1944, TNA, UKSSCV1.

234 *changed with the weather:* Pile, *Ack-Ack,* 290.

On June 20, an unnamed: Crossbow, Inter-Departmental Radiolocation Committee, "Minutes of the Fourth Meeting Held on Tuesday, June 20,

1944, in Church House, Dean's Yard, S.W.1," n.d., NARA, RG 165, E NM84 498.

a rounds-per-bird ratio: L. E. Bayliss, Crossbow, Inter-Departmental Radiolocation Committee, memo, "Use of Proximity Fuses with SCR. 584. Note by Dr. Bayliss, A.O.R.G.," June 21, 1944, NARA, RG 165, E NM84 498.

Pile had 376 heavy guns: Pile, *Ack-Ack,* 288.

only 9 percent: Ibid., 291.

Churchill disclosed: Associated Press, "Churchill Gives Grim Facts of German Robot Attacks," *Lafayette Journal and Courier,* July 6, 1944.

235 *"harder to bear than":* Longmate, *The Doodlebugs,* 129.

"thick and fast": E. Stern to S. Stern, July 3, 1945, available at london memorial.org.

"The effect of the new weapons": Longmate, *The Doodlebugs,* 129.

"a bigger mess": Regan-Atherton, *Heavy Rescue Squad Work on the Isle of Dogs,* 168.

"heard at a great distance": Haining, *The Flying Bomb War,* 16.

rate of destruction continued: Longmate, *The Doodlebugs,* 136.

236 *"Kids playing happily":* Regan-Atherton, *Heavy Rescue Squad Work on the Isle of Dogs,* 168.

"The stages were so prolonged": Longmate, *The Doodlebugs,* 198.

at least five minutes: Hölsken, *V-Missiles of the Third Reich,* 203.

"grisly transport service": Longmate, *The Doodlebugs,* 200.

"the world stood still": Ramsey, *The Blitz,* vol. 3, 386.

"As it stopped": Jacobsen, *The Deadly Fuze.*

Twenty thousand Morrison shelters: Longmate, *The Doodlebugs,* 134.

"sordid piles of bedding cluttering": George Orwell, *As I Please: 1943–1946* (Boston: Nonpareil, 2000), 196.

237 *Hatch rode the Underground:* Hatch, "Of Sorrow and Joy."

helped the British rescue squads: 130th Unit History, 10.

Monday, July 3: Hatch, "Of Sorrow and Joy"; Hatch, "Service Above All."

Eric Stern, the interpreter: E. Stern to S. Stern, July 3, 1945.

238 *"living in a play":* "Death Circles WAC in Havoc of Buzz Bomb," *Chicago Sunday Tribune,* July 1, 1945.

239 *a roll call was held:* Ibid.

sixty-two men of the 130th: There were at least sixty-six military casualties and nine civilian casualties. See the excellent data at www.londonmemorial .org; Jan Gore, *Send More Shrouds: The V1 Attack on the Guards' Chapel 1944* (Barnsley, UK: Pen and Sword, 2017), 26.

the incident marked: Ramsey, *The Blitz,* vol. 3, 408; 130th Unit History, 11.

28. "TURKEY SHOOT"

240 *new, high-ranking committee:* Associated Press, "Group Named to Plan Postwar Research, *Washington, D.C., Evening Star,* June 23, 1944.

"war seems to inspire": Kenneth E. Dutcher, "Science and War," *Washington, D.C., Times-Herald,* June 28, 1944, LOC, MAT.

"world-unifying force": Waldemar Kaempffert, "Science and a Lasting Peace," *Washington Post,* March 22, 1944.

241 *"Scientist does not":* Tuve, "Notebook PWP," June 16, 1944.

Merle solicited opinions: Merle Tuve to Deak Parsons, July 15, 1944, LOC, MAT; George E. Smith to M. A. Tuve, July 7, 1944, LOC, MAT; Lawrence Hawkins to M. A. Tuve, July 5, 1944, LOC, MAT; M. A. Tuve to E. O. Lawrence, June 19, 1944, LOC, MAT; Ernest C. Lawrence to M. A. Tuve, July 24, 1944, LOC, MAT.

third-largest research group: Per F. Dahl, *From Nuclear Transmutation to Nuclear Fission, 1932–1939* (Boca Raton, FL: CRC, 2002), chapter 13.

Section T employees now numbered: John R. McPhee to Chief, Bureau of Ordnance, Navy Department, memo, "Tabulation of Data Regarding Special Request for Draft Deferment Assistance," April 15, 1944, NARA, RG 227, E NC-138 8.

over six hundred and fifty people: Applied Physics Laboratory, "Employee List as of the Close of Business," June 30, 1944, NARA, RG 227, E NC-138 8.

a complete quadrangle: Section T, "Quarterly Report as of March 31, 1944," 3–4.

The telephone switchboard: D. L. Hopkins to A.P.L. Personnel, memo, "Telephone Service," February 9, 1944, NARA, RG 227, E NC-138 8.

A public address system: APL-JHU-Z News 1, no. 1 (1944): 4, APLA.

boss to hundreds: McPhee to Navy Bureau of Ordnance, "Tabulation of Data Regarding Special Request for Draft Deferment Assistance."

242 *its own fledgling newsletter:* The original name was "APL-JHU-Z," with a *z.*

a library filled: Memo, "APL Technical Library—Current Book List," December 1, 1943, NARA, RG 227, E NC-138 8.

guards needed a night registrar: D. Luke Hopkins to Unit, Group, and Project Supervisors, memo, "Night Register," May 22, 1944, NARA, RG 227, E NC-138 8.

four hundred cases of Coca-Cola: H. Moore, Jr. to C. L. Tyler, memo, "Procurement of Coca-Cola for Applied Physics Laboratory," August 6, 1943, NARA, RG 227, E NC-138 8.

For morale, movies: L. R. Hafstad to C. L. Tyler, memo, "Movies of Section T activities," April 15, 1944, NARA, RG 227, E NC-138 8.

the Silver Spring Giants: Floyd Morris, "J.H.U. Sports Round Up," *APL-JHU-Z News* 1, no. 1 (1944): 4, APLA.

Between April 1 and June 30: Section T, "Quarterly Report as of June 30, 1944," 35.

took on other projects: Ibid., 42–56.

On June 17, he arranged: M. A. Tuve to Franklin S. Cooper, June 17, 1944, NARA, RG 227, E NC-138 1.

"Great rearing clouds": Musicant, *Battleship at War,* 247.

243 *"The fate of the Empire":* James Campbell, *The Color of War: How One Battle Broke Japan and Another Changed America* (New York: Crown, 2012), 215.

the Japanese massed: R. G. Grant, *Battle at Sea: 3,000 Years of Naval Warfare* (New York: DK Publishing, 2008), 321.

the flagship: Foerstner, *James Van Allen,* 61.

75 percent of rounds: Applied Physics Laboratory, *The "VT" or Radio Proximity Fuze,* 8.

The gadget took on: Baldwin, *They Never Knew What Hit Them,* 152–53.

244 *The dry batteries:* L. R. Hafstad to M. A. Tuve and C. L. Tyler, "Progress Report on the Mk 32," January 26, 1944, NARA, RG 227, E NC-138 8.

reach 120 degrees: Van Allen, interview, 100.

Van Allen personally devised: Ibid., 101.

forty-five vicious Zero: Tillman, *Clash of the Carriers,* part 3.

saw three planes escape: Foerstner, *James Van Allen,* 50.

Kamikaze tactics hadn't yet: Van Allen recalled it as a kamikaze attack, probably erroneously, as kamikaze tactics were not put into effect until several months later. See Steven Zaloga, *Kamikaze: Japanese Special Attack Weapons 1944–45* (Oxford, UK: Osprey, 2011).

"seemed to come in ones": Musicant, *Battleships at War,* 248–49.

shooting down twenty-five planes: Ibid., 249.

245 *"most memorable scene":* Tillman, *Clash of the Carriers,* part 3.

Not one of the forty-four: Ibid. I'm referring to the first raid.

roast beef and strawberries: Foerstner, *James Van Allen,* 50.

The Japanese lost 92 percent: Not counting floatplanes. Ozawa had 430 carrier aircraft, and 395 were lost. Tillman, *Clash of the Carriers,* part 6; Phillips Payson O'Brien, *How the War Was Won: Air-Sea Power and Allied Victory in World War II* (Cambridge: Cambridge University Press, 2015), 72.

"a very boring existence": Musicant, *Battleships at War,* 254.

Construction began on two airfields: Baldwin, *They Never Knew What Hit Them,* 176.

246 *"Almost every night":* Ibid.

29. ACK-ACK GIRLS

247 *Ed Hatch was treated:* Hatch, "Service Above All."

A ghost plane detonated: 130th Unit History, 11–13.

"a number of bodies": "Raid Victims Rescued More Quickly," *Daily Telegraph,* July 4, 1944.

"waiting to go on rescue": Robert Wilson, "Robot Kills Yanks on Rescue Mission," *Boston Globe,* July 6, 1944.

248 *figures were redacted:* Austin Bealmear, "Whole Sections of London Affected by Robot Raiders," *Boston Globe,* July 6, 1944.

"a little help from outside": Longmate, *The Doodlebugs,* 157.

"The most harrowing part": Ibid., 241.

249 *Windows had to be kept:* D. G. Collyer, *Buzz Bomb Diary,* 64.

"like sheets of rain": Longmate, *The Doodlebugs,* 169.

The blasts blew tiles: D. G. Collyer, *Buzz Bomb Diary,* 9, 10, 14, 81, 119, 121, 130.

"suddenly wake, standing": Longmate, *The Doodlebugs,* 147.

already approaching three thousand: AP, "Churchill Gives Grim Facts of German Robot Attacks."

250 *Pile argued forcefully:* Pile, *Ack-Ack,* 291.

"All right": Ibid.

coastal gun belt stretching: Dobinson, *Operation Diver,* 58.

chronic staffing problems: Pile, *Ack-Ack,* 164.

Pile sought female soldiers: Ibid., 165.

251 *"The girls cannot be beaten":* Ministry of Information, *Roof Over Britain: The Official Story of Britain's Anti-Aircraft Defences, 1939–1942* (London: Her Majesty's Stationery Office, 1943), 64.

"the first battle of the robots": Pile, *Ack-Ack,* 286.

Some seventy-four thousand: Dobinson, *AA Command,* 313.

typical heavy mixed battery: Ibid., 312.

Under their charge were: Ibid., 433; Dobinson, *Operation Diver,* 27–30.

over twenty thousand pounds: I'm referring to the Mk 6.

"great trek southward": Pile, *Ack-Ack,* 294.

The transfer relocated: Ibid. 294–95; Dobinson, *Operation Diver,* 59–60.

"atop the furniture": Collyer, *Buzz Bomb Diary,* 29.

252 *Vee Robinson, who enlisted:* Robinson, *On Target,* 155; Vera (Vee) Robinson, "Life on the Guns!," WW2 People's War, BBC, November 28, 2003.

Within three days: Pile, *Ack-Ack,* 295.

"sky was alight": Robinson, *On Target,* 155–56.

there were forty-seven batteries: Dobinson, *Operation Diver,* 60.

253 *from 9 to 17 percent:* Dobinson, *AA Command,* 437. According to Salant et al., "Part Twelve," 16, the initial British figures for V-1 flying bomb statistics are slightly different than those generally cited today. "The British Information Services, in their pamphlet 'Flying Bombs,' published December 1944, state that the guns 17% of the flying bombs in the first week, 24% in the second, 27% in the third, 40% in the fourth, 55% in the fifth, and 74% in the sixth week."

British Lancaster bombers: Section T, "Quarterly Report as of September 30, 1944," 7.

"dainty sensitive fingers": Longmate, *The Doodlebugs,* 415; For a well-done study of gender norms in mixed AA batteries, see Gerard J. DeGroot, "Whose Finger on the Trigger? Mixed Anti-Aircraft Batteries and the Female Combat Taboo," *War in History* 4, no. 4 (1997): 434–53.

six hundred and eighty-four heavy guns guarded: Dobinson, *Operation Diver,* 60, 65.

"On a clear day": James Johnson, "Buzz Bomb Fighters Recall War: Group Gathers for Reunion," *Oklahoman,* October 3, 1987.

"use binoculars to tell": William Payne, ed., *The Screaming Eagle in Action: A History of the 127th AAA Gun Battalion Mbl.: 10 May 1943 to 21 March 1946* (Carlisle, PA: U.S. Army Military History Institute, 1987), 16.

106 times a day: I've divided Dobinson's total of 9,017 by the 85 days between June 13 and September 5, 1944; Dobinson, *AA Command,* 438.

"they'd look like stars": Joseph H. Gigandet, ed., "125 AA Battalion Unit History: A Straight Shooting Outfit, the History of Its Training, Travels, and Achievements," n.d., 82.

254 *"the stillness of the night":* Ibid., 41.

"There was never such": Ibid., 69.

dubbed Shrapnel Heights: Richard D. Flora, ed., "Madcap Memories: The History of Battery 'D' 134 AAA GN. BN.," n.d., 5. "Hell's Corner" was actually the wartime name for Dover's Shakespeare Cliff.

"But after we got": Johnson, "Buzz Bomb Fighters Recall War."

grew to 24 percent: Boyce, *New Weapons for Air Warfare,* 155.

practically all the heavy guns: Ibid.

255 *out of General Pile's headquarters:* Teas, "Trip Report," 8; Pile, *Ack-Ack,* 81.

arrived on July 30: Salant et al., "Part Twelve," 23.

lived among the Diver Belt: Chicago Tribune Press Service, "U.S. Defeated Flying Bomb, Briton Reveals," *Chicago Tribune,* April 6, 1946.

in an Army jeep: Mann, *Was There a Fifth Man?,* 36.

Salant saw a buzz bomb: Salant, interview.

some seventeen thousand times: Williams to Tuve, "Characteristics of Guns of Interest to Section T."

Each second, they flew: Ibid.

256 *which matched the dimensions:* Hölsken, *V-Missiles of the Third Reich,* 335. The wingspan of nearly all models was 18 feet, 9¾ inches. But there was an early version of the Fi 103, the A-1, with a shorter wingspan of 17 feet, 7¾ inches.

passed a triggering threshold: Applied Physics Laboratory, *World War II Proximity Fuze,* 173.

eighteen hundred pounds of explosives: Hölsken, *V-Missiles of the Third Reich,* 335. Hölsken has the exact weight at 1,830.

24 percent to 46 percent: Boyce, *New Weapons for Air Warfare,* 155.

"which is so secret that": F. A. Pile to George Marshall, August 12, 1944, NARA, RG 165, E NM84 421. As Pile noted, "Already we are far away ahead of the fighters."

Salant estimated that the gadget: Salant, "Status of VTF in the U.K."

A British analysis agreed: Teas, "Trip Report," 12.

776 heavy guns: Dobinson, *Operation Diver,* 65. Dobinson notes that by July 27, 1944, there were 124 HAA guns in the Diver Belt with plans in place to bring the total to 160; these extra 36 guns are included in my total.

257 *Ten times better:* Salant estimated that standard fuses took 500 or 600 per V-1. I've used a more conservative comparison with 400 RPB from Franklin, "Army Operational Research Group Memo No. 422."

radar and aiming devices: These were the SCR-584 and the T-10 electric director known in the Army as the M-9; Boyce, *New Weapons for Air Warfare,* x. At the heart of the SCR-584 was the cavity magnetron delivered by the British during the Tizard mission; Buderi, *The Invention That Changed the World,* 221.

V-1s was a stunning 8.1: Again, I've used the optimistic 400 RPB estimate for standard fuses from Franklin, "Army Operational Research Group Memo No. 422."

"clear for all to see": Soames, *A Daughter's Tale,* 276.

46 percent to 67 percent: Boyce, *New Weapons for Air Warfare,* 155.

downing some two-thirds: Dobinson, *Operation Diver,* 82.

planes claimed 17 percent: Ibid., 81.

"It was quite uncanny": Jacobsen, *The Deadly Fuze.*

fuse looked *different:* One of the reasons why data on the fuse is difficult to parse is that fuses were often shot in mixed combinations of regular and smart fuses, in part due to security concerns.

258 *"the enemy might have seen":* Salant, "Status of VTF in the U.K."

unleashed a "godsend": Ibid.

only two or three shots: Franklin, "Army Operational Research Group Memo No. 422."

"reached the stage where": Longmate, *The Doodlebugs,* 415.

soared to 79 percent: Boyce, *New Weapons for Air Warfare,* 155.

climbed up to 82 percent: Dobinson, *AA Command,* 437. The 82 percent figure is a daily, not weekly, average.

104 drones were fired: Boyce, *New Weapons for Air Warfare,* 155–56.

100 percent: Jacobsen, *The Deadly Fuze.*

not their most vibrant memory: See the unit histories of the 124th, 125th, 127th, 134th, and 601st AA Battalions.

"Any gum, chum": Gigandet, "125 AA Battalion Unit History," 40.

259 *"a good portion":* This and the following quotes come from ibid., 9, 22, 31, 54, 69.

wireless radio for entertainment: Dobinson, *Operation Diver,* 82.

By September, the V-1 attack: Dobinson, *AA Command,* 438.

downed 1,550 drones: Pile, *Ack-Ack,* 303.

"With my compliments": Boyce, *New Weapons for Air Warfare,* 156.

Churchill later asked Roosevelt: Baldwin, *They Never Knew What Hit Them,* 161.

"More was learned about": Dobinson, *AA Command,* 437.

thanked Ed Salant personally: Salant, in his September 5 memo to Hafstad, noted that Pile thanked him warmly. Also see Pile, "U.S. and Flying Bombs." Although Pile mistakenly describes Salant as the proximity fuse's "chief designer" and gets much of the fuse history wrong, in this article and in *Ack-Ack,* he noted that "due credit has not been given to the great part played by the United States" and that "American scientists, together with American production methods, and, above all, American generosity, gave us the final answer to the flying bomb."

"Our reputation in the": Salant, "Status of VTF in the U.K."

260 *too young to have:* AAP, "London 'Dim-Out' Hastily Turned Black," *Launceston Examiner,* September 19, 1944.

30. A COLD WINTER

261 *"You wait until this":* V. Bush to Bradley Dewey, September 28, 1944, LOC, VB.

a solar-powered pump: James F. Bell to Vannevar Bush, November 6, 1944, LOC, VB; V. Bush to James F. Bell, November 10, 1944, LOC, VB.

a cheap solar "collector": V. Bush to James F. Bell, September 29, 1944, LOC, VB.

"the knot book to end": V. Bush to Edith L. Bush, October 11, 1944, LOC, VB.

pondered a book deal: Paul Brooks to Vannevar Bush, November 30, 1944, LOC, VB.

now a national celebrity: J. D. Ratcliff, "War Brains," *Collier's,* January 17, 1942; Henry Gemmill, "Secret Weapons: Finding and Financing Them a $135 Million Job of Government OSRD," *Wall Street Journal,* July 17, 1943; "Honoring Dr. Bush," *New York Times,* December 28, 1943.

the cover of Time: "Yankee Scientist," *Time,* April 3, 1944.

262 *join him in prayer services:* Stephen Early to Vannevar Bush, February 26, 1944, LOC, VB.

"at the peak": Zachary, *Endless Frontier,* 184.

"the most extraordinary": V. Bush to the president, "Report to the President on Activities of the Office of Scientific Research and Development," August 28, 1944, 19, NARA, RG 227, E NC-138 2.

Over 1.75 million Section T: Brehon Somervell, "Memo for the Chief of Staff," October 4, 1944, NARA, RG 165, E NM84-421.

"logically untenable": Zachary, *Endless Frontier,* 155–56.

"You may be able": Ibid., 156.

263 *As Merle reported:* Memo, "Study of Factors Involved in Release of VT Fuzes: Report by Panel of Technical Experts, 1 September 1944," compiled in a report dated September 29, 1944, NARA, RG 165, E NM84-412.

"I have agreed to meet": Bush, *Pieces of the Action,* 110.

"It is a combined": Ibid.

the Joint Chiefs approved: Zachary, *Endless Frontier,* 181; C.F.F., memo, "Release of VT Fuses for General Use," October 27, 1944, NARA, RG 165, E NM84-412.

His itinerary took him: Donald David to Vannevar Bush, "Invitational Travel Order, Shipment IJ-847-YD," November 18, 1944, LOC, VB; memo,

"Vannevar Bush, Itinerary: November 19–November 29," LOC, VB; Bush OH, reel 7B, 451-1A. Bush mistakenly recalled this trip as occurring within twenty-four to forty-eight hours of the release of the fuse. He also could not recall the name of the French town, Vittel, that appears on his itinerary.

An atmosphere of ecstasy: Neiberg, *The Blood of Free Men,* 221–22.

264 *Cabinet settled on December 31:* Beevor, *Ardennes 1944,* 5.

entered Belgium in September: Ibid., 10.

Antwerp was seized: Headquarters Antwerp X, "The Story of Antwerp X," 50 AAA Brigade, Army, n.d., 11.

locked in empty animal cages: Beevor, *Ardennes 1944,* 11.

men of the 101st Airborne: Lennon, *Battle of the Bulge.*

265 *"There's nothing out there":* Ibid.

nineteen hundred German artillery pieces: Peter Schrijvers, *Those Who Hold Bastogne: The Story of the Soldiers and Civilians Who Fought in the Biggest Battle of the Bulge* (New Haven, CT: Yale University Press, 2014), 3.

"decapitated by an exploding": Caddick-Adams, *Snow and Steel,* 343.

"a bunch of wild cattle": Lennon, *Battle of the Bulge.*

seventy-five hundred American troops: Timothy J. Lynch, ed., *The Oxford Encyclopedia of American Military and Diplomatic History* (Oxford: Oxford University Press, 2013), 117.

output of wartime armaments: Charles Whiting, *Ghost Front: The Ardennes Before the Battle of the Bulge* (Cambridge, MA: De Capo, 2002), 167.

266 *authorized to start using:* Thompson, "Employment of VT Fuzes in the Ardennes Campaign," 10. Strangely enough, this extensive document, which consists mainly of frontline reports and interviews highly laudatory of the fuse, has been cited as a statistically "rigorous" study that stands against arguments for the fuse's overwhelming effectiveness during the Battle of the Bulge. More bizarre, Thompson clearly states that "effectiveness was difficult to evaluate objectively" because, while "the conviction is widespread that the fuze is of the highest value," statistically rigorous data does not exist. Of course such data would not exist; war is not a randomized control trial.

Beginning on December 18: Ibid., 2.

702 German troops: Ibid., 26.

the Nazi invasion receded: Ibid., 33; "It is very significant that the tip of the German salient began to wither rapidly within a day or two after the field of artillery fire of the 1st and 3rd Armies overlapped so that all supply roads were covered."

German snipers: Report, "OFS ETO Section T," 1, NARA, RG 227, E NC-138 177, 1.

"new secret American artillery": Associated Press, "New U.S. Weapon Shatters German Counterattacks," *Washington Post,* January 6, 1944.

267 *"that these shells drive"*: D. W. Hoppock to G. M. Barnes, memo, "Weekly Performance Report #22," May 10, 1945, NARA, RG 156, E NM26 646A.

"huge puffs of black": James W. Clyburn, memo, "AGF Report No. 597 —POZIT FUZE," January 31, 1945, 4, NARA, RG 227, E NC-138 7.

as low as ten feet: AFEO Training Circular, "VT Fuzes," April 1945, NARA, RG 156, E NM26 646A.

"PWs agree that it is": Thompson, "Employment of VT Fuzes in the Ardennes Campaign," 39.

"had never experienced": Ibid., 32.

They dubbed it "Doppelzünder": Ibid., 38.

"coming back in droves": Baldwin, *They Never Knew What Hit Them,* 187.

268 *"When all armies"*: Ibid., 26.

called it "unprecedented": J. P. Teas, "VT Fuzes in Action," n.d., NARA, RG 227, E NC-138 7.

lost over twelve thousand: Steven Zaloga, *Omar Bradley* (Oxford: Osprey, 2012), 42.

Artillery caused most: Donald L. Miller, *The Story of World War II* (New York: Simon and Schuster, 2001), 350.

was shot amidst: Details were drawn from Lennon, *Battle of the Bulge.*

"had a large share": D. W. Hoppock to G. M. Barnes, memo, "Weekly Performance Report #27," June 11, 1945, NARA, RG 156, E NM26 646A.

"won the Battle": Baldwin, *The Deadly Fuze,* 279.

Colonel Max Wachtel attacked: Wachtel, "Unternehmen Rumpelkammer," 118.

Of 2,394 V-1s: Robert J. Katz, *The Buzz Bomb Kings: A History of the 407th AAA GUNBN* (New York: Lewis Historical Publishing, 1945), 41–42.

269 *success rate of 97:* Headquarters Antwerp X, "The Story of Antwerp X," 23.

A month later: The defense of Antwerp ended on March 30, 1944; between the end of the defense of Antwerp and the end of the war was a month and a week.

31. TUG OF WAR

270 *he got hold of an Opel P4:* Wachtel, "Unternehmen Rumpelkammer," 118–19; Walpole, *Hitler's Revenge Weapons,* 198.

at a remote resort: Jacobsen, *Operation Paperclip,* 66.

While less deadly: I'm referring to total deaths caused by the V-1 and the

V-2 in England. The usual figures cited are 6,184 for the V-1, 2,754 for the V-2.

ten times faster: The V-2 flew at 3,580 miles per hour.

271 *Germany's top scientists:* Jacobsen, *Operation Paperclip,* 35, 66, 68 75.

shared with the Japanese: Jones, *Most Secret War,* 491.

dozens of intelligence programs: Jacobsen, *Operation Paperclip,* 37.

"Hap" Arnold suspected: Summary, "From CG, AAF to CG, STAF in Europe," NARA, RG 165, E NM84 421. According to a short summary dated October 1, 1944: "General Arnold indicates recent heavy losses of bombers apparently from anti-aircraft indicates Germans are using proximity fuzes and also have better fire control system for anti-aircraft guns. Requests reaction."

Salant began his new line: Roslyn K. Waksberg to Carroll L. Wilson, memo, "WL-168," February 23, 1945, NARA, RG 227, E NC-138 1.

Salant joined Pash in Paris: Pash, *The Alsos Mission,* 162–63.

272 *Salant interrogated hundreds:* APL-JHU-S 1, no. 20 (1946): 2, APLA.

Germans had perhaps fifty projects: Ibid. The full quote for "Every one was a fizzle" is "By and large, everyone was a fizzle." I felt the "By and large" could be excluded because every program, in fact, did fail. The quotes beginning with "Their only objection" also come from this source.

"Abstandzünder": E. O. Salant "Survey of German Program for Development of Proximity Fuze," June 18, 1945, NARA, RG 38, E MLR P5.

produced some rocket fuses: Report, "OFS ETO Section T."

an acoustic fuse at Peenemünde: R. D. Huntoon, memo, "Interrogation of Dr. Hans Kramer on 16 May 1945," n.d., NARA, RG 38, E MLR P5.

as early as 1937: Paul Ertsgaard, memo, "High Frequency Radio Bomb Fuze, 'Zünder 19' Rheinmetall Borsig A.G.," n.d., NARA, RG 38, E MLR P5.

microphonics problem had hampered: Ibid.

declared the task impossible: Salant, "Survey of German Program for Development of Proximity Fuze," 4.

"It must be apparent": Salant, "Summary of the Investigation of German Proximity Fuzes," August 20, 1945, 3, NARA, RG 38, E MLR P5.

"failed miserably in availing": Report, "OFS ETO Section T," 8.

273 *"gentlemanly collaboration among Allies":* Jacobsen, *Operation Paperclip,* 38.

described as "looting": Report, "OFS ETO Section T," 6.

Alsos Mission revealed: Pash, *The Alsos Mission,* 159.

American and British attitudes: Report, "OFS ETO Section T," 7–8.

"At least some of the prominent": Kart T. Compton to Vannevar Bush, May 25, 1945, NARA, RG 227, E NC-138 181.

as would Robert Lusser: Edward Jones-Imhotep, *The Unreliable Nation: Hostile Nature and Technological Failure in the Cold War* (Cambridge: MIT Press, 2017), 109.

274 *thousands of German experts:* Charles F. Pennacchio, "The East German Communists and the Origins of the Berlin Blockade Crisis," *East European Quarterly* 29, no. 3 (1995).

Russian officers toured the Diver: Baldwin, *The Deadly Fuze,* 6.

a Soviet spy: Sam Roberts, "Bad Cheer: The Holiday Exchange Precipitating 'Bridge of Spies,'" *New York Times,* November 25, 2015; Steven Usdin, *Engineering Communism: How Two Americans Spied for Stalin and Founded the Soviet Silicon Valley* (New Haven, CT: Yale University Press, 2005), 84; Michael Dobbs, "Julius Rosenberg Spied, Russian Says," *Washington Post,* March 16, 1997.

for rockets and bombs: Boyce, *New Weapons for Air Warfare,* 198, 202–6; *Radio Proximity Fuzes for Fin Stabilized Missiles* (Washington, D.C.: OSRD, 1946). The National Bureau of Standards, it should be noted, was also working on fuses before Tuve sent his rocket and bomb team over to them.

which in 1960: Roberts, "Bad Cheer."

a "year of madness": Jones, *Most Secret War,* 492.

275 *"An Improved Scientific Intelligence":* Goodchild, *Most Enigmatic War,* 560.

"little British service which has": Jones, *Most Secret War,* 497.

Jones left MI6 for good: Goodchild, *Most Enigmatic War,* 562. Jones was not finished with scientific intelligence work altogether; see Jones, *Reflections on Intelligence,* 7.

32. DOWNTIME

276 *On May 8, 1945:* Merle Tuve, memo, "V-E Day Remarks," May 1, 1945, LOC, MAT. Tuve had prepared these remarks along with a "Suggested Agenda for VE-Day Announcements" a week earlier. President Truman announced the end of the war via radio on May 8 at nine a.m. An addition to the speech, which is penned in Tuve's handwriting, is signed and dated at 9:30 a.m. that morning. I have assumed that he then delivered the speech to APL as planned in the attached agenda. The plan called for Tuve's agenda and speech to begin immediately following a radio announcement.

277 "EMERGENCY PEACE": Merle Tuve, memo, "EMERGENCY PEACE STATUS," May 23, 1945, LOC, MAT.

gained some fifteen pounds: Merle Tuve, accident policy, "Copy of Application," Union Mutual Casualty Company, September 27, 1928, LOC, MAT; Merle Tuve "Identification Card," Applied Physics Laboratory, LOC, MAT.

"personal life in all areas": Merle Tuve to Paul Kwei, September 27, 1944, LOC, MAT.

"grown into quite competent": Merle Tuve to Odd Dahl, July 13, 1945, LOC, MAT.

letter from an Italian physicist: Edoardo Amaldi to Merle Tuve, July 6, 1945, LOC, MAT.

278 *"SPLENDID NEWS THAT":* Merle Tuve to Odd Dahl, telegram, July 14, 1945, LOC, MAT.

"entirely out of touch": Merle Tuve to Odd Dahl, July 13, 1945, LOC, MAT.

return to their rightful jobs: Merle Tuve, "<u>EMERGENCY</u> PEACE STATUS."

the laboratory had: APL-JHU-S 1, no. 5 (1944): 1, APLA; *APL-JHU-S* 1, no. 11 (1945): 1, APLA. The dances were JHU dances held in Washington, D.C.

publish a drink recipe: APL-JHU-S 1, no. 15 (1945): 3, APLA.

279 *"war fever still in":* Foerstner, *James Van Allen,* 67.

an internal survey of one hundred and fifty: Memo, "Key and 'Essential' Technical Personnel as of Aug. 1, 1945," n.d., LOC, MAT.

Merle was reading passages: Vannevar Bush, "Science: The Endless Frontier. A Report to the President" (Washington, D.C.: United States Government Printing Office, 1945), LOC, MAT. Tuve's copy has his signature on the cover, dated July 16, 1945, in red pencil. The booklet is marked inside in the same red pencil.

Bush himself was: Zachary, *Endless Frontier,* 279.

Bush and Lawrence gawked: Rhodes, *The Making of the Atomic Bomb,* chapter 18.

280 *aboard the* Enola Gay: Christman, *Target Hiroshima,* 189.

equipped with a smart fuse: Brode's fuse wasn't a Section T radio proximity fuse but used, as a sensor, a modification of a British tail-warning radar.

thirteen hundred Section T workers: Baldwin, *The Deadly Fuze,* 308–16.

"Old Bastard Eddie": Mann, *Was There a Fifth Man?,* 42.

reported 162 inventions: D. Luke Hopkins, "Final Report on Inventions for OSRD Contract OEMsr-431," Applied Physics Laboratory, March 1, 1945, APLA.

awarded crisp one-dollar: Van Allen kept one of his dollars; John S. Lacey to J. A. Van Allen, "Nominal Compensation in Re: Patent Application Ser. No 555.148," UI, JAVA.

time in the national spotlight: All quotes were drawn from "News Clippings, V. T. Fuze Publicity, Sept—Dec., 1945," DTM. The booklet is an assemblage of clipped and photocopied news stories from across America on Section T's smart fuse. It also includes the newsreel ads from Universal, Paramount, and MGM. The quote "greatly shortened the war" is attributed to General Carl Spaatz. Also see "Secret Weapon War Capital's 'Brain Child,'" *Washington Post,* September 22, 1945.

EPILOGUE

282 *Section T's contributions gradually faded:* As of this writing, Section T, incredibly, does not have its own Wikipedia page. That many details circulating about the proximity fuse are false or misleading also speaks to the fading attention paid to the story.

British work on rocket fuses: As mentioned in the notes for chapter 6, much of the confusion stems from Brennan, "The Proximity Fuze." Brennan is referring to "unrotated projectiles," meaning rockets or bombs, not fuses for shells, which was the true puzzle. The real story, as I hope this book makes clear, is roughly as summarized here: "The British had been developing proximity fuses for use in bombs and rockets, but had decided that the technical problems in making such fuses rugged enough to be fired from a gun were, for the time being, unsurmountable. The USN [Tuve] pressed ahead with their development, solving many of the practical problems." Jon Robb-Webb, "New Tricks for Old Sea Dogs: British Naval Aviation in the Pacific, 1944–45," in *British Naval Aviation: The First 100 Years,* ed. Tim Benbow (Surrey, UK: Ashgate, 2011), 114.

Historians implied: For example: "U.S.-manufactured proximity fuzes, based on the British invention. . . ." Stewart Halsey Ross, *Strategic Bombing by the United States in World War II: The Myths and the Facts* (Jefferson, NC: McFarland, 2003), 171. Again, the confusion often stems from not distinguishing rocket and bomb fuses from fuses for shells. Even Buderi, *The Invention That Changed the World,* 221, suggests of the fuse that "the basic design had arrived in the United States as part of the Tizard mission in September 1940, with the Americans picking up development at a laboratory paid for by Vannevar Bush's National Defense Research Committee." This framing, a standard narrative, is at a minimum highly misleading.

hired a reporter: "Transcript of Meeting on VT (Proximity) Fuze," 1. The higher-ups at APL persuaded Luke Hopkins to employ the reporter. By an odd nominal quirk, Hopkins happened to be the APL liaison for Johns Hopkins University.

Van Bush, too: Bush OH, reel 3B, 185: "I asked him why he didn't put on something interesting, such as the story of the proximity fuse, which has never been well told."

An insider's history: Baldwin, *The Deadly Fuze.* The book does not have footnotes, endnotes, or even a bibliography, but Baldwin, *They Never Knew What Hit Them,* which is essentially a reprint of the original text, does have a small number of footnotes that make clear that he did not have access to the vast holdings now available at the National Archives in College Park, Maryland, many of which were only declassified in 1993, after *The Deadly Fuze* was published.

He returned home: Patrick Johnson, "Samuel 'Ed' Hatch, WWII Veteran, Purple Heart Recipient & Local TV Pioneer Recalled," *MassLive,* December 24, 2018. I confirmed these details with Alex Schneider, Hatch's grandson.

Wachtel was found in hiding: Wachtel, "Unternehmen Rumpelkammer," 118–19; Walpole, *Hitler's Revenge Weapons,* 198.

283 *Rousseau was close to death:* Grimes, "Jeannie Rousseau de Clarens."

professor at Aberdeen University: Weiner, "R. V. Jones, Science Trickster."

named the ribosome: Roy J. Britten, "Richard Brooke Roberts, 1910–1980," in *National Academy of Sciences Biographical Memoirs* 62 (1993): 327.

for General Motors: John D. Caplan, "Lawrence R. Hafstad, 1904–1993," in *Memorial Tributes: National Academy of Engineering* 7 (1994): 105.

crystals generated electric charges: P.C.Y. Lee and D. W. Haines, "Piezoelectric Crystals and Electro-Elasticity," in *R.D. Mindlin and Applied Mechanics,* ed. George Herrmann (Oxford: Pergamon, 1974).

studied lunar craters: Ralph Baldwin, *The Measure of the Moon* (Chicago: University of Chicago Press, 1963).

Ed Salant went on: "Dr. Edward Salant Dies," *Washington Post,* September 19, 1978; Seaburg, "Whatever Happened to Thelma Adamson?," 82; J. Butler Tompkins (Superintendent of Brattleboro Retreat) to Melville Jacobs, May 31, 1952, Melville Jacobs Papers, University of Washington.

married over fifty years: They married on October 13, 1945. Tuve received an invitation. Van Allen, "My Life at APL," 175.

"far and away": Van Allen, "What Is a Space Scientist?," 12.

Van Allen remained: The term "Altitude Research" is used in the booklet "Service and Emergency Call Listings," Applied Physics Laboratory, Johns Hopkins University, Silver Spring, Maryland, UI, JAVA. He discusses his involvement in the V-2 work in Van Allen, interview, 108. I am unsure as to the degree of Van Allen's involvement in testing the American version of the V-1, but in Van Allen's archives in Iowa there are a series of official photos, stamped CONFIDENTIAL, of firing tests of the LTV-N-2 Loon on the USS *Norton Sound* from January 16, 1950. At a minimum Van Allen was somehow looped in regarding these particular tests.

"psychosomatic difficulty": Zachary, *Endless Frontier,* 347. For "super-gad-geteer," see 311. For details of Bush's retirement, see 381.

284 *The military-industrial complex:* Ibid., 315. Zachary discusses the issue more broadly throughout the book.

Albert Speer warned: Stephen Holmes, "Why International Justice Limps," *Social Research* 69, no. 4 (2002): 1055.

CIA opened its own department: See Karl H. Weber, *The Office of Scientific Intelligence, 1949–68,* vol. 1 (Washington, D.C.: Directorate of Science and Technology, CIA, 1972).

agency honored Amniarix and Jones: Weiner, "R. V. Jones, Science Trickster."

Section T "graduates": For Salant, see the Melville Jacobs Papers related to Thelma Adamson at the University of Washington. Also see the correspondence between Samuel Goudsmit and Edward Salant, stamped CONFIDENTIAL, on October 14, 1948, available in the AIP archives, in which Goudsmit asks Salant to arrange a feeling-out meeting with a physicist in charge of all research in the Russian zone of occupied Germany. For Van Allen, see Tom Walsh, "Intelligence Work: CIA Was Regular Visitor to UI Physicist Van Allen," *Cedar Rapids Gazette,* February 26, 2002. For Bush, see his OH.

Roswell, New Mexico: Annie Jacobsen, *Area 51: An Uncensored History of America's Top Secret Military Base* (Boston: Little, Brown, 2011); Tony Long, "Sept. 24, 1947: MJ-12—We Are Not Alone . . . Or Are We?," *Wired,* September 24, 2007. According to UFO enthusiasts, Tuve was one of the scientists recruited by the "Majestic 12" group to investigate downed spacecraft. According to a document viewed as fraudulent, Tuve "was part of the Bell Laboratories team which developed the principles of laser light physics derived from study of systems found aboard the Aztec disc-craft."

Tuve became director: Brown, *Centennial History of the Carnegie Institution,* 117.

Merle finally bought a house: Comly, interview.

285 *He gave up cursing:* Ibid.

"five or six years": Tuve, interview by Christman.

the clear-eyed patriotism: One might compare, for example, the Christman interview with Tuve's inspirational letter to his staff in 1944, printed in *APL-JHU-Z* 1, no. 2, (1944): 2, APLA: "When we have carried our job through to a good smash finish (and now we know that Section T will be there, talking the only language they understand, in every battle on every front in a mighty blasting volume)."

SELECTED SOURCES

ARCHIVAL MATERIAL

Archambault, Bennett, ed. Report of OSRD Activities in the European Theater During the Period March 1941 through July 1945. Washington, D.C.: Office of Scientific Research and Development, 1946, MIT Archives, Cambridge, MA, Bennett Archambault Papers, MC 555.

Bureau of Ordnance. *VT Fuzes for Projectiles and Spin-Stabilized Rockets.* Washington, D.C.: Department of the Navy, 1946, UI, JAVA.

Bush, Vannevar. "Laboratory Notebook" (original manuscript, June 1942–April 1946, LOC, VB).

Franklin, R. A. "Army Operational Research Group Memo No. 422. Interim Note on Anti-'Diver' Firing with Bonzo, August, 1944," October 23, 1944, TNA, PRO WO 291/746.

Mitchell, B. I. "Special Mission: Instruction in Use of Fuze Shell T74E6 in ETOUSA, 3-15-44 to 7-1-44," August 1, 1944, NARA, RG 156, E NM26 646A.

Office of Scientific Research and Development. *Report of the National Defense Research Committee for the First Year of Operation: June 27, 1940 to June 28, 1941.* Washington, D.C.: OSRD, 1941.

Roberts, Richard. "Autobiography of Richard Brooke Roberts: World War II." Unpublished manuscript, June 5, 1979, DTM.

———. "Development of the Proximity Fuze." Unpublished manuscript, October 26, 1977, DTM.

Salant, Edward O. "Part Twelve: Report of OSRD London Mission Activities in the Field of Proximity Fuzes for Shells on Behalf of Section T, OSRD," in *Report of OSRD Activities in the European Theater During the Period March 1941 through July 1945,* ed. Bennett Archambault.

Scientific Sub-Committee (Flying Bomb) of Crossbow Committee: Letters and miscellaneous (German flying bomb, V1), January 1, 1944, to December 31, 1944, TNA, AVIA 6/25668.

Section T. *Progress Report: February 19, 1942*. Washington, D.C.: Department of Terrestrial Magnetism, 1942, DTM.

———. *Progress Report as of April 19, 1942*. Washington, D.C.: Department of Terrestrial Magnetism, 1942, DTM.

———. "Quarterly Report as of September 30, 1942." November 30, 1942, 16, NARA, RG 227, E NC-138 8.

———. "Quarterly Report as of December 31, 1942." January 30, 1943, NARA, RG 227, E NC-138 8.

———. "Quarterly Report: January 1, 1943-March 31, 1943." May 10, 1943, NARA, RG 227, E NC-138 8.

———. "Quarterly Report as of June 30, 1943." August 10, 1943, NARA, RG 227, E NC-138 8.

———. "Quarterly Report as of March 31, 1944," May 15, 1944, NARA, RG 227, E NC-138 8.

———. "Quarterly Report as of June 30, 1944." August 15, 1944, NARA, RG 227, E NC-138 8.

———. "Quarterly Report as of September 30, 1944." November 15, 1944, NARA, RG 227, E NC-138 8.

———. *Final Report on the Development of the Radio and Other Proximity-Fuzes*. Washington, D.C.: Department of Terrestrial Magnetism, 1944, DTM.

Teas, J.P. "Trip Report: European Theater of Operations, 16 March to 27 October 1944," March 2, 1945, NARA, RG 156, E NM26 646A.

Thompson, Royce L. "Employment of VT Fuzes in the Ardennes Campaign, European Theater of Operations, December 1944–January 1945," 1950, US Army Center of Military History.

Tuve, Merle. "1913 Boy Scouts' Diary" (manuscript entries, 1913, LOC, MAT).

———. "1914 Boy Scouts' Diary" (manuscript entries, 1914 LOC, MAT).

———. "Notebook II" (original notebook, September–October, 1940, DTM).

———. "Notebook IV" (original notebook, November–December, 1940, DTM)

———. "Notebook V" (original notebook, December 1940, DTM).

———. "Notebook VII" (original notebook, January 1941, DTM).

———. "Notebook X" (original notebook, March 1941, LOC, MAT).

———. "Notebook XII" (original notebook, April–May, 1941, DTM).

———. "Notebook XVI" (original notebook, July–August, 1941, LOC, MAT).

————. "Notebook XX" (original notebook, December 1941–January 1942, DTM).

————. "Notebook XXI" (original notebook, January 1941–April 1942, DTM).

————. "New Mexico Trip Notebook" (original notebook, August 1943, LOC, MAT).

————. "Notebook PWP" (original notebook, June 1944, LOC, MAT).

Van Allen, James. "Field Notes" (original memo book, November 1942–September 1943, UI, JAVA).

BOOKS AND DISSERTATIONS

Andrus, E. C., et al., ed. *Advances in Military Medicine.* 2 vols. Boston: Little, Brown, 1948.

Applied Physics Laboratory. *The First Forty Years: A Pictorial Account of the Johns Hopkins University Applied Physics Laboratory Since Its Founding in 1942.* Baltimore: Schneidereith and Sons, 1983.

————. *The "VT" or Radio Proximity Fuze.* Silver Spring, MD: Johns Hopkins University, 1945.

————. *The World War II Proximity Fuze: A Compilation of Naval Ordnance Reports.* Silver Spring, MD: Johns Hopkins University Applied Physics Laboratory, 1950.

Atkinson, Rick. *The Guns at Last Light: The War in Western Europe, 1944–1945.* New York: Henry Holt, 2013.

Avery, Donald H. *The Science of War: Canadian Scientists and Allied Military Technology during the Second World War.* Toronto: University of Toronto Press, 1998.

Baker, Richard Brown. *The Year of the Buzz Bomb: A Journal of London, 1944.* Whitefish, MT: Literally Licensing LLC, 2011.

Baldwin, Ralph B. *The Deadly Fuze: Secret Weapon of World War II.* San Rafael, CA: Presidio, 1980.

————. *They Never Knew What Hit Them: The Story of World War II's Best Kept Secret.* Naples, FL: Reynier, 1999.

Baime, A. J. *The Arsenal of Democracy: FDR, Detroit, and an Epic Quest to Arm an American at War.* New York: Mariner, 2014.

Baxter, James Phinney, III. *Scientists Against Time.* Boston: Little, Brown, 1946.

Beevor, Anthony. *Ardennes 1944: The Battle of the Bulge.* New York: Viking, 2015.

Bergen, Peter and Daniel Rotherberg, eds., *Drone Wars: Transforming Conflict, Law, and Policy.* Cambridge: Cambridge University Press, 2015.

Bourgeois-Doyle, Richard. *George J. Klein: The Great Inventor.* Ottawa: NRC Press, 2004.

Boyce, Joseph Cannon, ed. *New Weapons for Air Warfare*. Boston: Little, Brown, 1947.

Brown, Anthony Cave. *"C": The Secret Life of Sir Stewart Menzies, Spymaster to Winston Churchill*. New York: Macmillan, 1987.

Brown, Louis. *Centennial History of the Carnegie Institution of Washington*. Vol. 2, *The Department of Terrestrial Magnetism*. Cambridge: Cambridge University Press, 2005.

———. *A Radar History of World War II: Technical and Military Imperatives*. Bristol, UK: Institute of Physics Publishing, 1999.

Buderi, Robert. *The Invention That Changed the World: How a Small Group of Radar Pioneers Won the Second World War and Launched a Technological Revolution*. New York: Touchstone, 1996.

Burchard, John E, ed. *Rockets, Guns, and Targets*. Boston: Little, Brown, 1948.

Bush, Vannevar. *Modern Arms and Free Men: A Discussion of the Role of Science in Preserving Democracy*. New York: Simon and Schuster, 1949.

———. *Pieces of the Action*. New York: William Morrow, 1970.

Caddick-Adams, Peter. *Snow and Steel: The Battle of the Bulge, 1944–1945*. Oxford: Oxford University Press, 2015.

Campbell, Christy. *Target London: Under Attack from the V-Weapon*. Boston: Little, Brown, 2012.

Childs, Herbert. *An American Genius: The Life of Ernest Orlando Lawrence, Father of the Cyclotron*. New York: E. P. Dutton, 1968.

Christman, Al. *Target Hiroshima: Deak Parsons and the Creation of the Atomic Bomb*. Annapolis, MD: Naval Institute Press, 1998.

Churchill, Winston. *The Second World War*. Vol. 5, *Closing the Ring*. Boston: Mariner, 1951.

———. *The Second World War*. Vol. 2, *Their Finest Hour*. Boston: Mariner, 1949.

Collyer, D. G., ed. *Buzz Bomb Diary*. Kent, UK: Kent Aviation Historical Society, 1994.

Compton, Karl et al. *Scientists Face the World of 1942*. New Brunswick: Rutgers University Press, 1942.

Conant, Jennet. *Tuxedo Park: A Wall Street Tycoon and the Secret Palace of Science That Changed the Course of World War II*. New York: Simon and Schuster Paperbacks, 2002.

Cornell, Thomas David. "Merle A. Tuve and His Program of Nuclear Studies at the Department of Terrestrial Magnetism: The Early Career of a Modern American Physicist." PhD diss., Johns Hopkins University, 1986.

Cull, Brian with Bruce Lander. *Diver! Diver! Diver! RAF and American Fighter*

Pilots Battle the V-1 Assault over South-East England, 1944–45. London: Grub Street, 2008.

D'Albert-Lake, Virginia. *An American Heroine in the French Resistance: The Diary and Memoir of Virginia D'Albert-Lake.* New York: Fordham University Press, 2006.

De Maeseneer, Guido. *Peenemünde: The Extraordinary Story of Hitler's Secret Weapons V-1 and V-2.* Vancouver: AJ Publishing, 2001.

Dennis, Michael Aaron. "A Change of State: The Political Cultures of Technical Practice at the MIT Instrumentation Laboratory and the Johns Hopkins University Allied Physics Laboratory, 1930–1945." PhD diss., Johns Hopkins University, 1990.

Dobinson, Colin. *AA Command: Britain's Anti-Aircraft Defences of the Second World War.* London: Methuen, 2001.

———. *Operation Diver: England's Defense Against the Flying Bomb June 1944– March 1945.* York, UK: Council for British Archeology, 1996.

Donoughue, Bernard, and G. W. Jones. *Herbert Morrison: Portrait of a Politician.* London: Orion, 2001.

Duffy, Peter. *Double Agent: The First Hero of World War II and How the FBI Outwitted and Destroyed a Nazi Spy Ring.* New York: Scribner, 2014.

Earls, Alan R., and Robert E. Edwards. *Raytheon Company: The First Sixty Years.* Charleston, SC: Arcadia, 2005.

Evans, Margaret Carpenter. *Rosemond Tuve: A Life of the Mind.* Portsmouth, NH: Peter E. Randall, 2004.

Feigel, Lara. *The Love-Charm of Bombs: Restless Lives in the Second World War.* London: Bloomsbury, 2013.

Foerstner, Abigail. *James Van Allen: The First Eight Billion Miles.* Iowa City: University of Iowa Press, 2007.

Foot, M.R.D. *Six Faces of Courage.* London: Eyre Methuen, 1978.

Ford, Ken. *The Bruneval Raid: Operation Biting 1942.* Oxford: Osprey, 2010.

Fourcade, Marie-Madeleine. *Noah's Ark.* New York: E. P. Dutton, 1974.

Fritz, Stephen. *The First Soldier: Hitler as Military Leader.* New Haven, CT: Yale University Press, 2018.

Garliński, Jósef. *Hitler's Last Weapons: The Underground War Against the V1 and V2.* New York: Times Books, 1978.

Goodchild, James. *A Most Enigmatic War: R. V. Jones and the Genesis of British Scientific Intelligence, 1939–45.* Solihull, UK: Helion, 2017.

Goudsmit, Samuel A. *Alsos.* Woodbury, NY: AIP Press, 1947.

Greenberg, Daniel S. *The Politics of Pure Science.* Chicago: The University of Chicago Press, 1999.

Haining, Peter. *The Flying Bomb War: Contemporary Eyewitness Accounts of the German V-1 and V-2 Raids on Britain.* London: Robson, 2002.

Hartcup, Guy, and T. E. Allibone. *Cockcroft and the Atom.* Bristol, UK: Adam Hilger, 1984.

Haskins, Caryl P., ed. *The Search for Understanding: Selected Writings of Scientists of the Carnegie Institution.* Washington, D.C.: Carnegie Institution, 1967.

Hastings, Max. *Overlord: D-Day and the Battle for Normandy.* New York: Simon and Schuster, 1984.

Helm, Sarah. *Ravensbrück: Life and Death in the Hitler's Concentration Camp for Women.* New York: Anchor, 2015.

Herbert, A. P. *The Battle of the Thames: The War Story of Southend Pier.* London: County Borough of Southend-on-Sea, 1945.

———. *The Thames.* London: Weidenfeld and Nicolson, 1966.

Hershberg, James. *James B. Conant: Harvard to Hiroshima and the Making of the Nuclear Age.* New York: Alfred A. Knopf, 1993.

Hiltzik, Michael. *Big Science: Ernest Lawrence and the Invention That Launched the Military-Industrial Complex.* New York: Simon and Schuster, 2015.

Holand, Hjalmar Rued. *Norwegians in America: The Last Migration.* Translated by Helmer Blegen. Sioux Falls, SD: Augustana College, 1978.

Hölsken, Dieter. *V-Missiles of the Third Reich: The V-1 and V-2.* Sturbridge, MA: Monogram Aviation, 1994.

Howard, Michael. *Strategic Deception in the Second World War: British Intelligence Operations Against the German High Command.* New York: W. W. Norton, 1995.

Jacobsen, Annie. *Operation Paperclip: The Secret Intelligence Program That Brought Nazi Scientists to America.* Boston: Back Bay Books, 2015.

Johnson, David. *V-1 V-2: Hitler's Vengeance on London.* New York: Stein and Day, 1981.

Jones, R. V. *Most Secret War.* Hertfordshire, UK: Wordsworth Editions, 1998.

———. *Reflections on Intelligence.* London: Mandarin, 1990.

Keith, Phil. *Stay the Rising Sun: The True Story of USS* Lexington, *Her Valiant Crew, and Changing the Course of WWII.* Minneapolis: Zenith, 2015.

King, Benjamin, and Timothy Kutta. *Impact: The History of German's V-Weapons in World War II.* Rockville Centre, NY: Sarpedon, 1998.

Klee, Ernst, and Otto Merk. *The Birth of the Missile: The Secrets of Peenemünde.* New York: Dutton, 1965.

Klein, George J. *The Great Inventor.* Ottawa: NRC Press, 2004.

Klingaman, William K. *APL—Fifty Years of Service to the Nation: A History of the Johns Hopkins Applied Physics Laboratory.* Laurel, MD: Johns Hopkins University Press, 1993.

Lindqvist, Sven. *A History of Bombing*, transl. Linda Haverty. New York: New Press, 2001.

Longmate, Norman. *The Doodlebugs: The Dramatic Story of the Flying Bombs of World War II*. London: Arrow, 1981.

Mackay, Robert. *Half the Battle: Civilian Morale in Britain During the Second World War*. Manchester, UK: Manchester University Press, 2002.

Mann, Wilfrid Basil. *Was There a Fifth Man? Quintessential Recollections*. Oxford: Pergamon, 1982.

Martelli, George. *Agent Extraordinary*. London: Collins, 1960.

McClure, Rusty, David Stern, and Michael Banks. *Crosley: Two Brothers and a Business Empire That Transformed the Nation*. Cincinnati, OH: Clerisy, 2008.

McKinstry, Leo. *Operation Sea Lion: The Failed Nazi Invasion that Turned the Tide of the War*. New York: Overlook Press, 2014.

Merriam, Robert. *Dark December: The Full Account of the Battle of the Bulge*. Whitefish, MT: Literary Licensing LLC, 2012.

Middlebrook, Martin. *The Peenemünde Raid: 17–18 August, 1943*. London: Cassell Military Paperbacks, 1982.

Millar, George. *The Bruneval Raid: Stealing Hitler's Radar*. London: Cassell, 1974.

Mortimer, Gavin. *The Longest Night: The Bombing of London on May 10, 1941*. New York: Berkley Caliber, 2005.

Musicant, Ivan. *Battleship at War: The Epic Story of the USS* Washington. New York: Avon, 1986.

Neiberg, Michael. *The Blood of Free Men: The Liberation of Paris, 1944*. New York: Basic Books, 2012.

Neliba, Günter. *Kriegstagebuch des Flakregiments 155 (W) 1943–1945: Flugbombe VI*. Berlin: Duncker and Humblot, 2006.

Neufeld, Michael J. *The Rocket and the Reich: Peenemünde and the Coming of the Ballistic Missile Era*. Cambridge, MA: Harvard University Press, 1996.

Nijboer, Donald. *Meteor I vs. VI Flying Bomb, 1944*. Oxford: Osprey, 2012.

Noyes, W. A., Jr., ed. *Chemistry*. Boston: Little, Brown, 1948.

Olson, Lynne. *Citizens of London: The Americans Who Stood With Britain in Its Darkest, Finest Hour*. New York: Random House, 2010.

———. *Those Angry Days: Roosevelt, Lindbergh, and America's Fight Over World War II, 1939–1941*. New York: Random House, 2013.

Overy, Richard. *The Bombing War: Europe 1939–1945*. New York: Penguin, 2014.

Pash, Boris. *The Alsos Mission*. New York: Award Books, 1969.

Petersen, Michael B. *Missiles for the Fatherland: Peenemünde, National Socialism, and the V-2 Missile*. Cambridge: Cambridge University Press, 2009.

Pile, Frederick. *Ack-Ack: Britain's Defense Against Air Attack During the Second World War.* London: Panther, 1956.

Ramsey, Winston G., ed. *The Blitz: Then and Now.* Vol. 2. London: Battle of Britain Prints, 1988.

———. *The Blitz: Then and Now.* Vol. 3. London: Battle of Britain Prints International Limited, 1990.

Regan-Atherton, Ann. *Heavy Rescue Squad Work on the Isle of Dogs: Bill Regan's Second World War Diaries.* Scotts Valley, CA: CreateSpace, 2015.

Rhodes, Richard. *The Making of the Atomic Bomb.* New York: Simon and Schuster, 1986.

Robinson, Vee. *On Target.* Roberttown, UK: Verity, 1991.

Rosbottom, Ronald C. *When Paris Went Dark: The City of Light Under German Occupation, 1940–1944.* Boston: Little, Brown, 2014.

Rossiter, Margaret L. *Women in the Resistance.* New York: Praeger, 1986.

Russell-Jones, Mair, and Gethin Russell-Jones. *My Secret Life in Hut Six: One Woman's Experience at Bletchley Park.* Oxford: Lion Books, 2014.

Ryan, Cornelius. *The Longest Day: The Classic Epic of D-Day, June 6, 1944.* New York: Simon and Schuster, 1959.

Samuel, Wolfgang W. E. *American Raiders: The Race to Capture the Luftwaffe's Secrets.* Jackson, MS: University Press of Mississippi, 2004.

Schlesinger, Henry. *The Battery: How Portable Power Sparked a Technological Revolution.* New York: Harper, 2010.

Sebba, Anne. *Les Parisiennes: How the Women of Paris Lived, Loved, and Died Under Nazi Occupation.* New York: St. Martin's, 2016.

Sellier, André. *A History of the Dora Camp: The Story of the Nazi Slave Labor Camp That Secretly Manufactured V-2 Rockets.* Chicago: Ivan R. Dee, 2003.

Shirley, Craig. *December 1941: 31 Days That Changed America and Saved the World.* Nashville: Nelson Books, 2011.

Simons, Graham M. *Operation Lusty: The Race for Hitler's Secret Technology.* South Yorkshire, UK: Pen and Sword Aviation, 2016.

Soames, Mary. *A Daughter's Tale: The Memoir of Winston Churchill's Youngest Child.* New York: Random House, 2012.

Stargardt, Nicholas. *The German War: A Nation Under Arms, 1939–1945.* New York: Basic Books, 2015.

Stewart, Irvin. *Organizing Scientific Research for War: The Administrative History of the Office of Scientific Research and Development.* Boston: Little, Brown, 1948.

Suits, C. G., et al., ed. *Applied Physics.* Boston: Little, Brown, 1948.

Thiesmeyer, Lincoln Reuber. *Combat Scientists.* Boston: Little, Brown, 1947.

Tillman, Barrett. *Clash of the Carriers: The True Story of the Marianas Turkey Shoot of World War II*. New York: Nal Caliber, 2005.

Waller, Maureen. *London 1945: Life in the Debris of War*. New York: St. Martin's Press, 2004.

Walpole, Nigel. *Hitler's Revenge Weapons: The Final Blitz of London*. Barnsley, UK: Pen and Sword Aviation, 2018.

Ward, Laurence. *The London Country Council Bomb Damage Maps: 1939–1945*. London: Thames and Hudson, 2015.

Watson-Watt, Robert. *The Pulse of Radar*. New York: Dial Press, 1959.

Wegener, Peter P. *The Peenemünde Wind Tunnels: A Memoir*. New Haven, CT: Yale University Press, 1996.

Williams, Allan. *Operation Crossbow*. London: Arrow, 2013.

Young, Irene. *Enigma Variations: A Memoir of Love and War*. Edinburgh: Mainstream, 1990.

Zachary, G. Pascal. *Endless Frontier: Vannevar Bush, Engineer of the American Century*. Cambridge, MA: MIT Press, 1997.

Zaloga, Steven J. *German V-Weapon Sites 1943–45*. Oxford: Osprey, 2008.

———. *Operation Crossbow 1944: Hunting Hitler's Weapons*. Oxford: Osprey, 2018.

———. *V-1 Flying Bomb 1942–52*. Oxford: Osprey, 2005.

MAGAZINES, NEWSPAPERS, AND PERIODICALS

"Dover Sylvania Workers Built Anti-Kamikaze, Anti-Buzz Bomb Devices." *Portsmouth Herald*, September 28, 1945.

"Local Engineer Played Vital Role in Helping Shorten World War II." *Bridgeport Sunday Post*, February 19, 1967.

"In Memoriam: Merle Antony Tuve, 1901–1982." *Johns Hopkins APL Technical Digest* 3, no. 2 (1982): 214.

"Mill Hall Sylvania Plant Wins Highest Navy Production Flag." *Lock Haven Express*, October 11, 1945.

"New Products, Moderns Plastics Helped Forge Secret Weapons." *Benton Harbor News-Palladium*, December 31, 1945, 16.

"Our Scientific High Command." *Reader's Digest*, June 1944, 37–39.

"Proximity Fuse An Iowa Product: Staff of 300 Engaged in Research at I. U." *Council Bluffs Nonpareil*, October 5, 1945.

"Role of Proximity Fuse, No. 2 Secret of War, Told at Chamber Dinner Meet." *Annapolis Evening Capital*, April 30, 1946.

"Secret Proximity Fuse Vital to U.S. Victory." *Cullman Democrat*, December 6, 1945.

Associated Press. "'Radio Fuse' Was One of Scientific Marvels of War." *Titusville Herald,* September 21, 1945.

Associated Press. "7 Titusville Men Worked On Radio Fuse." *Titusville Herald,* September 21, 1945.

Brennan, James W. "The Proximity Fuze: Whose Brainchild?" *United States Naval Institute Proceedings* 94, no. 9 (1968): 73–78.

Burke, Alan. "Ipswich Woman Was a Mom in Proximity." *Salem News,* September 12, 2011.

Cornell, Douglas B. "On Trail of Spy Stole Top Secret U.S. Proximity Fuse." *Alberta (Canada) Medicine Hat News,* December 3, 1953.

Dillman, Grant. "Proximity Fuse Had Big Part in Sending Axis Down to Defeat." *Sandusky Register Star News,* December 11, 1945.

Dobbs, Michael. "Ex-Agent: Rosenberg Was Spy, but Not Atomic Spy." *Santa Fe New Mexican,* March 16, 1997.

Gurnett, Donald A., and Stamatios Krimigis. "The Life and Accomplishments of James A. Van Allen." *IEEE Transactions on Plasma Science* 35, no. 4 (2007) 745–47.

Krimigis, Stamatios. "James A. Van Allen, "From Iowa to APL to Iowa to Space." *Johns Hopkins APL Technical Digest* 33, no. 3 (2016): 162–64.

International News Service. "Medal Awarded to Man Who Produced Proximity Fuse." *Lebanon Daily News,* October 26, 1950.

Poole, Lynn. "The Proximity Fuse Helped Win Historic 'Bulge' Battle." *Cedar Rapids Gazette,* December 20, 1959

Roberts, Sam. "A 'Bridge of Spies' Back Story," *New York Times,* November 26, 2015.

Sacks, H. William. "Local Man Who Helped Develop 'Best Kept Secret' Relives WW II Memories." *Gettysburg Times,* May 23, 1992.

Science Service. "Proximity Fuse Revealed as Our No. 2 Secret Weapon." *El Paso Herald Post,* February 23, 1946.

Seaburg, William. "Whatever Happened to Thelma Adamson? A Footnote in the History of the Northwest Anthropological Research." *Northwest Anthropological Research Notes* 33, no. 1 (1999): 73–84.

Sullivan, Walter. "James A. Van Allen, Discoverer of Earth-Circling Radiation Belts, Is Dead at 91." *New York Times,* August 10, 2006.

Tomlin, George. "Shells that Know When." *Guardian,* August 10, 1992.

United Press. "Radio Fuzes Protect U.S.: New Weapons Ready, Patterson Says." *Nevada State Journal,* October 24, 1945.

Van Allen, J. A. "What Is a Space Scientist? An Autobiographical Example." *Annual Review of Earth and Planetary Sciences* 18, no. 2 (1990): 1–26.

———. "My Life at APL." *Johns Hopkins APL Technical Digest* 18, no. 2 (1997): 173–77.

Wachtel, Max. "Unternehmen Rumpelkammer." *Der Spiegel,* December 1965, 99–119.

Walsh, Tom. "Pioneer of a New Frontier: UI Physicist's Breakthroughs Shattered Scientific Thinking at Start of Space Age." *Gazette,* February 24, 2002.

———. "Work for the War Effort." *Iowa Today,* December 7, 2003.

INTERVIEWS, ORAL HISTORIES, AND TRANSCRIPTS

Bush, Vannevar. Oral history interviews by Eric Hodgins, 1964, MIT Archives, MC-0143.

Comly, Lucy Tuve. Interview by Paul Ertsgaard, October 11, 1987, DTM.

de Clarens, Jeannie. Interview by Austin Mitchell, in P. Jones, *The Secret War of Dr. Jones,* 1977.

———. Interview by David Ignatius, 2011, International Spy Museum Archive, Washington, D.C.

Salant, Edward O. Interview by Paul Sulds for "Science in War and Peace," Mutual Broadcasting System, November 12, 1945, 8, JAVA UI.

"Transcript of Meeting on VT (Proximity) Fuze." New York City, New York, annual conference of the Radio Club of America, 1978. Navy Yard Archives, Washington, D.C.

Tuve, Merle. Interview by Thomas D. Cornell, January 13, 1982, session 1. Niels Bohr Library and Archives, American Institute of Physics, College Park, Maryland, www.aip.org/history-programs/niels-bohr-library/oral-histories/4921-1.

———. Interview by Thomas D. Cornell, February 5, 1982, session 2. Niels Bohr Library and Archives, American Institute of Physics, College Park, Maryland, www.aip.org/history-programs/niels-bohr-library/oral-histories/4921-2.

———. Interview by Charles Weiner, March 30, 1967. Niels Bohr Library and Archives, American Institute of Physics, College Park, Maryland, www.aip.org/history-programs/niels-bohr-library/oral-histories/4920.

———. Interview by Albert Christman, May 6, 1967. Niels Bohr Library and Archives, American Institute of Physics, College Park, Maryland, www.aip.org/history-programs/niels-bohr-library/oral-histories/3894.

Van Allen, James. Interview by David DeVorkin and Allan Needell, Washington, D.C.: Smithsonian National Air and Space Museum Oral History Program, 1981, JAVA, IU.

Whitman, Winifred Gray. Interview with Paul Ertsgaard, September 8, 1987, DTM.

FILMS AND TELEVISION

Battle of the Bulge: The Deadliest Battle of World War II. DVD. Directed by Thomas Lennon. Boston: American Experience, 2004.

The Deadly Fuze: The Story of WWII's Best Kept Secret. DVD. Directed by Maurice Jacobsen. Grand Rapids, MI: WGVU Productions, 1995.

The Secret War: The Complete Original 1977 Series. DVD. Directed by Glynn Jones. London: BBC and the Imperial War Museum, 2014.

The Secret War of Dr. Jones. Directed by Peter Jones. 1977. Yorkshire, UK: Yorkshire Television and ITV Studios.

INDEX